W9-AWI-625

Landmass

Continent	Land Area (Sq. Mi.)	Percentage of Earth's Total Land Area
Asia	17,300,000	30
Africa	11,700,000	20
North America	9,500,000	16.5
South America	6,900,000	12
Antarctica	5,400,000	9.5
Europe	3,800,000	6.5
Australia	3,300,000	5.5

Highest Point

Continent	High Point and Country	Elevation Above Sea Level (in Feet)
Asia	Everest, China–Nepal	29,028
South America	Aconcagua, Argentina	22,831
North America	McKinley (Denali), AK, U.S.A.	20,320
Africa	Kilimanjaro, Tanzania	19,340
Europe	El'brus, Russia	18,510
Antarctica	Vinson Massif	16,066
Australia	Kosciusko	7,310

Lowest Point

Continent	Low Point	Elevation Below Sea Level (in Feet)
Antarctica	(Ice-covered)	–8,327
Asia	Dead Sea	–1,339
Africa	Lake Assal	–512
North America	Death Valley	–282
South America	Valdes Peninsula	–131
Europe	Caspian Sea	–92
Australia	Lake Eyre	–52

Population

Continent	Total Population	Population Density (People per Sq. Mi.)
Asia	3,547,788,000	205
Africa	730,952,000	62.5
Europe	697,013,000	183
North America	462,161,000	49
South America	323,554,000	47
Australia	18,311,000	5.5
Antarctica	0	0

(summer population of research scientists not counted)

tear here

alpha
books

Trivial Stuff: The World

World Extremes

Highest mountain: Mount Everest, China–Nepal	29,028 feet
Deepest ocean point: Challenger Deep, Pacific	–35,826 feet
Highest waterfalls: Angel Falls, Venezuela	3,212 feet
Deepest lake: Lake Baikal, Russia	–5,371 feet
Largest canyon: Grand Canyon, U.S.A.	95,144 feet (width); 5,249 feet (depth)
Longest reef: Great Barrier Reef, Australia	1,250 miles
Largest desert: Sahara, Northern Africa	3,474,918 sq. mi.

Earth Stats

Mean diameter	7,917.52 mi.
Polar circumference	24,855.34 mi.
Mean distance to the sun	93,020,000 mi.
Mean distance to the moon	238,857 mi.
Total area	197,000,000 sq. mi.
Land area	57,900,000 sq. mi.
Ocean area	139,100,000 sq. mi.

Largest Islands

Greenland	840,004 sq. mi.
New Guinea	309,000 sq. mi.
Borneo	287,300 sq. mi.
Madagascar	226,658 sq. mi.
Baffin	195,928 sq. mi.

Largest Oceans

Pacific	63,800,000 sq. mi.
Atlantic	31,800,000 sq. mi.
Indian	28,900,000 sq. mi.
Arctic	5,400,000 sq. mi.

Largest Seas

Caribbean Sea	970,000 sq. mi.
Mediterranean Sea	969,000 sq. mi.
South China Sea	895,000 sq. mi.
Bering Sea	875,000 sq. mi.
Gulf of Mexico	600,000 sq. mi.

Largest Natural Lakes

Caspian Sea (saltwater)	143,243 sq. mi.
Lake Superior (freshwater)	31,820 sq. mi.
Lake Victoria	26,724 sq. mi.
Lake Huron	23,010 sq. mi.
Lake Michigan	22,400 sq. mi.

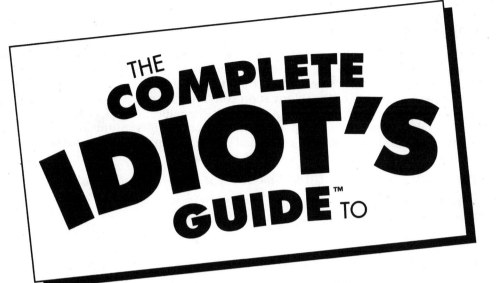

THE COMPLETE IDIOT'S GUIDE™ TO

Geography

by Thomas E. Sherer, Jr.

alpha books

A Division of Macmillan General Reference
A Simon & Schuster Macmillan Company
1633 Broadway, New York, NY 10019-6785

THE COMPLETE IDIOT'S GUIDE name and design are tademarks of Macmillan, Inc.

International Standard Book Number: 0-02-861955-2
Library of Congress Catalog Card Number: 97-073162

99 98   8 7 6 5 4 3 2

Interpretation of the printing code: the rightmost number of the first series of numbers is the year of the book's printing; the rightmost number of the second series of numbers is the number of the book's printing. For example, a printing code of 97-1 shows that the first printing occurred in 1997.

Printed in the United States of America

ALPHA DEVELOPMENT TEAM

Brand Manager
Kathy Nebenhaus

Executive Editor
Gary M. Krebs

Managing Editor
Bob Shuman

Senior Editor
Nancy Mikhail

Development Editor
Jennifer Perillo

Editorial Assistant
Maureen Horn

PRODUCTION TEAM

Director of Editorial Services
Brian Phair

Development Editor
Sharron S. Wood

Production Editor
Rebecca Whitney

Cover Designer
Michael Freeland

Illustrator
Judd Winick

Designer
Glenn Larsen

Indexer
Chris Barrick

Production Team
Aleata Howard, Mary Hunt, Scott Tullis, Megan Wade

Contents at a Glance

Contents

Part 2: A Regional Look: The Developed World   63

6 North America: Land of Opportunity   65

7 The British Isles   77

8 Western Europe: The Heart of the Continent   91

9 Northern Europe: Land of Ice   105

12 The New Russia — 143

13 Japan: The Pacific Dragon — 155

Part 3: A Regional Look: The Developing World   193

16 Middle America: A Bridge in Time   195

Foreword

In the days when foreign travel scarcely existed—when people vacationed only in the next town or camped at a nearby lake—there was perhaps an excuse for being ignorant of geography. When the world of other nations seemed so remote that it scarcely entered into our journalism or conversations, it was no great thing never to have heard of Serbia or Chechnya, the El Niño current, or Mount Saint Helens.

Today, not to understand geography is not to have lived, not to understand most of the issues that concern us, not to have opinions or make choices in important daily matters. Think of how the scope of our interests has expanded in the past century.

Today, we travel as casually to other nations as we once boarded a train to a neighboring state. At parties, when someone asks us about our next trip, we answer that we are undecided between Australia or Costa Rica, between Europe or Hong Kong. At night, some of us tune our television sets to British newscasts or view a program that includes correspondents reporting from around the world. In our business dealings, we discuss exports to places thousands of miles away.

And all of this is new. We are the first generation in human history to enjoy foreign travel as a casual part of life, to communicate by direct-dialing with remote continents, to receive instant news of world happenings, to send salespeople to other lands on behalf of the most mundane products, to contemplate even the exploration of the heavens.

A knowledge of geography aids us to enjoy the fullness of our contemporary lives. And none of us knows enough. Turning the pages of Thomas Sherer's lucidly written, enthusiastic guide to geography, I've picked up a much newer understanding of our world, a totally fresh awareness of many formerly unknown lands and regions that will not only enhance my future travels but also enliven my daily reading and discussions.

Apart from everything else it provides, the *Complete Idiot's Guide to Geography* is fun to read, fun to return to. Even later dipping into its pages, after reading it only once, has reawakened my thoughts of the wonderfully varied world in which we live.

Arthur Frommer

Introduction

Thank you for purchasing *The Complete Idiot's Guide to Geography.* Please don't be put off by its title: We're all idiots when it comes to geography! As a geographer, I'm supposed to be somewhat of an expert, but when the crossword puzzle clue asks for a tributary of the Rhone River, I have to bury myself in an atlas just like everyone else does! There's just too much to geography to know it all—but that doesn't mean, of course, that you can't work at it.

This book is designed to give you the basics. It tells you what geography encompasses and how it applies to your life. You will learn a little about the natural world and get a feel for the people who inhabit the many environments of the earth. Best of all, you won't feel like an idiot the next time your game piece lands on the geography category of your favorite trivia board game.

I hope that this book is a stepping-stone toward sending you on a lifelong quest for geographic knowledge. You can check out a few of the books listed in the bibliography in Appendix B, subscribe to a geographic magazine, buy yourself a good atlas, but most of all—travel! The experiences people have as they venture to foreign lands brings geography to life.

When I teach a geography class, I always ask first whether anyone has been to the place the class is preparing to study. It seems that a good story is always waiting to be told or an exploit to be relived. These tales help bring the world to life and elevate it from the arid text pages into the light of reality. The stories are also a great deal of fun, which is what geography should be!

This book has drawn on many excellent geographic texts for its information. Please remember that it provides only a quick overview of the world's geography. For additional information or to investigate a region in more depth, consult one of the sources listed in Appendix B, "Select Bibliography."

What You Learn About in This Book

The Complete Idiot's Guide to Geography divides your geographic journey into four parts.

Part 1, "An Overview of Geography," defines geography and gets at some of the nuts and bolts of the discipline. It helps you become familiar with some terminology and principles addressed later in the book. Part 1 gives you the foundation on which your geographic knowledge will be built.

In Part 2, "A Regional Look: The Developed World," you begin your epic journey around the globe, starting with the developed nations of the world and exploring ten regions and dozens of countries. This part hops from region to region, describing a little history, nature, culture, and perhaps a geographic oddity or two. If you're a place-name person, you see the names of a few tall mountains, long rivers, big cities, capitals, and countries. Don't get overwhelmed or feel obliged to memorize it all—that's what atlases and almanacs are for.

Part 3, "A Regional Look: The Developing World," is virtually a clone of Part 2, except that it focuses on the developing countries of the world. Part 3 describes ten more regions, dozens more countries, and lots of other stuff—plenty to keep you busy!

The shortest part of the book, Part 4, "A Global Overview," looks at problems that concern every region on the earth—global concerns that threaten us all and call for universal cooperation to resolve.

Extras

As you read through the four parts of this book, you see aids to help you along the way. First, a number of pictures and simplified maps provide a graphical reference. Although you won't want to navigate by these maps, you should get at least a feel for what goes where! To spice things up and to provide clarifying information, each chapter is loaded with extra tidbits of information:

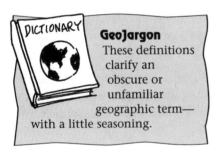

GeoJargon
These definitions clarify an obscure or unfamiliar geographic term—with a little seasoning.

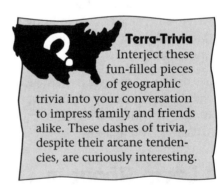

Terra-Trivia
Interject these fun-filled pieces of geographic trivia into your conversation to impress family and friends alike. These dashes of trivia, despite their arcane tendencies, are curiously interesting.

> **Geographically Speaking**
> These boxes describe a geographic concept or an aside in more depth. They might talk about the nature of earthquakes or the exploits of pirates and buccaneers or any other interesting geographic tidbit.

Acknowledgments

The materials in this book were compiled with help from two skilled researchers, Susan Calhoun Heminway and Thomas E. Sherer, Sr. Both contributed long hours of research, fact finding, and trivia hunting that are reflected in the chapters in this book. Because both are avid geography buffs, I drew heavily on their extensive travel experience to fill in those spots on the earth I've missed!

The maps in this book were created from MapArt, from Cartesia Software. Macmillan has been granted permission to use these maps. For information about MapArt, call 800-344-4291 or visit www.map–art.com.

Special Thanks to the Technical Reviewer

The Complete Idiot's Guide to Geography was reviewed by an expert who not only checked the technical accuracy of the information in this book but also provided valuable insight to help ensure that it tells you everything you need to know about geography.

Our special thanks are extended to Geoff Golson. He grew up in Europe and had traveled to 27 countries by the time he settled in the New York area. With a keen interest in maps and travel, Geoff has worked as executive editor at Rand McNally, in charge of new-product development, and as editorial director at Macmillan Publishing USA, in charge of the world atlas products. He is now publisher at Macmillan Library Reference.

Part 1
An Overview of Geography

One of the most difficult challenges geographers face is defining exactly what geography means. Although that process might be similar to trying to hold the wind in a box, Part 1 of this book tries to flesh out the term geography *and gives you a little background in physical, cultural, and regional geography. It might sound like a great deal of ground to cover in just a few chapters, but try not to let those practical concerns daunt you.*

Along the way, you even get to refresh your map skills. Although some of this stuff may seem sort of heavy-duty, it's useful as you set forth on your global explorations later in this book. With maps in hand, your newfound knowledge of physical and cultural geography, and your grasp of the regional concept well entrenched, you'll be ready to take on the world.

What Is Geography?

You've probably picked up this book because you have some interest in geography. Perhaps you're a geography buff and you grab any material that has to do with the subject or perhaps you simply want to brush up your geography skills to enhance your reputation as a Trivial Pursuit ringer.

More likely, you believe that you know nothing about geography and have decided to rectify your hideous state of geographic illiteracy! If that's your situation, you may not even know what an understanding of geography entails—to be truthful, most people don't. Geography is a difficult discipline to define because it doesn't fit neatly into some academic box. This chapter describes what geography is all about, what it is not, and why the subject should matter to you.

The Mother of All Sciences

Geography has been called "the mother of all sciences" and "the science of place." However you describe geography, it involves the examination of the physical and cultural factors that interact to make up the diversity of the earth.

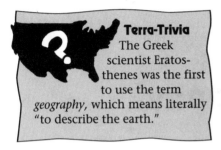

Terra-Trivia
The Greek scientist Eratosthenes was the first to use the term *geography*, which means literally "to describe the earth."

Is this scope somewhat broad? Yes, it is. Almost anything can fall into the realm of geographic study. If you want to learn about the earth, its peoples, and why the two interact as they do, you have the makings of a geographer. Although this book doesn't really restrict *what* geographers study, it's more specific about the *way* geographers look at things.

Geography is a *spatial* discipline, which means that geographers are concerned not only with what something is but also with the way it is distributed in space. Although geographers don't yet have the medical expertise to develop a cure for cancer, for example, they can still study the distribution of cancer cases and suggest possible causes based on that spatial information.

My favorite definition of geography is "anything that can be mapped." I use this definition again later in this book because it gets at the heart of geography. If something can be mapped, it has a spatial component. Maps are the tools of geographers: If something can be mapped, it's geography.

What Geography Is Not: The Dread Place-Name Quiz

Many people, because of their experiences in school, incorrectly believe that the study of geography involves nothing more than studying for and taking place-name exams. If you mention geography to most people, you get a sour look. To them, geography means no more than an outline map and the memorization of country names and capitals. When I tell people that I have a graduate degree in geography, I can almost see them thinking, "Wow—this guy must memorize more than the names of cities and rivers; he must also memorize village names and small streams!" Geographers can never get away from the place-name outline map.

Why is the place-name map so entrenched? Because, for decades, geography has been taught that way. Spatial thinking and investigation have been left in the shadows as teachers have focused on the place-name map. Although outline maps are easy to display in a classroom, they present only the most superficial geographic knowledge and leave the more important issues unchallenged.

The importance of knowing where things are in the world can't be denied. As the world increasingly becomes a smaller place, you have to know the names of the places around you, although geography consists of much more than this single component. Would you grant a medical doctor a license based on his or her ability to name the parts of the body? Of course not. In addition to knowing medical terminology, doctors must be able to prescribe medication, diagnose illnesses, and complete successful surgery.

If you like the use of place-name maps, don't worry—I've put a few in this book. I hope that this book gives you more than just maps, though—I hope that it gives you a whole new spatial sense of how the people and places of the earth interact.

Despite what I've said about place-name maps, after reading this book, you probably will gauge your newfound geographic literacy by your ability to identify places on a map. Because most people accept this convention, I might as well give you a yardstick by which to measure. It's time for you to take the dread place-name quiz.

This ten-question quiz is designed to test your place-name knowledge. Don't worry: It doesn't ask you to locate the third-largest city in Togo. Each country in the quiz has been in the news in the past ten years or contains significant physical features on the earth's surface. (That's right—you have to be familiar with mountains and rivers, too.)

Check your answers at the end of this chapter. If you missed several questions, try this quiz again after you finish reading this book to see whether you fare any better.

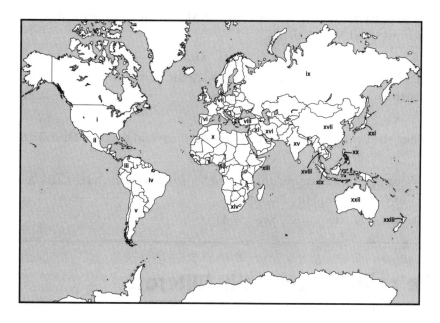

The Dread Place-Name Quiz

1. This country used to comprise the bulk of the Soviet Union: Find Russia.

   A. xvi   B. xvii   C. ix   D. xxii

2. This country has the second-largest population on earth: Find India.

   A. iv   B. xv   C. xi   D. xx

3. This country was the stomping ground of Evita Perón, the subject of a recent hit movie: Find Argentina.

   A. ii   B. iii   C. v   D. xii

4. When a United States teenager was caned for vandalizing a car, this country was in the news: Find Singapore.

   A. ii   B. iv   C. xviii   D. xix

5. This country's invasion of Kuwait resulted in Operation Desert Storm: Find Iraq.

   A. xi   B. xvi   C. ix   D. xiii

6. Much of the world's rain forest and its largest river by volume are located in this country: Find Brazil.

   A. xii   B. iv   C. xviii   D. iii

7. Those pesky Kiwis have helped to smash U.S. dominance of the America's Cup in yachting: Find New Zealand.

   A. v   B. xix   C. xxii   D. xxiii

8. The United Nations recently took on warlords in this impoverished country: Find Somalia.

   A. xiii   B. xvi   C. xviii   D. xii

9. When the Berlin Wall finally came down, the world rejoiced: Find Germany.

   A. iii   B. vi   C. vii   D. viii

10. A volcanic eruption caused the United States to close the huge Clark Air Force Base in this country: Find the Philippines.

    A. v   B. xiv   C. xviii   D. xx

Why We're Geographically Illiterate

If you live in the United States and took a geography class in school, you're in the minority. Most people in this country got their limited dose of geography in a social studies class because most states ditched geography as a distinct course decades ago. Geography was absorbed into the more general social studies curriculum along with history, economics, sociology, and anthropology.

One result of this amalgam called social studies is that Americans don't know their geography. Article after article in the press cites the woeful state of our country's geographic education. Although a "geographically illiterate" nation is good news when you're writing a book like this one, it's bad news for U.S. business.

Virtually every other industrialized nation on earth features geographic education as a primary component of its academic instruction. In most countries, geography is taught—from elementary school through high school—in a separate class. When graduates in these countries enter the global marketplace, they're familiar with resources, transportation systems, landforms, foreign languages, religions and customs, climate, demographics, and political systems. How many U.S. graduates can say the same?

Terra-Trivia
Did you know that the highest wind gust ever recorded was 231 m.p.h.? It wasn't recorded in Antarctica or at the top of Mount Everest or in a Philippines typhoon. No, it was recorded atop Mount Washington in quiet New Hampshire, U.S.A., by a small weather station anchored to the bedrock with steel rods.

Educators are coming to terms with the problem, however. They're again acknowledging the importance of geography as a vehicle for studying the environment, multiculturalism, and human–environment interaction.

Ways to Understand Geography

You can learn about the topic of geography in an infinite number of ways. Entire geography courses have been structured around newspaper clippings, for example. Go through your local daily newspaper or a weekly magazine and you can find stories of interest from almost every corner of the globe. (That expression has some geometric inconsistencies!)

Geographically Speaking

Do you remember working on a book report back in middle school and asking one of your parents to spell a word for you? I'm sure that you received the same retort kids everywhere receive: "Look it up in the dictionary, and then you'll always remember it." Although I hated that response, our parents were right.

The process of learning geographic place-names is similar: Every time you hear the name of a place on a news program or in daily conversation, look it up. If you use an atlas in the same way you use a dictionary, you soon will have an expansive place-name repertoire—without ever having to fail the dread blank map place-name quiz.

You can also use literature as the basis for geographic learning. Some geography courses use *Around the World in Eighty Days* or other books as the framework for their exploration. Others use history as a guide, such as following the expedition of Lewis and Clark.

Although all these approaches are valid, two main lines of teaching geography have emerged: introductory and world regional.

The *introductory* approach, which is a great place to start, simply describes the techniques and terminology geographers use. This type of course helps distinguish between a meridian, a lingua franca, and a tsunami, for example. Here are some topics discussed in an introductory approach:

➤ Maps and globes

➤ The earth–sun relationship

➤ Weather and climate

➤ Landforms

➤ Soils and vegetation

➤ Demographics (populations)

➤ Culture

➤ Political systems

➤ Economic systems

➤ Settlements and urban studies

➤ Resources

➤ Regions

➤ Human–environment interaction

Terra-Trivia

Do you know what the words *tsangpo, sungai, stroom, song, shatt, myit, nahr, kong, fluss, fleuve, flume, batang,* and *alf* mean? They're all foreign words for *river.* You might encounter these words as you peruse maps in an atlas. Most atlases have a section describing foreign words they use—check it out the next time you find yourself up a kei without a paddle.

The other major line of inquiry, the *world regional* approach, focuses not so much on techniques and terminology (although they inevitably creep into the discussion) as on describing the specific geography of world regions. This method examines each of the earth's regions and a wide range of geographic topics that help to describe them.

This book uses a little of each of these methods. Part 1 explains a few key geographic terms, techniques, and concepts in the standard introductory approach; it also provides an overview to prepare you for the regional sections later in the book. Part 2 takes a world regional approach and helps you begin to look at the world's developed regions. In more of the world regional approach, Part 3 describes the developing regions of the earth. Both Parts 2 and 3 consider the physical and human attributes of each region on earth in some detail.

Breaking away from regional concerns to consider the earth as one entity, Part 4 returns to the introductory approach to discuss global concerns and global solutions for the entire earth. In this part of the book, you begin to understand the geography of the whole earth and realize that the actions of one person, one country, or one region are closely tied into an integrated web of systems that forms the complex planet earth.

Why Study Geography?

One of the most frequent complaints from kids in school is, "Why do I have to learn this stuff? I'll never need to use it." We never know what life will bring, of course, nor what we may someday need. But try to tell that to a 15-year-old! Nonetheless, people seem to focus better when they feel that what they're doing has some purpose.

So what purpose does geography have in your life— why learn about this stuff? To answer that question, I've compiled a top-ten list of why geography is important. Although you might not find every point important, you should relate to enough of them to give your quest for geographic knowledge some meaning:

Terra-Trivia
Most of the maps people use in their lives are similar to roadmaps—they show a small amount of area (as with a state or town). An example of a *really* small-scale map shows the solar system. The earth is the third planet from the sun (of nine orbiting planets); only Mercury and Venus are closer. The mean distance from the earth to the sun is 93,020,000 miles.

Top Ten Reasons Geography Is Important to You

1. **It helps you enjoy the great outdoors.** Planning a successful camping trip or an invigorating day hike can be a pleasure as you apply your newly acquired map skills. With map and compass in your hand, no overgrown trail is too daunting. You will no longer pitch your tent in dried river beds prone to flash floods or choose the trail with the 45-degree incline. The primary tool for mastering geography—the map—can make your annual outdoor adventure a confident pleasure.

2. **It gives you a marketing perspective.** Suppose that you're sick of that nine-to-five job and are ready to set off on your own enterprise. You even have the perfect start-up idea: the Mouse-Cozy for those cold-weather computer users. But where should you open a Mouse-Cozy store, and where should you send your mass mailings? Thank goodness for your knowledge of demographics and geographic information systems. As you enter age profiles, income levels, and climatic data—behold!—Point Barrows, Alaska, beckons, and financial success is ensured.

3. **It can serve as a guide for planning.** Do you want to start a new career as a city or town planner or a zoning official? Perhaps you just want to plan an addition to your family room. Either way, geography can be your guide. Are you encroaching on wetlands, crowding your property line, or exceeding local height restrictions? You don't have to worry because your newfound geographic confidence makes zoning maps as easy to read as Dick and Jane.

> ## ! Geographically Speaking
>
> In a typical geography classroom in which a particular topic, such as Mexico City's overpopulation, is being discussed, some students are clearly interested and concerned—perhaps the teacher has truly reached them.
> Invariably, however, a student (generally in the last row) raises his or her hand and asks, "Do we have to know this stuff for the final?"
>
> This question tears at the heart of most geographic learning. Much of what I discuss in this book is not information you need in order to do your work. This book isn't a how-to guide that enables you to replace the sink trap in the bathroom. Do you need to know this information for the final? No, you need to know it because you're a citizen of the planet earth.

4. **It can make you the life of the party.** Have you ever embarrassed yourself at a party by displaying geographic ineptitude? Perhaps someone has said something like, "That nuclear disaster at Chernobyl has threatened the health of hundreds of thousands of people," and you responded, "Yes, that's true. I just thank goodness I don't live in Ohio."

   Humiliating, yes. Uncorrectable, no. After reading this book, you'll be able to dazzle your friends with comments such as, "Yes, that's true, and the danger is not simply limited to the Ukraine. Dangerous levels of radiation were detected even north of the Arctic Circle in Scandinavia."

5. **It reinforces your spatial thinking.** This catch-all reason affects almost every aspect of life. When I did some consulting work for an engineering company, I was asked to be part of a team that inspected a huge chemical plant. The plant's concrete floors were rapidly deteriorating, and I was asked to investigate the cause. The plant was six floors high and filled with hundreds of vats, chests, and vessels, each filled with a different toxic brew.

   Not sure exactly where to start, the team suggested that the company map out the topsides and undersides of each floor slab and indicate levels of deterioration (geographers call this process "choropleth mapping"). Although the plant operators assumed that spills from the vessels had been the cause of the deterioration, our team found something different. The mapping showed that the slab deterioration had been caused by rising vapors from vats on the floors below. Using the geographic approach, the company arrested the vapors, stopped the deterioration, and solved the problem.

6. **It can help save the earth.** People are increasingly concerned about the plight of this planet. If you want to tread a little more softly on the earth and to teach children to be gentle stewards of the environment, where do you start?

You start with geography. When kids ask you about deforestation in the Amazon Basin, just show them a map. Investigate the geography of the place, its vegetation and soils, the indigenous people who call the forest home, and the settlers who are encroaching on the forest. Point out which animals live there, how trees affect the climate, and how trees contribute to the oxygen we breathe. Use geography to add depth and meaning to the issue. You might be surprised at the direction in which a child's questions can lead.

7. **It helps you choose where to set up house.** So you want to move. Do you want a neighborhood that's rural or urban? Do you need to be near public transportation? Are you looking for an ethnically diverse spot with superior schools? What about taxes and the cost of housing in the area? If you ski, for example, you had better not stray far from the mountains. If you sail, you probably won't move too far from the sea.

   Perhaps you thought that geographic questions would involve only certain houses located on fault lines, situated on flood plains, or skirting national borders. Prospective home buyers have to ask lots of questions, many of them about geography.

8. **It helps you understand politics.** The union of geography and politics is not a new marriage. Struggles for territory, disputes about borders, and fights over resources are as old as humankind. If gerrymandering is occurring within local congressional districts, that's a part of geography. If the former Yugoslavia is being partitioned into new nations, that's also a part of geography. If the United States and Canada are sending aid and troops to central Africa, that's geography. Politics and geography are inextricably intertwined. If you like your politics, know your geography.

9. **It's a critical tool for business.** Do you want to get a job with a huge multinational corporation so that you can travel? Or, if you work for a multinational corporation, do you want to find a job with a small local business so that you don't have to travel? Because the world is quickly becoming one giant marketplace, your knowledge of other countries, foreign customs, and global resources can determine your success. This knowledge is geography!

   Suppose that the president of the Thai division of your company takes you for a tour of the local Buddhist temples, and you forget to take off your shoes, so you lose your job. Or you send your best client in Saudi Arabia a canned ham for Christmas, so you lose your sale. Or you establish a tea plantation in Norway, so you go bankrupt. With a little geographic understanding, you can avoid all these international faux pas.

10. **It can help you get away from it all.** Are you sick of that same old summer lake rental? How about using your geographic awareness to plan something a little different? You can make the bed-and-breakfast circuit around England with stops at every castle and pub you encounter. Or how about a spartan ride across the steppes

of Asia on the Siberian Express (a railroad that runs across Russia)? If that's not spartan enough for you, how about a three-week backpacking trek through the Southern Alps of New Zealand? If you want more luxury, consider the sunny beaches of the French Riviera.

Whether it's Mayan ruins or African safaris, the world is out there waiting for you. You just need enough geographic knowledge to be able to know what you're missing. At some point, the rut you've dug becomes a chasm. It's time to break out, explore, and experience. Geography is the vehicle. Get in and drive!

The Least You Need to Know

➤ Geography is a spatial discipline that encompasses "anything that can be mapped."

➤ There's more to geography than place names.

➤ Social studies is an amalgam of disciplines, including history, economics, sociology, civics, anthropology, and—geography.

➤ Two of the most common ways to learn about geography are the introductory approach and the world regional approach.

➤ Geography is important in your life in ways you may not guess. Knowledge of geography opens your life to new frontiers.

Answers to the Dread Place-Name Quiz

1. C   2. B   3. C   4. D   5. A   6. B   7. D   8. A   9. C   10. D.

Chapter 2

Physical Geography: A Look at Mother Earth

In This Chapter

➤ How the sun is responsible for day and night and the seasons

➤ How geography affects the climate

➤ Whether the earth is really on solid ground

➤ The earth's moving surface

➤ How plate movement and volcanoes build up the earth's surface

➤ How water and ice wear down the earth's surface

Physical geography is the study of the earth. Understanding the earth has always been a primary focus for geographers. In the study of physical geography, geographers often step on other scientists' toes. Geologists, hydrologists, meteorologists, biologists, and virtually every "ologist" you can name has some interest in the subject of geography. Because geographers are concerned with the earth primarily as it affects the human experience, they spend less time on isolated details; they're the "big picture" people when it comes to earth study.

This chapter addresses the geographical big picture (think broad strokes on an extremely large canvas). Although entire college courses are dedicated to the subject of physical geography, even a college course is the short version of physical geography: Many people spend their entire lives trying to understand the subtleties of the earth and its systems. After a lifetime of effort, these people would report that they had only begun to scratch the surface. Complex? Yes. Daunting? Yes. Intimidating? No.

It All Revolves Around the Sun

The source for all energy on earth is the sun. Without it, the earth would have no day and night and no seasons. *No* life would exist, in fact. Throughout human history, as people have observed the sun's path through the sky, apocalyptic fears have swept across nations during solar eclipses, and societies have even worshipped the sun as a god. Just how does the sun influence people's lives now, though?

Spin Control: What Causes Day and Night

Most people are aware that the sun is responsible for day and night and the changing seasons. When it comes to explaining the nuts and bolts of how it all happens, though, our knowledge begins to fade.

As shown in the figure, a flashlight aimed at a tennis ball sums up the relationship between the sun and the earth. When the light (the sun) shines on one side of the tennis ball (the earth), the other side of the ball is dark. Therefore, half the earth is always illuminated, and half the earth is always in the dark (although a look at some politicians might lead you to believe that more than half of us are in the dark).

How the sun causes day and night.

I know what you're thinking: "Thank goodness I live on the illuminated side!" Everyone has a little of the light and a little of the dark, of course. The reason that one side isn't always in the light or always in the dark is that the earth rotates; it always spins one full rotation every 24 hours. The result is that about half the time you're in the light (daytime) and half the time you're in the dark (nighttime).

Naturally, this process is a little more complicated than my simple description. You know that during the summer the days are longer and the nights are shorter and that during the winter the days are shorter and the nights are longer. You may even know that at certain times of the year, near the North Pole or the South Pole, the sun never rises nor sets. I talk more about this subject later in this book, when I describe Northern Europe and the North and South poles. For now, however, remember that day and night are caused by both the sun's illumination of half the globe and the spinning of the earth.

Why the Seasons Change: Talking About a Revolution

The seasons also depend on the earth's relationship to the sun. In this case, the spinning of the earth is not the important factor; rather, it's that the earth *revolves* around the sun.

Although Copernicus had a tough time selling the idea to the Pope, the earth does orbit the sun. Because the earth's axis is inclined, or tilted (23½ degrees, to be exact), sometimes the Northern Hemisphere tilts toward the sun (in June), and sometimes the Southern Hemisphere does (in December).

As the figure shows, the hemisphere tilted toward the sun experiences summer, and the hemisphere tilted away from the sun experiences winter. Although residents of the Northern Hemisphere assume that Christmas and New Year's are synonymous with winter, half the world would disagree. While New Englanders are bundling up their kids for a sleigh ride to grandmother's house for a Christmas ham, Australians are throwing a shrimp on the barbie for a Christmas dinner on the beach.

GeoJargon
Rotation is the spinning of the earth on its axis; a complete rotation occurs once every 24 hours. *Revolution* describes the earth's orbit around the sun; the earth completes a full revolution once every year (365 days, 5 hours, 49 minutes, and 12 seconds)—except in leap years, but that's another matter.

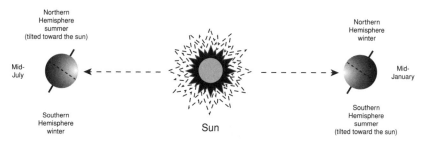

Northern Hemisphere summer (tilted toward the sun)

Mid-July

Southern Hemisphere winter

Sun

Northern Hemisphere winter

Mid-January

Southern Hemisphere summer (tilted toward the sun)

How the earth's tilt causes the seasons.

Considering Climate

Although people usually use the terms *weather* and *climate* interchangeably, they mean different things. Weather refers to the temporal, or the here and now. When people speak about the weather (which many people do frequently), they're talking about the temperature, wind, and moisture of a specific place at a specific time. Weather is localized, it changes, and (as your local meteorologist often has to admit), it can be unpredictable.

Climate refers more to the long term, or to average weather conditions over an extended period for large regions of the earth. The world can be divided into generalized climatic zones that correspond roughly to the distance from the equator. The map shows where climates typically are located:

➤ **Tropical, rainy:** Near the equator

➤ **Dry, desert:** In bands just north and south of the equator

➤ **Humid:** Farther to the north and south than the dry, desert areas

➤ **Polar:** Toward each pole

World climatic zones.

Cold/Polar

Humid

Dry

Tropical/Rainy

Why Is It Hot at the Equator and Cold at the Poles?

Even during the middle of winter at the equator, it's hot. Accordingly, even during the balmiest summer day at one of the poles, it's cold. The primary reason is that the earth is hottest at the point where the sun's rays strike it most directly. At the equator, the sun is always close to being directly overhead. At the North and South poles, the sun never is directly overhead. During the winter months, the sun does not even appear. Even at the height of a polar summer, the sun remains close to the horizon.

The air temperature you feel when you step outside is caused only indirectly by the sun's rays. The process of heating the air is called *reradiation.* As the sun's rays penetrate the earth's atmosphere and strike the earth's surface, the surface absorbs the light energy from the sun and reradiates it as heat energy that heats the air.

Some surfaces are more effective than others at absorbing and radiating heat. The earth's vast oceans effectively absorb heat energy, as do dark surfaces, such as asphalt, soil, and tree leaves. Smooth and light-colored surfaces, such as ice and snow, are less effective in absorbing energy from the sun.

Because of this process, you probably don't want to sell air conditioners in Antarctica. There, the sun's rays are never direct, the surface does a poor job of absorbing what little energy it receives, it's always cold, and you always crave hot cocoa. Conversely, at the equator, the sun's rays are always relatively direct, the surface does a good job of absorbing the sun's energy, it's always hot, and you always need sunscreen.

Why Does the Wind Blow?

To understand this question, you have to understand air pressure. As you stroll along the beach, you feel some pressure. Although your walk may seem relaxing, you support a column of the earth's atmosphere that surrounds the earth and extends upward about six miles, as shown in the figure.

Terra-Trivia
The average high temperature in July in Singapore—located almost directly on the equator—is 88 degrees F. Even in January, it's 86 degrees F! The average high temperature in July in Barrow, Alaska—located well within the Arctic Circle—is 45 degrees F. In January, its average high temperature is –7 degrees F.

Terra-Trivia
The earth's lowest temperature was recorded at Vostok Station, Antarctica. On July 21, 1983, the temperature dropped to a remarkable –129 degrees F. (If you spit while you're there, it will freeze before it hits the ground!)

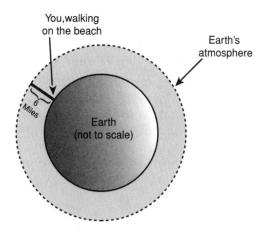

You, walking on the beach

Earth's atmosphere

6 Miles

Earth (not to scale)

The weight of the atmosphere is about 14.7 pounds on every square inch of your body. If this pressure is wearing on you, head for the mountains. The higher the elevation, the smaller the column of air you're supporting and the lower the pressure. (The downside is that you have to follow those special baking instructions on the cake-mix box.)

Temperature also affects air pressure. As you've witnessed if you've ever watched a hot-air balloon drift gracefully overhead, the warmer the air, the lighter it becomes (and the higher the balloon rises). Warm air means lighter air, which means lower air pressure; conversely, cold air means heavy air, which means higher air pressure.

Nature follows predictable patterns. Water seeks its own level, mountains rise and valleys fall, and lawyers eventually move to the D.C. beltway. So it is with air pressure (the equalizing part, not the move to the beltway). The air in high-pressure areas moves toward areas of low pressure, resulting in wind. The greater the difference between the high pressure and low pressure, the faster the wind blows. Extreme differences in air pressure result in tornadoes and hurricanes.

They Call the Wind Many Things

Many people use the terms cyclones, hurricanes, monsoons, tornadoes, twisters, and typhoons interchangeably to refer to any blast of windy, wet weather. Although some terms have similar meanings, they're not all synonymous and can refer to different phenomena.

Twisters, tornadoes, and *whirlwinds* all refer to isolated points of extreme low pressure (warm, moist air) surrounded by rapidly spinning columns of wind. Although these roaring, funnel-shaped vortexes are not generally more than a couple hundred yards wide, they leave terrible destruction in their path when they touch down on the ground surface (often at a wind speed of as much as 300 m.p.h.). The central plains states are the most tornado-prone.

Waterspouts and *dust devils* are miniature and often short-lived versions of tornadoes. Waterspouts, as their name implies, are misty columns of spray over bodies of water. Dust devils are similar to waterspouts, except that they occur over land. Dust devils are often so small that they would be unnoticeable if not for the tiny column of sand and dust they swirl into the air. (Don't confuse dust devils with dust bunnies, the reclusive lint puffs found under beds and in the corners of rooms with hardwood floors.)

Hurricanes and *typhoons* are severe tropical storms with sustained winds of 74 m.p.h. or higher. These massive, circular- or oval-shaped storms can measure more than 300 miles across. A peculiar characteristic of these storms is their *eye,* an area of calm (often complete with blue sky) directly in the center of the storm. As with other types of storms, hurricanes and typhoons are associated with extreme low pressure. Hurricanes are storms that occur in North or Central America; typhoons, in the western portions of the Pacific Ocean. Heavy rain, destruction, and coastal flooding are all elements associated with these huge storms.

Cyclones include all the storms, of varying intensities, in this list (except for monsoons). Even low-pressure centers that do not develop into one of these types of severe storms are considered cyclones. Interestingly, the circulation of winds associated with a cyclone varies its direction depending on the hemisphere in which it occurs. In the Northern Hemisphere, cyclonic winds spin counterclockwise. In the Southern Hemisphere, they circulate in a clockwise direction.

Monsoons: Monsoons differ from the rest of the terms in this list in that they're not a particular type of storm. Instead, they refer to a seasonal shifting of winds. Generally associated with southern Asia and depending on the season, monsoons may be "wet" or "dry." During a wet monsoon, large amounts of precipitation fall, and the weather can remain rainy for months.

Terra-Trivia
On November 13, 1970, a tropical cyclone moved north through the Bay of Bengal and struck the low-lying delta that comprises the country of Bangladesh. The high winds, flooding, and resulting exposure and disease killed 300,000 people.

The Earth's Moving Surface

When geographers become concerned not only with the sky above their heads but also with the ground beneath their feet, they encroach on the territory of geologists. Geographers don't delve deeply into their discipline, however, because they're interested primarily in the uppermost layer of the earth's crust, where humans walk and live.

Solid Ground—or Not?

The earth's crust is good old terra firma—except that it's not so firma. What people like to think of as their anchor, as the bedrock on which they build, is really the equivalent of many giant rafts moving and bumping around on an orb floating in space. So much for stability.

If you were to assemble a jigsaw puzzle in which the pieces represent the seven major continents, you might notice something that's crucial to explaining the movement of the earth's surface. If you jockey the pieces around a little, you would see that the shapes of the continents, especially South America and Africa, all fit together in one huge, interlocking piece—and for good reason: 225 million years ago, all the continents were once part of a huge supercontinent called Pangaea.

The theory of *plate tectonics* suggests that this supercontinent eventually broke up into a dozen or so major crustal plates, as shown on the following map. These rigid plates "float" around on the earth's surface, in a process called *continental drift*.

Summarizing Subduction

In some places, such as a ridge on the floor of the Atlantic Ocean, continental plates are forced apart by upheavals of molten rock from beneath the surface. When the plates are forced apart and then together again, the area of the collision is called a *subduction zone.*

In a subduction zone, one plate is typically forced down below another. The impact of these land masses colliding causes a folding of the plates that results in the upheaval of large mountain chains. When the plates collide and crack (fault) in these subduction zones, earthquakes and volcanoes are created. An example of a subduction zone is the west coast of North America, with its great folded mountains of the coastal range, frequent earthquakes, and volcanic activity, such as the 1980 eruption of Mount Saint Helens.

Subduction zones exist across the earth, the most significant of which is around the perimeter of the Pacific Ocean. This vast ring at the edge of the Pacific Plate is the *Pacific Ring of Fire,* as shown in the figure. Aptly named, this ring is marked by frequent earthquake activity and more than 75 percent of the world's volcanic eruptions.

Tectonic plates and the Ring of Fire.

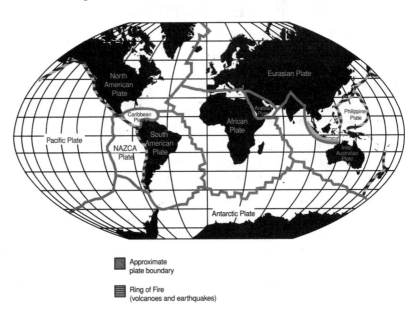

North American Plate

Eurasian Plate

Caribbean Plate

Arabian Plate

Philippine Plate

Pacific Plate

African Plate

NAZCA Plate

South American Plate

Indo Australian Plate

Antarctic Plate

■ Approximate plate boundary

▤ Ring of Fire (volcanoes and earthquakes)

The Ring of Fire, as shown on the preceding map, stretches from New Zealand north through Indonesia, Japan, and the Pacific coast of Russia. Continuing across the north Pacific, the ring crosses Alaska's Aleutian Islands and continues south along the west coast of North and South America. Some of the earth's largest population clusters are also located along this treacherous ring.

The Great Shake: Earthquakes and ✳ Faults

If you were to draw a map with dots placed on the world's earthquake *epicenters* (focal points or starting points) and then connected the dots, you would have almost an exact map of the earth's tectonic plates. The vast majority of the world's earthquakes occur in and around subduction zones. When the massive force of two or more plates pressing against one another eventually gives—crunch!—an earthquake occurs.

Terra-Trivia
When one plate dips below another in a subduction zone, huge trenches are formed, such as around the periphery of the Pacific Ring of Fire, which has some of the world's deepest ocean trenches. The Mariana Trench in the western Pacific is the deepest spot on earth. There, in an abyss called Challenger Deep, the earth's surface plunges to 35,826 feet below sea level.

The movement generated by earthquakes causes *faults,* or cracks, in the earth's surface. Faults can result from either up-and-down movement or side-to-side movement, as is the case with California's famous San Andreas Fault.

The severity of quakes is usually measured on the *Richter scale.* On this logarithmic scale, every whole-number increase is equal to ten times the earthquake's magnitude, or intensity. A seven on the Richter scale, for example, represents an earthquake ten times as powerful as a six. Property damage and loss of life are a result of earthquake intensity and the degree to which the area around the earthquake epicenter is populated. By the time an earthquake hits seven or eight on the Richter scale, the damage is extensive. A nine on the scale has never been recorded—an earthquake of that magnitude would result in virtually total destruction.

Tsunami or Tidal Wave?

Earthquakes that occur underseas can generate *tsunamis* (destructive ocean waves), a term often incorrectly used interchangeably with *tidal waves.* Tidal waves have nothing to do with ocean tides.

Out at sea, a tsunami is not the huge wall of water you may remember from *The Poseidon Adventure.* A tsunami is barely noticeable, in fact, on the ocean's surface. When these fast-moving waves reach the shallow water of the shoreline, however, they build up to enormous heights and can cause extensive coastal damage. Seismic monitoring stations strung around the Pacific Ocean help warn vulnerable areas, such as Hawaii, of potential tsunamis.

Geographically Speaking

Earthquakes in populated areas have historically caused a catastrophic loss of life. A 16th century quake in Shaanxi, China, was estimated to have caused 830,000 deaths. The famous San Francisco quake of 1906 killed "only" 700 people, a modest number by world standards. In recent history, a 1976 Chinese quake registered 8.2 on the Richter scale and resulted in nearly 250,000 deaths.

In addition to intensity and population density, time of day and local construction techniques are factors in an earthquake's death toll. A nighttime quake in an area of masonry construction can have terrible consequences. The recent quake in Kobe, Japan, demonstrated that even stringent building codes are no insurance against destruction and loss of life in a powerful quake.

Building Up the Earth's Surface

The movement of the earth's plates is one mechanism for the land-building forces of the earth. As the continental plates press against one another, the earth's crust folds, warps, and faults. This surface-building activity, called *diastrophism,* is to the earth's surface as a fender bender is to the bumper of your car.

Terra-Trivia
The world's tallest mountains are also its fastest-growing. The Himalayas are folded mountains, formed by the bending of the earth's crust, located in the subduction zone between the Eurasian Plate and the Indo-Australian Plate. (This zone falls along the border between India and China.) As the plates press on each other, the Himalayas continue to grow, at a rate of about a half-inch per year.

Another primary land-building force is *volcanism,* which occurs most commonly along the fractured edges of continental plates. *Magma,* or molten rock, is forced up through cracks in the earth's crust to emerge at the earth's surface as *lava.* The build-up of lava and ash forms the familiar volcanic cones that soar majestically on the landscape. Some of the world's most famous mountains (Mount Fuji, Mount Kilimanjaro, Mount Rainier, and Mount Etna, for example) are volcanoes.

Wearing Down the Earth's Surface

Nature's penchant for equilibrium is both minutely complex and beautifully simple. Just as nature's forces build up, they inevitably also tear down. These *gradational forces,* which eventually combine to level even the highest mountains, are evident in many forms. Huge rocks are reduced to small stones by prying plant roots and the freeze–thaw action. Acids and oxidation (rust) chemically break down the earth's minerals. The wind sculpts sand dunes, and windborne abrasives rasp away at exposed rock surfaces. Although all these actions are agents of erosion (wearing away), water is by far the most potent.

Follow the Flowing Water

As a gentle rain falls to the earth's surface, the water runs off into countless tiny creeks that in turn feed into streams and eventually into rivers. Ultimately, the rush of water makes its way to the sea. At every stage of this cycle, water is eroding, wearing, and carving away at the earth's surface.

Sometimes the action of water on the landscape is dramatic, as with raging floods that shift the course of rivers. Severe down-cutting into the earth's crust, as with the Colorado River's Grand Canyon, can leave a spectacular tribute to the power of moving water. Rocky coastal cliffs slowly yield to the ceaseless pounding of the ocean's waves.

Water not only cuts away at the earth's surface but also moves around great quantities of its material. Streams typically originate in highlands and mountain areas. In this steep terrain, streams drop quickly and are characterized by rapids and waterfalls. The quickly moving water is capable of moving stones, sand, and silt (the *load* of the stream) as it descends through sharply cut gorges and ravines.

When fast-moving streams join to form a river in the lower flatlands, the flow begins to slow. The slower-moving water then spreads out to meander through large, level *floodplains,* or wide valleys formed by rivers. No longer capable of moving large stones, it carries primarily sand and silt in its path to the sea.

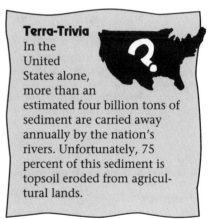

Terra-Trivia
In the United States alone, more than an estimated four billion tons of sediment are carried away annually by the nation's rivers. Unfortunately, 75 percent of this sediment is topsoil eroded from agricultural lands.

As the widening river finally reaches the ocean, its flow has slowed to a crawl. Tremendous amounts of sand and silt are deposited at the mouth of the stream, often forming deltas. The deltas grow outward from the mouth of the stream, stretching their fingers into the sea. Deltas are often characterized by rich alluvial (nutrient-rich) soils renewed by the continual deposit of silt carried by the river from the land's interior.

The Crushing Effects of Ice

Water can also carve away effectively at the earth's crust in another form: ice. Many landforms in the Northern Hemisphere are the result of *glaciation:* During the ice ages, the last of which occurred between 8,000 and 15,000 years ago, huge sheets of ice spread southward from the North Pole. These massive 10,000-foot-thick glaciers scoured the land surface and left glacial lakes and deposits in their wake.

The remnants of these *continental glaciers* are still found today on the island of Greenland and on Antarctica. Smaller glaciers, called alpine glaciers, are also found in high mountain areas. *Alpine glaciers* advance and retreat seasonally, grinding the mountain rock into a fine mineral dust called *glacial till* that causes the spectacular aqua blue hues of many high mountain lakes.

Landforms created by glaciers have many names. In fact, the names constitute their own geological language. Here are a few of the better-known terms:

Alpine glaciation:

➤ Aretes: Sharp mountain ridges

➤ Cirques: Mountain valleys

➤ Cols: Mountain passes

➤ Horns: Three- or four-sided pointed peaks

➤ Tarns: Mountain lakes

Continental glaciation:

➤ Drumlins: Rounded (egg-shaped) hills

➤ Eskers: Long serpentine ridges

➤ Kettles: Lakes

➤ Moraines: Hills of glacial debris

The Least You Need to Know

➤ The earth's rotation on its axis once every 24 hours causes day and night.

➤ The earth's revolution around the sun once a year causes the changing seasons.

➤ The earth is hottest at the equator because that's where the sun's rays strike the earth most directly. It is coldest at the earth's poles because these areas receive the most indirect light from the sun.

➤ Wind blows from areas of high atmospheric pressure to areas of low atmospheric pressure.

➤ The earth's crust is composed of slowly moving plates that grind and bump against one another.

➤ The earth's surface is built up primarily by folding, warping, and volcanism; the surface is worn down primarily by running water and ice.

Chapter 3

Cultural Geography: The Hand of Humankind

In This Chapter

➤ Pick a language—any language!

➤ Religious persuasions

➤ The distinction between race, ancestry, ethnicity, culture, and nationality

➤ The political landscape

➤ The business of business

Cultural geography, or perhaps more accurately, human geography, is a fairly self-descriptive term. This chapter focuses on the people of the earth. If that statement sounds somewhat general, just compare this chapter to a winery tour: You get at least a taste of the wide selection offered.

The human experience incorporates language, belief systems, racial groups, political systems, customs, livelihoods, and interaction with the environment. Throw in a dash of history, and you're looking at just a few facets of this broad enterprise. This chapter briefly discusses political, economic, and urban geography as it relates to culture; these areas are covered in more detail in Part 2 and Part 3 of this book. First, you must understand a little of the terminology and some classifications that cultural geographers use.

Language: It's All Greek to Me!

I have spent many evenings eating and laughing with a wonderful family who immigrated to the United States from Sicily. The family was blessed with five girls (all adults now), ranging from infancy to middle-school age when they arrived in the United States. Although they all speak English now (the parents still have some difficulty), none of them spoke one word of English when they arrived. Interestingly, the younger sisters speak English without a hint of their native Sicilian dialect, although the elder sisters still have a distinct Italian lilt.

All the sisters have their own children now, and their children speak only a few words of Italian. As a language is lost through the generations, so too is a strong tie to the original culture. (I can only hope that the Sicilian recipes are being passed down.) Language can unify people, and it can divide.

Parlez-Vous—More than 3,000 Languages?

If you ever have been in a situation in which everyone is speaking a language you don't understand, you know the feeling of isolation it creates. If you travel frequently, you may eventually be in that situation, forced to adopt the "point and grunt" system of communication. This awkward situation is bound to occur because the world is home to more than 3,000 different languages (and some experts place the total closer to 4,000). Regardless of the true number, if you speak only one language, you have a way to go to reach linguist status.

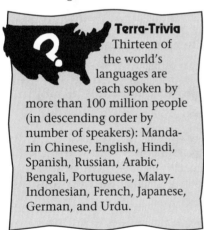

Terra-Trivia
Thirteen of the world's languages are each spoken by more than 100 million people (in descending order by number of speakers): Mandarin Chinese, English, Hindi, Spanish, Russian, Arabic, Bengali, Portuguese, Malay-Indonesian, French, Japanese, German, and Urdu.

Thousands of languages are spoken in the world, grouped by linguists into a few *language families* consisting of languages believed to have had a common historic origin. The largest of these families is Indo-European, which includes languages spoken by about half the world's people.

Within language families are *language branches* (or *subfamilies*). The Indo-European family, for example, includes the Germanic branch and the Romance branch. Although languages within each branch might still be mutually unintelligible, they share common origins.

Branches are additionally broken down into *language groups*. They share origins from more recent history and share much in both vocabulary and grammar. Within language groups are individual languages, such as English. It's a language in the West Germanic group of the Germanic branch of the Indo-European language family. Complicated? It doesn't even stop here. An example of how English and German are related is shown in the similarity of words such as *white* and *witte*—they have similar roots. Conversely, *white* and the French *blanc* indicate different roots.

It's the Same but Different: Dialects, Pidgins, Creoles, and Lingua Francas

You might think, after whittling this language jumble to a single language, that you would be pretty much home free. Not really. Although a resident of Chicago and a resident of London might both speak English, they might have a hard time carrying on a conversation because they speak different dialects. *Dialects* reflect regional differences in vocabulary, pronunciation, and syntax that sometimes hinder fluid communication. The dialectic differences between a lobsterman in Maine and a catfish farmer in Mississippi, for example, might prove to be conversationally humbling.

A *pidgin* is a blend of different languages. Not the primary language of any of its speakers, a pidgin evolves so that speakers of different languages can converse. If, over time, the primary languages become lost and the pidgin language becomes the primary language, it's called a *creole*.

Another means by which people of different tongues converse is in a *lingua franca,* a language used by peoples whose primary languages are unintelligible to each other. As an example, ambassadors from Iran and Germany might converse in the language of diplomacy—which is now English. For ages, Roman Catholic priests could converse in Latin, no matter how different their principal languages.

Languages, Officially

An official language is one that has been designated by law as the tongue of the country. This type of decree might specify that the language (or languages) be used in schools, courts, businesses, and government. Mastery of the official language might even be a requirement for immigration into the country.

Some countries have not designated an official language, and others have several official languages. Areas within countries may promote languages different from the official language of the country.

Terra-Trivia
The United States has no official language. Although the majority of the population is English-speaking, other languages (Spanish is the most common) are spoken primarily in localized pockets in the United States. At the other end of the official-language scale, South Africa has 11 designated official languages.

A Look at the World's Religions: The Biggies

Because about six billion people live on the earth and because religion is so intensely personal, you probably could make the argument that six billion religions exist on the earth. As with other geographic topics discussed in this book, some lumping is in order here. This book defines a *religion* as any value system that incorporates worship and faith in a divine creative force.

GeoJargon
Geographers classify religions into two primary types: A *universalizing religion* is open to all human beings and attempts to spread its faith through missionary activities. (The three largest are Buddhism, Christianity, and Islam.) An *ethnic religion* generally incorporates specific groups of people in a particular location on earth. (Hinduism is the largest.) Missionary activity is not a factor because this type of religion tends to be site-specific.

Although a description of all the world's religions wouldn't fit in this chapter, nor would the intricacies of the ones briefly described, the following list does include the most influential religions with the greatest number of adherents. You should have a basic understanding of these religions because of their huge impact on the world's geography. (Although this list is incomplete, it does cover the religions practiced by more than 75 percent of the world's population.)

Christianity (1.9 billion adherents): This universalizing religion springs from Judaic roots. Centered on the life and teachings of Jesus, who lived in the first century A.D., Christianity's principal religious text is the Bible. Although this religion originated in the Middle East, today it's widely dispersed throughout much of the world.

Primary divisions: Roman Catholics, Protestants, Orthodox

Place of worship: Church

Islam (1 billion adherents): Also a universalizing religion, Islam descends from the Judaic–Christian faiths and is centered on the seventh century A.D. life and teachings of the prophet Mohammed. Its principal religious writings are in the Koran. Islam is now most prevalent in the Middle East, North Africa, and Southeast Asia.

Primary divisions: Sunni Muslims, Shiah Muslims

Place of worship: Mosque

Hinduism (750 million adherents): Perhaps the world's oldest religion, the origins of Hinduism date back more than 5,000 years. Linked closely with the people and culture of India, this major religion is the world's largest ethnic religion. Not tied to any one prophet, it incorporates many doctrines and deities.

Place of worship: Temple, shrine

Buddhism (335 million adherents): The universalizing religion of Buddhism had its genesis as a Hindu reform movement in the sixth century B.C. The reformer was Sidhartha Gautama, the Buddha (meaning the "Enlightened One"). Although Buddhism arose in India, today it's focused primarily in Southeast Asia, East Asia, and Japan.

Primary divisions: Mahayana Buddhists (including Zen), Theravada Buddhists, Tantrists

Place of worship: Stupa, temple or pagoda, monastery

Chinese religions (145 million adherents): The three major Chinese religions (Three Teachings of Mahayana Buddhism, Confucianism, and Taoism) are not always distinct. *Confucianism* is more of an ethical code than a religion; the founder of the movement was the sixth century B.C. philosopher Confucius (K'ung Fu-tzu). *Taoism* also arose in China in the sixth century B.C. Its teacher, Lao Tsu, based his philosophy on a simple life in harmony with the way of nature. The Chinese religions are concentrated primarily in China, Korea, and Southeast Asia.

Place of worship: Temple, shrine, stupa

Judaism (18 million adherents): Rooted in the teachings of Abraham and the Torah of Moses, Judaism is thought to be between 3,000 and 4,000 years old. Although it's not one of the world's largest religions, it has global importance politically and as the wellspring of both Christianity and Islam. Judaism is concentrated primarily in Israel and the United States and is in urban areas throughout the western world.

Place of worship: Synagogue, temple

Terra-Trivia
Although this chapter has identified the world's major religions, another major group is the *nonreligious*. People of the world who don't adhere to any specific religion, including atheists (those denying the existence of God), total more than 1.1 billion people.

Race, Ancestry, Ethnicity, Culture, or Nationality— What Does It All Mean?

Cultural diversity, ethnic cleansing, and racial tension are in the news on almost a daily basis. Although these terms are often used somewhat interchangeably, they have specific meanings to geographers. This section sorts through them and provides a few geographic examples.

The Diversity of Race

Race usually denotes a part of the population that is distinct based on inherited biological characteristics. These differing characteristics typically represent a physical response to the environment evolving over long periods. Physical differences can include hair texture, height, facial characteristics, skin pigmentation, and even variations in blood.

To try to draw exact lines distinguishing between racial groups is absurd. The distinctions may be extremely subtle, and variations within a racial group can be extremely wide. Nonetheless, for purposes of classification and study, geographers tend to generalize groups of human races. Although each geographer has a slight variation of this list, these nine large racial groups are often identified:

➤ African

➤ Asian

➤ Australian (aboriginal, not *"Crocodile" Dundee*)

➤ Caucasian

➤ Indian (from South Asia, not American Indian)

➤ Indigenous American

➤ Melanesian

➤ Polynesian

➤ Micronesian (not always included in this list)

} island groups of Pacific

> ## ! Geographically Speaking
>
> The Lakota people (Western Sioux), as is typical of many Native American groups, associate colors with the four cardinal directions and with the races of humankind. The use of these symbolic colors dates into distant Lakota history, long before contact with other racial groups. Interestingly, the colors red, yellow, white, and black were used by the Lakota to describe different people even before they had experiential knowledge of other races.
>
> In addition to the colors of the races of humankind, the colors have many symbolic meanings for the Dakota. Red is associated with the north, the bald eagle, and health; yellow represents east, the golden eagle, and wisdom; white symbolizes south, the white crane, and destiny; and black signifies west, the black eagle, and renewal.

Ancestry: Climbing the Family Tree

Ancestry is a cloudy term that simply refers to one's line of descent, or lineage. More and more, people attempt to piece together "family trees" and spend a great deal of time probing back through the generations. The term is cloudy from a geographical standpoint because it can encompass just about anything to different people, including race, nationality, social status, and economic status. In census tables, ancestry typically refers to a connection to a particular country. In the U.S. census, a leading response was not a foreign country, but rather simply "American."

The Nuances of Nationality

Although nationality is often confused with issues involving inherited traits, it is, strictly speaking, independent. A nation can be thought of as a unified group of people sharing customs and ethnicity. Nationality can also simply refer to citizenship in a nation or country. Considering this definition, you may realize that it's possible to change your nationality. After an immigrant is naturalized in a new country, he or she can become officially one of its citizens.

Coping with Changing Culture

Culture is a broad term that incorporates a wide array of social attributes. Technology, art, symbols, music, belief systems, and laws are all aspects of culture. As these elements are passed down from generation to generation, a culture becomes richer and more complex.

An interesting example of the genesis of culture is Los Angeles, California. It has emerged as the capital of United States pop culture. Trends, fashions, and fads emanate from Los Angeles (Hollywood and the film industry serve as a mouthpiece) and gradually diffuse to the rest of the country and the world.

Terra-Trivia
In 1993, the United States admitted about 900,000 immigrants. The countries representing the top five sources of U.S. immigration were (from largest to smallest) Mexico, China, Philippines, Vietnam, and the former Soviet Union.

An Eye on Ethnicity

Ethnicity includes culture but can be based on other criteria. It can incorporate racial characteristics, language (or dialects), religion, and nationality. Any of these factors can identify an ethnic group as separate from a larger population.

An example of the misunderstanding of ethnicity arises with the misuse of the term *Hispanic.* It's often listed in population tables as a race, although people of Hispanic origin can be of any race, and it does not refer to any one ancestry or nationality. Hispanics may be from Mexico, Puerto Rico, Cuba, Central or South America, or a number of other countries. Perhaps the most unifying element connected with the term is use of the Spanish language, although even Spanish use within this group is subject to extreme dialectic differences.

Terms of Settlement: The Political Landscape

Geographers often use the term *political* to describe the human-made lines and dots on a map (as opposed to *physical* maps that show mountains, rivers, and other features). Countries, states, territories, counties, and cities often have prescribed outlines that appear on a map but do not appear on the physical landscape. This section skips over political systems (governments and laws) because they're addressed in Part 3; the chapter focuses instead on a variety of human constructs and sorts through a confusing array of terms and misconceptions.

Not Bound by Boundaries

Boundary lines surround us. If you build a fence that crosses the invisible boundary surrounding your yard, your neighbor might take offense (or take off fence, as the case may be). The situation wasn't always like this, though. In some cultures (many Native American groups, for example), lines on maps were never used to demarcate one group's land from another. "No man's lands" were often established in which both parties could

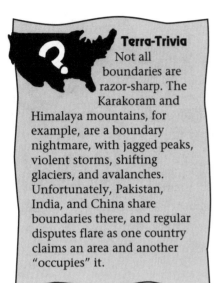

Terra-Trivia
Not all boundaries are razor-sharp. The Karakoram and Himalaya mountains, for example, are a boundary nightmare, with jagged peaks, violent storms, shifting glaciers, and avalanches. Unfortunately, Pakistan, India, and China share boundaries there, and regular disputes flare as one country claims an area and another "occupies" it.

jointly use the middle ground without fear of conflict. With today's laser surveying equipment, those fuzzy boundaries are now often razor-sharp.

Geographers classify boundaries according to these types of distinctions:

➤ **Natural versus artificial boundaries:** Boundaries based on landscape features (such as mountains, lakes, and rivers) are called *natural,* or *physical, boundaries.* The following map shows how easily they can be recognized (even on a map that doesn't show physical features) because they're rarely straight lines. The boundary between Texas and Mexico, for example, follows the Rio Grande and is therefore a physical boundary. If you see straight lines dividing places on a map, you're probably looking at *geometric,* or *artificial, boundaries,* which are usually surveyed lines often based on a parallel or meridian (a line of latitude or longitude). The straight boundary between Arizona and New Mexico is an example of an artificial boundary.

Natural versus geometric boundaries

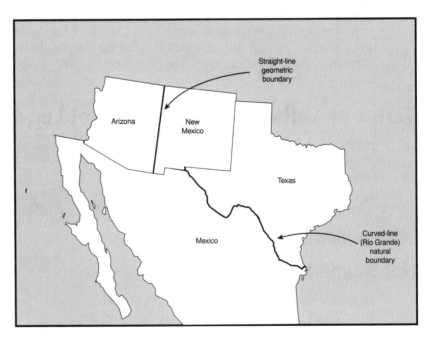

➤ **Antecedent versus subsequent boundaries:** Boundary lines drawn before an area has been well populated are *antecedent boundaries.* The western border between the United States and Canada is an antecedent border, drawn before "the West was won." (Native Americans were not factored into the governments' decision.) Borders drawn after an area has developed cultural characteristics are called *subsequent borders,* such as new boundaries drawn after the break-up of the former Yugoslavia.

➤ **Consequent versus superimposed subsequent boundaries:** Boundaries drawn after an area has well-established cultural characteristics can either respond to those characteristics or ignore them. *Consequent,* or *ethnographic, borders* are drawn in an attempt to respond to cultural dictates, as with the former Soviet republics, which have now become independent countries. *Superimposed boundaries* are borders drawn rather arrogantly that "do their own thing" regardless of any preexisting cultural patterns. Superimposed boundaries are placed by colonizing European powers, for example, typically divided indigenous peoples. Africa is laced with these arrogant boundaries. Tribal areas were simply cut in half as European countries divvied up the land of Africa by imposing boundary lines on a map.

Terra-Trivia
Geographers call the forces that tend to pull a country apart *centrifugal forces.* Differences of religion, language, or ethnicity all may act as centrifugal forces that cause a country to fracture. Forces that help to bind a country together are called *centripetal forces.* The sharing of religion, language, and ethnicity, in addition to forces such as nationalism and transportation systems, can act as centripetal forces.

Countries, Nations, and States

The distinction between countries, nations, and states can be somewhat confusing. First, throw out the use of the word *state* as it applies to political units that form a federal government (such as New York state or the state of California). Now you have to deal with the word *state* only as it's used to describe a country or nation (that's difficult enough):

➤ **State:** An area that's a political unit with an established population and government controlling its own internal and foreign affairs

➤ **Country:** Much the same as state

➤ **Nation:** A closely linked group of people that occupies a territorial area

A stateless nation can exist (the Palestinians of the Middle East or the Cree of Canada, for example), and a distinct nation of people can be the primary occupants of a state (Italy, France, and Japan, for example). Many examples exist of multinational states in which several nations of people occupy one state (the Flemish and Walloons in Belgium, for example). And, after reading this paragraph, it's entirely possible to be in a state of confusion.

Sorting Through Settlements

The people of the earth are becoming more urbanized. As global economies shift away from an agricultural base, more and more people are flocking to urban areas. Many of these areas are growing at an alarming rate (Mexico City grows by more than one-half million people per year). In this chapter, *urban* refers to a centralized area regardless of size.

The following terms describe urban areas, in order from smallest to largest:

➤ **Hamlet:** Tiny; maybe just one or two dozen buildings (this one sounds quaint!)

➤ **Village:** Larger than a hamlet but not as large as a town

➤ **Town:** Bigger than a village and generally centered on a nucleated business area

➤ **City:** Same concept as a town, only larger and more complex

➤ **Conurbation:** A large, built-up area in which two or more cities have coalesced into one huge unit

➤ **Megalopolis:** The huge, built-up area in the United States extending from Virginia to New Hampshire, where several conurbations have coalesced

Getting Down to Business

As you surely already know, vast economic differences mark the countries of the world. What may be surprising however, is just how vast these differences are. One gauge of economic status is per capita *gross domestic product* (or GDP), which is how much money a country produces per person living in the country. Countries such as Mozambique (in Africa) have a per capita GDP of less than $120. At the other end of the scale are countries such as Switzerland, with a per capita GDP of more than $22,000.

Why do such huge disparities exist? The answer, as with most in geography, is often complex. History, resources, education, and economic systems all factor into the mix. Three primary types of economies now exist in the world. Although the economies of most countries are blends of two or more types, only three basic types exist.

A *subsistence economy* is based in rural areas and around family units. In this survival economy (existing in the world's developing nations), people are barely able to produce enough to sustain themselves, to subsist. The exchange of currency and goods is minimal, and people are always just one natural disaster away from famine. Typical of this economic type is *slash-and-burn* agriculture, which involves cutting and burning forested areas, nomadic herding, and intensive small-plot agriculture.

In a *commercial economy*, the most prevalent in the world, people produce goods and services for sale. It is controlled (often operating today on a global scale) by competition and the law of supply and demand. It's the dominant type of economy in developed countries and also operates to lesser degrees in developing countries.

A *planned economy* is associated with communist and socialist governments. It involves central government mandates regarding areas and amounts of production. As the number of communist governments worldwide has drastically declined in recent years, so has this type of notoriously inefficient economy.

The complex links between economic status and humans' relationship with their environment is discussed in Part 4, "A Global Overview."

GeoJargon

Slash-and-burn agriculture involves the cutting and burning of forested areas as a preparation for planting crops. Because the soil in many tropical forests is poor, soil nutrients decline after just a few years of planting. Typically, the farming family simply moves to another area of forest, and the cycle is repeated. This type of agriculture is also called *shifting cultivation,* or *swidden.*

The Least You Need to Know

➤ Although thousands of languages are spoken on the earth, they are classified into just a few language families.

➤ Although thousands of religions exist on earth, Christianity, Islam, Hinduism, Buddhism, Chinese religions, and Judaism are practiced by more than 75 percent of the world's population.

➤ Race, ancestry, nationality, culture, and ethnicity are all closely related words that have subtle differences many people misunderstand.

➤ *State* is virtually synonymous with *country* because both refer to a political unit. Nation is more of a reference to a people than to a specific area.

➤ Urban areas range in size from a tiny hamlet to a sprawling megalopolis.

➤ The three major types of economies are subsistence, commercial, and planned.

Chapter 4

Maps: A Geographic Love Affair

In This Chapter

➤ Identify different types of maps

➤ Know why a map can never be perfect

➤ Let symbols on the legend help you read a map

➤ Locate a map position

➤ Measure distance on a map

➤ Read the ups and downs of the earth's surface

➤ Develop a sense of direction

Ask a geographer, "What is geography?" and you might get the response, "Geography is anything that can be mapped." Maps are the basic tools of geography. Even people who don't know much about geography are drawn to the beauty and usefulness of a map. This chapter provides all the basics you need in order to own and operate your map.

Maps contain enormous amounts of information because of their graphical nature (a picture really is worth a thousand words). This book uses maps throughout to help you locate places around the world. The maps give you an idea of the relative size of places and the distances between them, and you can determine directions from the maps. The map skills described in this chapter will prove to be useful tools in your quest for geographic mastery.

Millions of Maps

Maps exist for virtually every type of information under (and including) the sun. Maps show the ocean bottom, temperatures in India, and the best route to get to a baseball game. Some examples of maps are wall maps, route maps, and globes. Maps are in the newspaper, in atlases, and on your computer screen.

One way you can identify maps is by the type of information they contain. This list shows a few common map types:

➤ **Weather:** On the TV news or in the newspaper, for example

➤ **Road:** Impossible to refold in its original form while you're in the car

➤ **Political:** Newspaper favorites typically filled with donkeys and elephant icons

➤ **Coastal chart:** Familiar primarily to old salts and weekend sailors

➤ **Star chart:** Charts the heavens (not to be confused with map of the stars)

➤ **Map of the stars:** Points out the homes of Hollywood luminaries

Pondering the Projection Dilemma

A map represents the ins and outs of the earth's surface on a flat piece of paper. If you've ever given a friend directions by scrawling a map on the back of an envelope, you know the basic process. Map makers (cartographers) attempt to do the same thing for the entire earth. Although this process might not seem difficult, there's a fly in the proverbial cartographic ointment.

When you peel an orange and attempt to lay the peel flat on a table, you see a glimmer of the problem cartographers face: The earth is not flat (if only Columbus's critics had been right). The spherical earth—an oblate spheroid, to be exact—poses a dilemma to cartographers: No matter how hard a mapmaker tries, the rounded surface can't transform 100 percent accurately into a flat map.

Cartographers call their attempts to create flat maps from the spherical earth map *projections*. Creating a perfect projection is impossible. If you're a doubting Thomas, try to draw every side of a baseball accurately on a piece of paper. You might end up with a pile of sketches that look vaguely like torn-off baseball covers, a pile of eraser shards, and a bad headache.

GeoJargon
The term *map projection* comes from the old practice of placing a light source inside a transparent globe and projecting the image on to a piece of paper.

The impossibility of a perfect projection.

The only truly accurate representation of the earth is a globe because it preserves all the map characteristics of interest to a cartographer: area, shape, direction, and distance. A map projection can preserve only one or two of these characteristics. If you focus on getting an area's size exact, for example, its shape may be distorted, and vice versa. Cartographers have names for projections that preserve each of these characteristics:

➤ **Equal area:** Area

➤ **Conformal:** Shape

➤ **Azimuthal:** Direction

➤ **Equidistant:** Distance

Sorting Through the Symbols

Because maps represent a reduced part of the earth's surface on a small piece of paper, certain symbols (points, lines, and patterns on a map) are often necessary to represent landscape features. To help map users recognize symbols, cartographers often use color:

➤ **Blue:** Water features (rivers, lakes, and ponds, for example)

➤ **White, purple, yellow:** Height and depth (mountains and valleys, for example)

➤ **Green:** Vegetation features (woods, orchards, and scrub brush, for example)

➤ **Yellow, tan, brown:** Roads and urban development (important roads and built-up urban areas, for example)

Smart cartographers consider who will use a map when they decide which symbols to use. If you design a map to direct wedding guests from a church to a reception, you should include symbols for landmarks, traffic lights, and street names. Most guests are unlikely to show up if you hand out a map with symbols representing industrial zoning, storm sewers, and tectonic plates.

When a map has many symbols or its symbols are tricky, it usually includes a *legend* (sometimes referred to as a *map key*) for explanation, as shown in the following simple map legend.

A map legend.

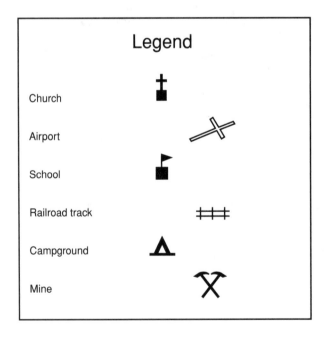

Now, Where Was I?

After you have your map in hand, you can locate a position on the map in two basic ways: relative location and absolute location.

Relative Location

You generally use relative location when you're giving someone directions. To direct a person to the local convenience store, you might say something like this: "Go to the airport, and continue through three lights; the Trendy Mart is on your right." These directions give the position of the Trendy Mart *relative* to the airport.

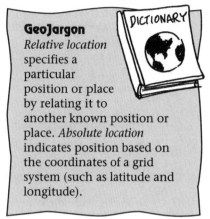

GeoJargon
Relative location specifies a particular position or place by relating it to another known position or place. *Absolute location* indicates position based on the coordinates of a grid system (such as latitude and longitude).

Absolute Location

The absolute location method specifically identifies the position of a place by means of a grid system. Although absolute location is a useful way to locate something in an atlas, it's not often used in everyday speech. The person who wants to find Trendy Mart might give you a strange expression if you say, "Proceed on a 232-degree azimuth to coordinates 41.45 degrees north latitude and 72.40 degrees west longitude."

The old geographic standbys of latitude and longitude, often used to determine absolute location, are difficult to work with because they're somewhat complicated to understand. Many atlas publishers have given up hope that readers will ever master that system and no longer use latitude and longitude to locate positions on their maps. The National Geographic Society's *Atlas of the World,* for example, uses letters down the sides and numbers across the top and bottom of each map to help readers locate places. To locate Ulan Bator, Mongolia, for example, you would first look in the *index* of the atlas and see that it's on page 83, grid square D14. (The grid square system is similar to playing Bingo, but without the potential cash rewards.)

Despite the difficulties of working with latitude and longitude, no discussion of absolute location would be complete without an explanation of the basics of these elements.

Learning to Live with Latitude and Longitude

Latitude and longitude are stated in degrees (a complete circle has 360 degrees), with latitude first and then longitude. Latitude indicates a location's north or south position on a map or globe, and longitude indicates its east or west position. Using both latitude and longitude, you can find any location on the earth.

Lines of latitude, or *parallels,* run east–west; lines of longitude, or *meridians,* run north–south. If you follow a meridian, you eventually run into the North or South Pole. If you follow a parallel, you head east or west and travel parallel to the equator.

This list shows the basics of working with latitude:

➤ Latitude measures position north or south of the equator.

➤ The equator is 0 degrees latitude.

➤ The North Pole is 90 degrees north latitude.

➤ The South Pole is 90 degrees south latitude.

➤ Latitude ranges from 0 degrees to 90 degrees north or south.

➤ Always state direction (north or south) when you're specifying latitude.

And here are the basics of working with longitude:

➤ Longitude measures east or west of the *prime meridian,* a north–south line that runs through Greenwich, England.

➤ The prime meridian is 0 degrees longitude.

➤ Longitude ranges from 0 degrees to 180 degrees east or west.

➤ Always state direction (east or west) when you're specifying longitude.

Terra-Trivia
The earliest known system of latitude and longitude dates back to Hipparchus in the year 150 B.C.

Lines of latitude and longitude.

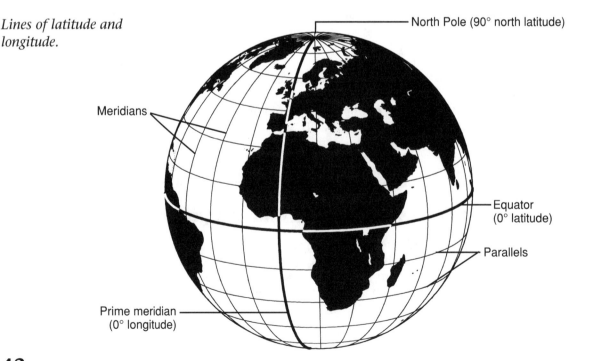

North Pole (90° north latitude)

Meridians

Equator (0° latitude)

Parallels

Prime meridian (0° longitude)

Sizing Up Scale and Distance

A map is generally created according to some scale because a measured distance between two features on a map is, obviously, smaller than the measured distance between the same two features on the earth's surface. You can think of a map as a reduction of the surface it represents.

If you have ever drawn up a plan for remodeling your kitchen, you know how scale works. Suppose that you draw your plan on ¼-inch grid paper and decide that each ¼-inch grid square on the paper equals a one-foot square in your kitchen. When you plot a refrigerator on your plan, you draw it as three grid squares across by three grid squares deep (or ¾-inch X ¾-inch). This *scale* drawing of the refrigerator in your kitchen measures three feet across by three feet deep.

The scale used on a map is generally much smaller than the scale used on an architectural plan because the geographic distances are so much larger. In contrast to a typical house-plan scale, in which ¼-inch equals 1 foot, a typical map scale inch equals one mile. (Because 63,360 inches are in a mile, the scale could be written as $\frac{1}{63,360}$). One inch of map, therefore, is equal to one mile of the earth's surface.

> **Terra-Trivia**
> The circumference of a globe in a typical classroom is only 48 inches; the equatorial circumference of the earth is 24,902 miles.

What Are Those Numbers at the Bottom of the Map?

On a map with a scale written as 1:24,000 or $\frac{1}{24,000}$, the numbers are called *representative fractions*. The first number in the ratio (the numerator in the fraction) represents a map distance; the second number (the denominator in the fraction) represents a ground distance. As long as you use the same unit for both the map and the ground distance, you can plug in any unit: 1 inch on the map equals 24,000 inches on the ground, 1 centimeter on the map equals 24,000 centimeters on the ground, and one pencil length on the map equals 24,000 pencils on the ground.

Bellying Up to the Bar Scale

All those numbers in the preceding section can lead, of course, to a discussion of fractions, ratios, and proportions—the kind of stuff many people like to avoid! If you're one of those people, you may still be in luck. On some maps, a *bar scale* is printed in the margin.

To use a bar scale (sometimes called a graphic scale) on a map to calculate the distance between two features on the earth's surface:

1. Locate two features on the map.
2. Lay a piece of paper along the two features.

3. Use a pencil to place a tick mark on the paper at each feature.

4. Move the paper to the bar scale, and compare the distance between the tick marks to the bar scale.

5. Notice that the distance on the scale represents the distance between the features on the ground.

6. Be sure to note the measurement used in the bar scale (feet, miles, or kilometers, for example).

Using a bar scale.

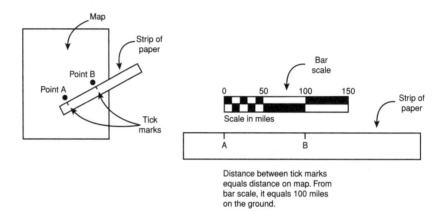

Remember that a map is a *reduced* picture of the surface it represents. You can use the numerical scale (representative fraction) or the bar scale on a map to determine actual ground distances, which is especially important if you're using a map to plan a backpacking adventure, canoe trip, or bike route. Be sure to measure along the route you intend to follow—don't just measure the straight-line distance between your starting and ending points.

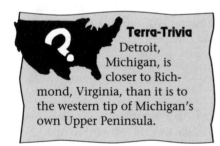

Terra-Trivia
Detroit, Michigan, is closer to Richmond, Virginia, than it is to the western tip of Michigan's own Upper Peninsula.

If you're estimating a drive you intend to make in your car, you have some simpler options. Many road maps specify the mileage along the major roads that connect towns and road junctions. Road maps and road atlases also often provide mileage charts that enable you to determine quickly the driving distances between large cities.

Distancing Yourself from Distance

I realized while traveling in Scotland how lackadaisical about distance Americans have become. When I told a Scottish gentleman that I live in Hartford, Connecticut, he asked how far it is from New York. I responded, "It's about two hours away."

The man seemed quite amused as he said, "Typical American response—I asked you for a distance, and you responded with a driving time." He was right, of course. Although I'm a geographer, I didn't know the distance from Hartford to New York without calculating it backward from my driving time. Most people are much more aware of the *time* it takes to drive between two places than of the distance between them.

The Rise and Fall of Elevation and Relief

This chapter has already examined the problems mapmakers face when they try to map the spherical earth on a flat piece of paper. The dilemma has another component because, as everyone knows, the earth's surface is far from smooth. Soaring mountain peaks reach more than five miles above the surface of the sea, and plunging ocean abysses dip almost seven miles deep. The earth is spherical and its surface irregular—a cartographer's nightmare.

Two terms are often confused in describing the ups and downs of the earth's surface. *Elevation* refers to the height (or depth) of the earth's surface above or below a baseline (usually sea level). *Relief* refers to the variation in elevation from one place to another. An area of extreme relief is an area with dramatic fluctuations in elevation.

Cartographers indicate elevation and relief on maps in many ways. Usually, shading and hatching (small parallel lines) show major mountain ranges or river valleys, colored tinting shows different elevational zones, and elevations printed at the tops of hills and mountains indicate their height.

Perhaps the most accurate—and difficult to understand—method of indicating elevation and relief is with *contour lines*. These lines of equal elevation (everywhere the same line touches is at the same elevation) trace the "contours" of the earth as they thread their way across a map. To understand how a contour line works, picture a map of a mountain lake. If you trace on your map the point at which the water meets the shore around the lake, you trace a line of equal elevation (because water always serves as a level surface). If you trace the shoreline again after the lake floods and the water level rises ten feet, the contour line of equal elevation is ten feet higher than the original line.

> **Terra-Trivia**
> The highest point on earth is the top of Mount Everest, located between China and Nepal in the Himalaya Mountains. Mount Everest's elevation is 29,028 feet (plus or minus ten feet for snow) above sea level—maybe. Although that height was calculated in 1954, many people accept the 1850 height of 29,002 feet. In 1987, satellite measurements set the height at 29,864 feet (go figure).

If you continue this process, you eventually create a contour map of the area around the mountain lake. A *topographic map* uses contour lines to show elevation. Hikers, developers, planners, and civil engineers use them because they show surface relief with a high degree of accuracy. The United States Geological Survey produces topographic maps in a variety of scales for most parts of the United States. You can obtain them from many public and private agencies or order them directly from the U.S.G.S. (call 1-800-USA-MAPS).

A topographic map.

Section of 1:24000 scale, New Britain Quadrangle, Connecticut-Hartford Co., 7.5 Minute Series (topographic) United States Geological Survey, Reston, VA 22092, in cooperation with Connecticut Highway Department.

Setting Off in the Right Direction

Some people are blessed with an innate sense of direction. They can quickly ascertain the lay of the land, find their way back to a friend's house they visited three years ago, and orient their tent so that they can watch the sun rise through the front flap. These people bother me.

I'm one of those people who spends a great deal of time backing out of dead-end alleys and searching tree trunks for moss to find north on camping trips. Although you can study the nuances of direction and learn to use it on a map, a good sense of direction is innate. Nothing is more humiliating than to be a geographer with a poor sense of direction!

The Cardinal Directions

Everyone has worked with direction from time to time. Perhaps you shop on North Main Street or have spent a romantic vacation evening watching the sun slowly set in the west. The terms *north, east, south,* and *west* conjure up visions of an elaborate compass rose (as shown in the following figure) surrounded by writhing serpents on sixteenth century mariners' charts. These *cardinal* directions are the most common form of direction.

Compass rose.

North refers to the direction you follow to get to the North Pole. No matter where you are on earth, if you set off toward the North Pole, you are heading due north. If you're directionally savvy, you may have already guessed that to head toward the South Pole is to travel due south, or directly opposite of due north. At a right angle to this north–south line lie east and west. On the face of a clock (nondigital, of course), if north is at 12 o'clock, then east is at 3 o'clock, south is at 6 o'clock, and west is at 9 o'clock.

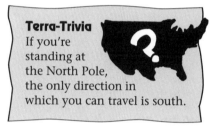

Terra-Trivia
If you're standing at the North Pole, the only direction in which you can travel is south.

Northeast is located directly between north and east, at 1:30 on the clock face (you can figure out southeast, southwest, and northwest). As shown in the compass rose, directions can be further subdivided, such as north northeast and east northeast. Although these subdivisions can go on ad nauseum, you rarely have to be that specific in working with basic directions.

Magnetic Attraction

In addition to looking at lines on a map, gazing at the stars, or observing the sun rise or set, people rely on another indicator of north: the magnetic compass. Perhaps you have mounted on your car's dashboard a compass orb that spins wildly as you career around city streets. Or maybe you remember that ten-mile hike you made with the scout troop, relying on only your trusty compass (and the knowledge that your scout leader was trailing 50 yards behind). In either case, you rely on the magnetic needle in your compass to point north.

This method has just one minor problem: In most places on earth, the magnetic compass needle points to a different north than the geographic North Pole. A magnetic compass points to the earth's *magnetic* north pole, as shown in the figure. The North Pole through which the earth's rotational axis passes (where meridians converge and where Santa Claus lives) is a different place from the magnetic north pole.

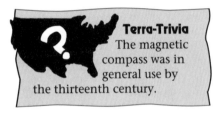

Terra-Trivia
The magnetic compass was in general use by the thirteenth century.

For many uses, the difference between the true north of the North Pole and magnetic north (called *magnetic declination*) is not significant. If you were to set off on a long hike, embark on a transoceanic voyage, or orient the photovoltaic cells on your solar home, however, you would have to consider this deviation. You can make the correction by using the *declination diagram* provided on most detailed (topographic) maps or with declination charts that show the magnetic variance over time. (These charts can be important because, in addition to not coinciding with true north, magnetic north moves around slightly over time.)

Magnetic north.

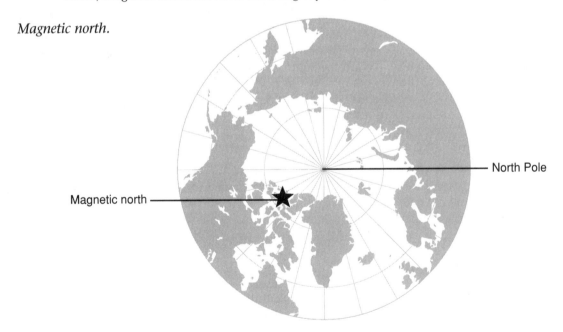

North Pole

Magnetic north

Technical Directional

Before finishing the subject of direction, you should be familiar with another term. *Azimuth* is a more specific way to state direction than to use the points of a compass rose. Again, visualize a clock face.

An azimuth, as shown in the figure, measures direction clockwise from a north baseline (12 o'clock on the clock face). Using this technique, the clock face contains 360 degrees starting at 12 o'clock and moving as the clock hands do, clockwise around the dial. Three o'clock, or due east, is a 90-degree azimuth; six o'clock, or due south, is a 180-degree azimuth.

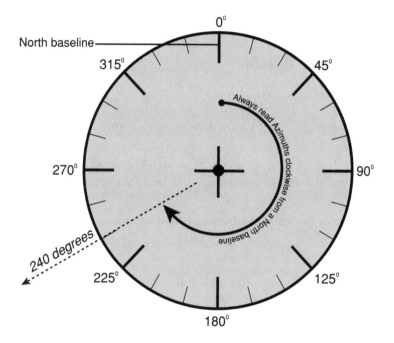

Azimuth.

New Directions in Mapping

In the second century A.D., the Roman geographer Ptolemy's atlas was the cutting edge of technology. Since that time, the field of geography has continued to evolve, and today technology breaks new ground at a blistering pace. You can access maps and atlases from a computer with the click of a mouse and transform data into maps with the press of a key.

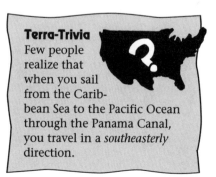

Terra-Trivia
Few people realize that when you sail from the Caribbean Sea to the Pacific Ocean through the Panama Canal, you travel in a *southeasterly* direction.

As you drive a car, your position is indicated on a small map screen on the dashboard. The Jetsons? The year 2020? No, this technology, the *global positioning system* (GPS), is available today. Using this system (look for it at your local car dealership), a small computer can track your position through coordination with satellite signals. U.S. soldiers used GPS to navigate the featureless sands of the Arabian Desert during Operation Desert Storm.

Another development getting lots of press is *geographic information systems* (GIS), in which map information can be recorded digitally in computer files. This technology has been a boon for environmental scientists, city and town planners, and marketing consultants—the list continues to grow. Geographers can use this digital information to determine the impact of local development on wetlands, track delivery of an Acme widget, or choose a prime location for the next Trendy Mart.

The Least You Need to Know

➤ No map can represent the earth's surface without distorting either size, shape, distance, or direction.

➤ Symbols and a symbol key (a legend) are used on maps to represent the landscape's features.

➤ Relative location is useful for locating one position with respect to another. Absolute location is used to describe a position by means of a grid (latitude and longitude, for example).

➤ Maps are drawn to scale in a reduction of the surface they represent.

➤ Elevation refers to a feature's height above or below sea level. Relief is the variation in elevation.

➤ The cardinal directions north, south, east, and west correspond to the directions of meridians and parallels.

Here We Go! A World of Many Regions

You're about to go on a geographic journey around the globe, to places you know well and places that may seem strange and exotic. Prepare to witness a world filled with diversity: mountains, deserts, great rivers, and teeming deltas. As you encounter indigenous people who have adapted to the land in which they live and people who rarely step from a paved surface, how do you make sense of all the variety?

The key is to divide the world into parts called regions or subregions (and also known as areas, territories, realms, ranges, fields, and domains). Although the terms are not important, you should understand each region's similarities and differences. This chapter identifies the regions and discusses why they qualify as regions.

The Hemispheres: Half a World Away

If you choose the equator as the earth's dividing line (midway between the North Pole and South Pole), you can effectively divide the earth into two halves. A hemisphere is simply half a sphere. The northern half of the earth is called the Northern Hemisphere; the southern half, the Southern Hemisphere.

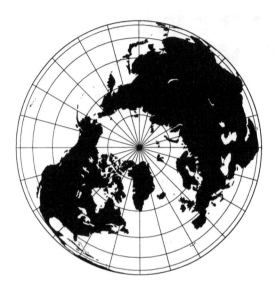

Northern Hemisphere   Southern Hemisphere

Another way to divide the earth is into an eastern half and a western half. Unlike with the equator, this more arbitrary division has no clear-cut dividing line. Nonetheless, the Atlantic Ocean and the Pacific Ocean are the de facto dividing lines for the Eastern and Western hemispheres.

The Americas and lands bordering the Atlantic Ocean generally comprise the Western Hemisphere; most everything else comprises the Eastern Hemisphere. If you're in the Eastern Hemisphere, you can get to the Western Hemisphere by traveling east or west. If you're in the Western Hemisphere, you can reach the Eastern Hemisphere by traveling east or west. (Just remember that this division is somewhat arbitrary.)

Western Hemisphere

Eastern Hemisphere

The Seven Continents—or Maybe It's Five

The process of distinguishing all the continents is even more arbitrary than for the hemispheres. No matter which definition you choose for "continent," you wind up with problems. A continent is generally considered to be a huge landmass surrounded entirely by water, a definition that works well for only about three of the seven continents.

! Geographically Speaking

Don't confuse the seven continents with the Seven Natural Wonders of the World. Traditional lore records these natural wonders as *Mount Everest,* the world's highest mountain, in the Himalaya chain between China and Nepal; *Victoria Falls,* one of the world's largest waterfalls, in eastern Africa; the *Grand Canyon,* the Colorado River's spectacular mile-deep chasm in Arizona; the *Great Barrier Reef,* the huge ocean reef skirting Australia's east coast; the *northern lights* (or aurora borealis), the colorful curtains of light in the northern skies; *Paricutin,* a beautiful volcano in west–central Mexico that is perhaps the least known of the natural wonders; and the *Rio de Janeiro harbor,* Brazil's gorgeous natural harbor surrounded by dramatic mountains.

People traditionally speak of the earth's seven continents (as shown on the following map): North America, South America, Europe, Asia, Africa, Australia, and Antarctica. Some experts argue, however, that North and South America comprise only one continent (because they're connected by the Middle American land bridge) and that Europe and Asia comprise one continent because they're not separated by an intervening body of water. If you accept these arguments, the world has only five continents.

The seven continents.

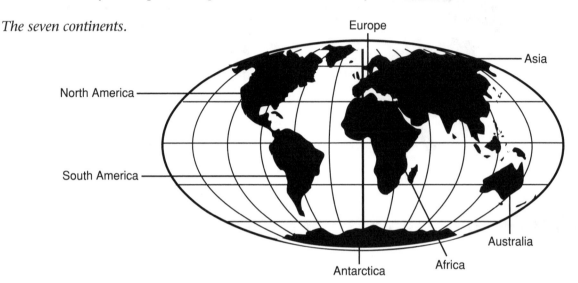

The Eurasia Distinction

Because the continental distinction between Europe and Asia is somewhat lame, you can throw that phrase "entirely surrounded by water" out the window. Most of the boundary between Europe and Asia is not water but dry land—the boundary runs through mountain chains, in fact. The following map shows the traditional dividing line between Europe and Asia (as you can see, it's a tortured distinction).

The line between Europe and Asia uses the physical features of the Ural Mountains, the Caspian Sea, the Caucasus Mountains, the Black Sea, and the Bosporus and Dardanelles straits. Because the distinction between Europe and Asia is difficult to justify using any definition of the word *continent*, many geographers simply refer to the continent of Eurasia.

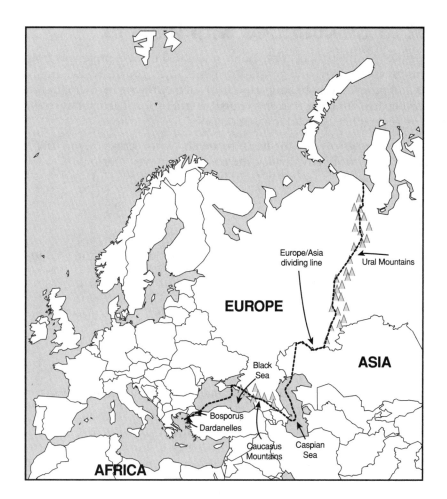

The Europe–Asia boundary.

Australia: Island or Continent?

Earlier, this chapter defined a continent as a huge landmass surrounded by water. Although islands are also surrounded by water, most are not huge landmasses. (Unfortunately, "huge" in this context has no clear distinction.)

Greenland is relatively huge (more than 840,000 square miles), but not quite huge enough for continent status. To be considered a continent, an area must be at least as huge as Australia (nearly three million square miles). If it's any smaller than Australia, it's just an island.

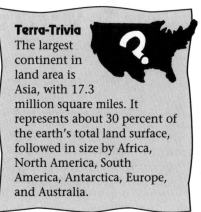

Terra-Trivia
The largest continent in land area is Asia, with 17.3 million square miles. It represents about 30 percent of the earth's total land surface, followed in size by Africa, North America, South America, Antarctica, Europe, and Australia.

Continents—Not to Be Confused with Regions

The business department of a university near my home once tried to have professors from three departments (business, sociology, and geography) teach a new course by dividing the earth into regions and progressively relating aspects of each of the three disciplines to each region. (I told the department chair that this course, world regional geography, was already being taught by geographers around the world.)

Unfortunately, the business professor used the terms *region* and *continent* synonymously in his segment of the class. He incorrectly divided the course into seven "regional" sections, each actually a continent. As I stood in front of 45 students who had been taught incorrect terminology, I was expected to describe the geography of Asia in one 50-minute period and cover all these regions:

➤ **The Middle East:** One of the most volatile and strategic areas on earth

➤ **The Soviet Union** *(it was the Soviet Union back then)*: The world's largest and second most powerful country

➤ **South Asia:** Home of half the world's population

➤ **China:** The country with the largest population and one of the world's fastest-growing economies

➤ **Southeast Asia:** With a rapidly growing economy, the site of protracted U.S. involvement over the past decades

➤ **Pacific Rim states:** Powerful and rapidly growing players on the world business scene

➤ **Japan:** One of the most powerful economic forces on earth

The lack of basic geographic understanding made the course a terrible failure. Continents are too big and too diverse to study as one unit, and they're not necessarily regions.

Okay, So What's a Region?

Regions are hard to define exactly—five geographers might identify the world's regions in five different ways. When you're identifying a region, you look for some degree of sameness. What physical, cultural, economic, or historical attributes make one area distinct from another? Geographic proximity is a factor, as political boundaries often are. Depending on the weight you assign to any particular factor, regions can and do vary.

Even if everyone could agree on a set number of world regions, the number eventually would change. People move, ideas diffuse, economies change, and even climate and vegetation can shift. What's considered a region today might not be so clearly a region 25 or 50 years from now.

Although regions may coincide with an entire continent, that's not always the case. Regions usually encompass much less area than continents do. Remember that the definition of a continent, as you read earlier in this chapter, has nothing to do with culture, history, economics, climate, or vegetation—all elements that factor strongly in the determination of a region.

The Idiot's Approach

In this book, I've used the homogeneous factors mentioned in the preceding section to identify the geographic regions of the earth. In addition, I've drawn on the experience of many noted geographers in making these selections. No other book outlines these regions in exactly the same way as this one does. In fact, rarely do any two geography books agree precisely. Nonetheless, the regions outlined in this book roughly parallel regional descriptions you might see elsewhere.

For the sake of clarity, this book does not divide countries between regions. Country lines are used as regional limits, though many countries are not quite culturally and physically diverse. Often, two countries abutting (touching) one another are in different regions.

Lines on a map also serve as regional divisions in this book. Remember that these lines, in addition to their political significance, generally represent subtle geographic transitions (variations) in the earth's surface.

As with continents, the larger the area you attempt to describe, the more general your description must be. Because it's almost impossible, for example, to make any universal geographic assertions about an area as large and diverse as Asia, this book divides the earth into regions small enough, and similar enough, that you can thoughtfully consider their geography.

Terra-Trivia
Don't confuse the division of continents with the *Continental Divide* (or *Great Divide*). It runs north–south through North America, roughly along the Rocky Mountain chain. Rivers to the east of the divide ultimately drain into the Atlantic Ocean; rivers to the west of the divide drain into the Pacific.

Terra-Trivia
Even Africa does not fit perfectly this book's definition of a continent. In the northeast, Africa is separated from Asia by the Gulf of Aden and the Red Sea. The Red Sea tapers into the Gulf of Aqaba and the Gulf of Suez. Only the thin, man-made Suez Canal connecting the Gulf of Suez with the Mediterranean Sea severs Africa from continental Asia.

On the following map, the earth is divided into these 20 world regions:

A. North America

B. The British Isles

C. Western Europe

D. Northern Europe

E. Southern Europe

F. Eastern Europe

G. Russia

H. Japan

I. Australia and New Zealand

J. Pacific Rim

K. Middle America

L. South America

M. Saharan Africa

N. Sub-Saharan Africa

O. The Middle East

P. South Asia

Q. Southeast Asia

R. East Asia

S. The Pacific

T. The Poles

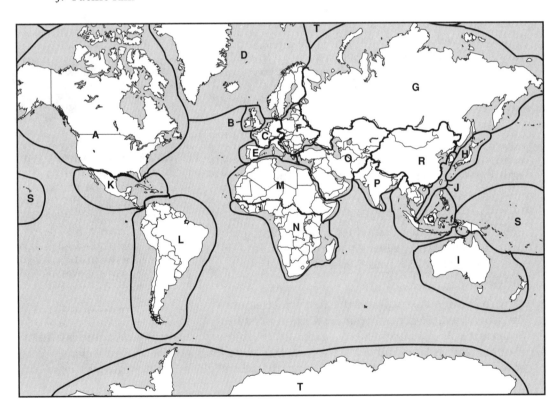

The 20 regions of the world.

The size of the regions varies greatly: Some are as small as the British Isles, and some are as large as Russia. The populations in different regions also fluctuate wildly: East Asia and South Asia have more than a billion people between them; at the North and South poles, only a few scientists are hunkered down in Quonset huts.

Similarly, the physical attributes of each region are quite diverse, from the tropical rain forests of Southeast Asia to the icy mountains and fjords of Northern Europe to the arid stretches of desert in the region of Saharan Africa. By discussing each region in detail, this book gradually paints a portrait of the entire earth.

Highlighting the Essentials

Parts 2 and 3 of this book describe the essential geography of each of the world's regions, often with a different focus in each region. In some places, the cities are the most important; sometimes, a river defines a region; in other places, such as at the poles, the climate is the most striking geographic feature.

For each region, you get an overview of several key geographic components. Although the focus changes from chapter to chapter, each one mentions the region's historical highlights, countries, and physical geography. The chapters also mention major cities and characteristics of the people and briefly explain why the region might be in the news.

Stepping Back through History?

Although this book is intended to broaden your knowledge of geography, it also dabbles in history. Understanding the geography of a place is often impossible without a glimpse of its history. Most of the New Zealand population, for example, is of European ancestry, although that wasn't the case only a couple hundred years ago. You can't find out what caused this dramatic change in cultural geography without peeking at New Zealand's history.

Although the discussion of regional history is brief and incomplete in this book, it helps flesh out subtleties in a region's geographic composition. History is also loaded with fun little quirks that can add interest and life to the character of each place you read about.

The Countries

Each region is composed of a mosaic of individual countries. Some regions, such as Japan, aren't much of a mosaic because they have only one country. Other regions, such as sub-Saharan Africa, have dozens of countries. This book obviously has to treat the approach to each region's countries in slightly different ways.

If a region described in a chapter has only one country, or just a few, I describe its workings in some detail. When the number of countries is large, space considerations allow little more than a list of the countries and a brief description of each one. Regional maps in each chapter also help you sort out the countries.

Unlike the place-name map quiz traditionally used in schools, the objective in this book is not to make you memorize. Rather, you should get a feel for the geography of each region. Identifying, grouping, and describing countries within a region is one way to help you make connections between them. If you think of sand, the Nile River, Arabs, nomads, and Islam when you think of the region of Saharan Africa, this book has done its job. Don't be upset, for example, if the name of Tunisia slips your mind—you will have the essentials.

Physical Attributes

Terra-Trivia
India has tropical rain forests, glaciers, deserts, and monsoon rains. In addition to high plateaus and two of the world's greatest rivers, it has low-lying swamps and deltas. India's soaring mountains rise more than 28,000 feet above sea level.

Each region described in this book has its own unique physical characteristics. Some regions are fairly homogeneous: The poles are covered with ice and snow, it's cold everywhere in the region, and winters don't have lots of light. Although it's relatively easy to generalize about that particular region, some regions feature an incredible range of physical characteristics.

This book describes such features as mountains, rivers, vegetation, and climate. For some regions, I focus on earthquakes and volcanoes; for others, I focus more on monsoon rains. Again, don't worry about memorization: Think about how the physical characteristics of each place help to shape the region's unique geography.

The Cities

The method for choosing which of the world's cities is important is not cast in stone. One way to decide a city's importance is by considering its population. This means of selection presents several problems: Census data is not entirely reliable (especially in developing countries); data is often wrong and quickly becomes outdated; and different population tables use different ways of estimating city populations.

What extent of the outlying city population (the metropolitan area) is included in an estimate of a city's population? In one table, Tokyo, Japan, might be listed as a city with a population of slightly more than 8 million people. In another table that estimates metropolitan areas (Tokyo's metropolitan area includes the cities of Yokohama and Kawasaki), Tokyo's population is listed at more than 30 million.

The size of a city's population generally should not serve as the sole indicator of the city's global importance. Jerusalem, for example—the center of three of the earth's largest religions—has a long and rich history and is a hotbed of political turmoil and cultural conflict. The city is in the news almost daily, yet it has only slightly more than 500,000 inhabitants and is considered only moderately sized by global standards. If population were the sole criteria for inclusion in this book, Jerusalem would not even be mentioned.

Although you could just stick with national capitals, this plan also has its flaws. Washington, D.C., would get some mention, for example, but New York, Chicago, and Los Angeles wouldn't make the cut. It would also be strange to include Belmopan, Belize, a capital city with a population of less than 25,000 people, and omit São Paulo, Brazil, a noncapital city with a population of more than 17 million people.

Historical importance, economic importance, and current events are all additional barometers to add to the selection criteria for mentioning cities in this book. I've selected cities that are geographic "need to know" locations. They have some significance whether you're a tourist, a diplomat, or just a guidebook adventurer.

Terra-Trivia

Many cities remain obscure until dramatic events make them household words. Warlords and United Nations troops pushed Mogadishu, Somalia, suddenly into the world's spotlight. Kobe, Japan, came under international scrutiny as the epicenter of a destructive earthquake. The Winter Olympics placed Lillehammer, Norway, on the map. (If your own city doesn't eventually get mentioned here, just wait.)

The People

In profiling the people of each region, Parts 2 and 3 describe geographic concepts, such as population density, distribution, and urbanization. They also describe cultural characteristics, such as religion, race, language, and customs. As with other considerations, the extent to which each of these topics is discussed depends on the individual region.

Some regions, such as Japan, have a high degree of cultural uniformity within their populations. In other regions, such as South America, the people form a varied and complicated cultural mosaic. Despite differences among people, you can generally find cultural ties and similarities.

Generalization is most difficult when it comes to the people of a region. Remember that no single group or individual perfectly fits the profile you construct for a region. The important task is not assigning labels to people; rather, it's understanding how different cultures adapt to and modify their environment.

In the News: A Case Study

The end of each chapter in Parts 2 and 3 of this book focuses on a particular issue of importance, typically something that has made some regional or international stir in recent years. This focus also helps to flesh out some geographic concept in much the same way as case studies are used in management seminars.

The sections at the end of each chapter are also interesting because they often involve developing issues that have not yet been resolved. You often can follow these issues in your daily newspaper or on the evening news. Keep an atlas at hand, and, as these stories develop, consider how these issues might affect the geography of the region.

The Least You Need to Know

➤ A hemisphere is simply one half of the earth.

➤ Geographers traditionally have recognized seven continents: North America, South America, Europe, Asia, Africa, Australia, and Antarctica.

➤ Regions are areas of the earth that share a variety of geographic characteristics that make them distinct.

➤ Twenty regions are described in this book: North America, the British Isles, Western Europe, Northern Europe, Southern Europe, Eastern Europe, Russia, Japan, Australia and New Zealand, the Pacific Rim, Middle America, South America, Saharan Africa, sub-Saharan Africa, the Middle East, South Asia, Southeast Asia, East Asia, the Pacific, and the North and South poles.

Part 2
A Regional Look:
The Developed World

It's time to set sail for the unknown—or maybe you're just setting sail for places less familiar than others. This part of the book doesn't represent the really unknown because it examines the developed regions of the world. If you're reading this book, you're probably from one of the world's developed regions, so the places described in Part 2 might be somewhat familiar to you. Because these places are also the world's primary economic players, they're typically not far from the headlines.

The phrase "developed world" implies a degree of bias because "developed" is often in the eye of the beholder. The ten regions grouped into Part 2 have a number of factors that generally fit into the geographic class called "developed." Although this class is defined largely by economic status, technology and demographics also factor into the equation.

The regions in Part 2 are the world's more economically and technologically advanced. Don't assume that you can extend the "advanced" label to anything beyond this phrase: The more I travel and the more time I spend with the people of the world's less developed regions, the more I question the common definition of "developed." If you spend all your time becoming economically and technologically developed, many of the more important human attributes just might go undeveloped.

Chapter 6

North America: Land of Opportunity

In This Chapter

➤ Recognizing that the new world isn't so new

➤ Focusing on physical features

➤ Poring over population

➤ Examining ancestry, religion, and language

➤ Investigating the issues of indigenous people

North America is a place of immense variety, as reflected in its people and the geography of the land. Within this region, you can see both palm trees in southern Florida and polar bears in Canada's Northwest Territories. You can dine on enchiladas rancheros in Albuquerque or pâté de foie gras in Quebec. Brokers in pinstripe suits hail taxis in the canyons of Wall Street, and ranchers drive their pickups down the dusty roads of Eagle Butte.

Perhaps this book will stir wanderlust in your spirit and you will someday travel to distant lands. If so, you will discover many beautiful places and exotic cultures. Yet the most common sentiment expressed by those who have traveled the world is a renewed love and appreciation for America.

If you try to describe the North American landscape and its people, you're doomed to failure: They change by the day and season. The more you experience the world, however, the more you realize that although North America defies description, it is certainly unique.

As a geographer, you must endure a bit of generalization. This chapter helps you step back and look at the big picture to find similarities in a sea of differences. Although it's a big task to accomplish in a small chapter, it's a good place to start.

The New World That Isn't

That North America is often called the New World reflects a cultural bias, considering that it was new only to the Europeans who "discovered" it. It was not so new to the indigenous people who had inhabited it for thousands of years, however. To lay claim to discovering an area when people who already live there walk to the beach to greet you when you arrive is also somewhat ethnocentric. Still, to the old world, the Americas were certainly new.

Although much debate takes place over which discoverer was the first to make safe passage to North America. Christopher Columbus first initiated sustained contact between the New World and the Old.

> ### Geographically Speaking
>
> When the Greek geographer Eratosthenes calculated the circumference of the earth in about 247 B.C., he came remarkably close to the true circumference of 24,900 miles. Later, when the Roman geographer Ptolemy recorded these calculations in his book *Geography,* he unfortunately "corrected" Eratosthenes' figures and incorrectly recorded the earth's circumference as considerably smaller than its true dimension.
>
> Christopher Columbus consulted Ptolemy's work in planning his famous trip. Unaware of the existence of the New World and based on the smaller circumference recorded by Ptolemy, Columbus figured that he could reach Asia by sailing west from Spain. Had he known the true circumference of the earth, it's unlikely that he ever would have attempted such a long ocean voyage and would never have "discovered" the Americas.

Two Countries but Still a Full House

As North America was settled, two nations emerged: the United States and Canada. These countries share many cultural and economic similarities and share the earth's longest international open border (a political border that doesn't inhibit crossing). Both highly industrialized countries are among the world's leaders in the standard of living their people enjoy. Because of these similarities and geographic proximity, the two countries form the region of North America.

Canada: A Look at the Provinces and Territories of the Great White North

The Great White North, as Canada is popularly called, has vast amounts of land (divided into ten provinces and two territories) yet few people. Although the country shares a common border with the United States (not to mention a host of professional sports teams), few people in the United States can name Canada's provinces and territories.

In the spirit of promoting geographic understanding between neighbors, I'll name these divisions (shown on the preceding map). The four small eastern provinces bordering the Atlantic Ocean are the Maritime Provinces: New Brunswick, Prince Edward Island, Nova Scotia, and Newfoundland. Moving west, you can see Canada's two largest provinces (in land area and in population): Quebec and Ontario. Farther west are the provinces known as the Prairie Provinces: Manitoba, Saskatchewan, and Alberta. In the far west, bordering the Pacific Ocean, is the tenth province, British Columbia.

Canada's territories lie to the north of the provinces and are large in land area and short on people. To the far northwest are the Yukon (not to be confused with the UConn that is the University of Connecticut), which shares a long border with Alaska. East of the Yukon are the huge Northwest Territories. In 1999, a third territory called Nunavut (an Inuit word meaning "our land") will be carved out of the Northwest Territories to serve as a homeland for the Inuit.

The United States: Fifty-Plus Possessions

If you received any geographic training during your school years, you likely had to take a place-name quiz covering the states and capital cities of the United States. Because there are so many of them (50) and they're on the map at the beginning of this chapter, I don't list them in this section. (I don't want to stir up dark fears of blank maps and panicked memorization, either. If it's any consolation, college geography students have asked me about the three countries of North America—Canada, the United States, and New Mexico!

! Geographically Speaking

The United States owns or administers many outlying areas that are not among the 50 states. These areas often have congressional representatives who cannot vote except in committees. Although some of the areas are tiny, uninhabited islands, others (such as Puerto Rico, with a population of more than 3.5 million) are larger than some states.

In the Pacific Ocean are American Samoa, Guam, the Commonwealth of the Northern Mariana Islands, the Republic of Pail, and many tiny islands, including Wake and Midway. The Caribbean Sea has the Commonwealth of Puerto Rico, the American Virgin Islands, and the tiny, uninhabited island of Novice.

What Are the "Lower 48"?

Geographers call the United States a *fragmented* country because it's interrupted by another country or by international waters. The United States is interrupted by both.

Alaska is separated from the bulk of the United States by Canada, and Hawaii is separated from the other states by the Pacific Ocean. The remaining 48 states are known as the lower 48. They're also called the 48 *contiguous* states or the 48 *conterminous* states. *Contiguous* means to be in contact with, or touching, each other; *conterminous* refers to a shared boundary.

Purple Mountains Majesty and Fruited Plains

Extensive plains and great mountain chains dominate North America's landscape. North America's mountain chains, shown on the following map, all run north–south, which can present an obstacle to anyone traveling east–west. (Early settlers undoubtedly wished that North America's mountains ran east–west.)

In eastern North America, the dominant mountain chain is the Appalachians. Although these old, eroded mountains no longer reach to great heights, they still stretch from Alabama to the Maritime Provinces in Canada. In the west, the higher Rocky and Pacific Mountains run the length of the United States and Canada.

GeoJargon
Avoid the term "continental United States" because it leads to all sorts of confusion. The continental United States generally refers to the lower 48 states but technically includes Alaska because it's also on the continent of North America.

Along the Atlantic Ocean and the Gulf of Mexico are the coastal plains on which the largest masses of North America's people live. Stretching north from the coastal plains and between the great mountain chains of the east and west are the vast central plains. These plains reach northward past the Great Plains of the Midwest to the icy tundra of the far north. North America's farmlands serve as the world's breadbasket because prolific farmers annually produce huge yields of grain.

The Mighty Miss

If you can remember only one river, make it the Mississippi, by far the largest river in North America. Feeding the Mississippi River are its two largest *tributaries* (feeders), the Missouri and Ohio rivers. The tributaries of the Mississippi reach to the Rocky Mountains in the west, the Appalachian Mountains in the east, and just below the Canadian border in the north. It all drains south to the Mississippi's great delta and the Gulf of Mexico. The Mississippi has nurtured westward expansion, the riverboat era, historic expeditions to the interior (such as Lewis and Clark), and the adventures of Huckleberry Finn.

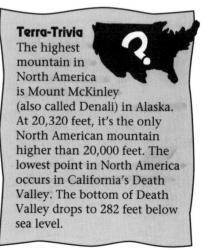

Terra-Trivia
The highest mountain in North America is Mount McKinley (also called Denali) in Alaska. At 20,320 feet, it's the only North American mountain higher than 20,000 feet. The lowest point in North America occurs in California's Death Valley. The bottom of Death Valley drops to 282 feet below sea level.

Physical features of North America.

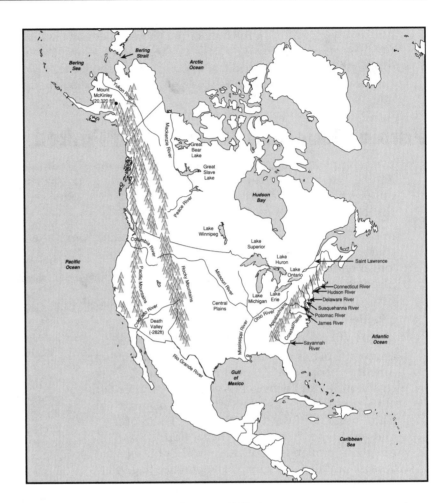

In addition to the Mississippi, noteworthy rivers in North America include the Columbia, Colorado (creator of the Grand Canyon), Mackenzie-Peace, Rio Grande, Saint Lawrence, and Yukon. Many smaller, well-known rivers lace the eastern part of the United States, including the Connecticut, Delaware, Hudson, James, Potomac, Savannah, and Susquehanna. (If you feel overwhelmed by this list of names, just pick two from the preceding map of North America and remember the Mississippi and the river closest to your hometown.)

Water, Water, Everywhere

The North American region is surrounded by oceans, seas, gulfs, and bays. Within the landmass of North America are several of the world's largest lakes. (Look at the preceding map to get yourself hydrographically oriented.) The three oceans in this region are the Pacific to the west, the Atlantic to the east, and the icy Arctic to the north. The Bering Sea is notable because it and the narrow Bering Strait were all that separated the world's two greatest superpowers, the United States and the Soviet Union, during the Cold War. Hudson Bay is memorable because of its vast size and the historical importance of Hudson's Bay Company and the fur trade. Many smaller bays in North America, including the Chesapeake Bay, have great local importance.

This region's primary gulf, the Gulf of Mexico, is formed by the Caribbean Sea and the Atlantic Ocean. The gulf borders five states in the southern United States and, as mentioned, is the exit point for the mighty Mississippi River. The Gulf of Mexico is also important for its oil reserves.

North America also has 8 of the world's 12 largest freshwater lakes. Three of them—Great Bear Lake, Great Slave Lake, and Lake Winnipeg—are located entirely within Canada. The other five lakes all comprise the Great Lakes, which are located along the border of the United States and Canada and drain primarily into the Atlantic Ocean via the Saint Lawrence River.

Terra-Trivia
One way to keep track of the names of the five Great Lakes is to remember the acronym HOMES: Huron, Ontario, Michigan, Erie, and Superior. Lake Superior is the largest freshwater lake on earth. All the lakes straddle the United States–Canada border except Lake Michigan, which is located entirely within the United States.

Weathering the Climate

North America stretches from near the tropics in the south to well within the Arctic Circle in the north, and these latitudes define the region's varied climate. The region is surrounded on three sides by oceans, which restrain its extreme climate in coastal areas. North America's location in the northern hemisphere leads to June–September summers and December–March winters.

The far north of Alaska and Canada have cold polar climates. These areas typically have short, cool summers and long, cold winters (not the ideal destinations for sunbathing). The huge mass of North America experiences humid climates, colder in the north and warmer in the south, with year-round precipitation. Southern Florida has a humid tropical climate, with typically dry winters. The southwestern United States and the south–central plains of Canada have a dry climate, with arid conditions prevailing in the far southwest.

From Tundra to Mangrove

North America's vegetation generally parallels its climatic areas (as happens worldwide). Tundra (a treeless expanse) in the extreme northern reaches of the region gives way to pine forests as you move south. As you move farther south, the pine forests yield to deciduous forests of broadleaf trees that seasonally lose their foliage and in the southeast return to pine forests.

In the drier central areas, tallgrass prairies dominate. The prairies, which once supported vast herds of bison and the Plains Indian culture, are now primarily used for ranching. In the driest areas, desert vegetation with cacti, shrubs, sage, and chaparral growth is the norm.

Another type of vegetation that's specific to North America is found in the Pacific Northwest. There, in old-growth temperate rain forests, are the largest trees on earth. Redwoods, Douglas firs, sitka spruces, and giant sequoias reach more than 300 feet tall in this area. The tallest tree on record is the National Geographic Society Tree, a California redwood that has been measured at 365 feet, 6 inches tall. California also boasts the largest tree on record, the General Sherman Sequoia, with a trunk circumference of 101.6 feet. In California's Inyo National Forest, you can also see the world's oldest living tree, a 4,600-year-old bristlecone pine.

Terra-Trivia
The number of bison (or buffalo) roaming the prairies in the early 1800s numbered more than 40 million. As western expansion reached the great plains, wholesale buffalo slaughter ensued, and by 1890 the number of bison was reduced to less than 200. Today the numbers of buffalo, raised largely on private ranches, have begun to rebound.

Terra-Trivia
The two countries of North America are both among the world's largest in total land area (Russia is largest). Canada is the world's second largest country, with 3,849,674 square miles of land area. The United States is the world's third largest country, with 3,787,425 square miles of land area.

Black Gold

Because fossil fuels are so important to modern economies, you should know about North America's reserves. The United States has larger coal reserves than any other country on earth. Despite coal's importance for energy production and heavy industry, its environmental cost has caused its value to decline in recent years. Despite the large oil and gas deposits in the south central and southwestern United States and in northern Alaska and Alberta, Canada, North America is a major importer of petroleum.

Mother Nature's Dark Side

Although North America is blessed with natural resources, it also has its share of natural disasters. Far northern seas are subject to temporary ice packs that wreak havoc on shipping much of the year. Also in the north, *permafrost* (permanently frozen ground) hinders permanent settlement in many areas.

Because the west coast of North America lies along the Pacific Ring of Fire, it is subject to volcanic eruptions and frequent earthquakes. A notable eruption occurred in Washington state in 1980. In May of that year, Mount Saint Helens exploded with the force of 500 atomic bombs, blowing away the entire north face of the mountain. California, infamous for its San Andreas Fault, is subject to the continual threat of earthquakes. The 1989 "World Series quake" in San Francisco dramatically illustrated this threat to a worldwide television audience.

If you're ready to pack up your possessions and move east, you might want to consider a few more natural disasters before you leave. Droughts ravage the southwest, devastating tornadoes frequently plague the central plains states, the Mississippi River and its tributaries are subject to severe flooding, and the Gulf of Mexico and eastern coasts of the United States are vulnerable to tropical storms and hurricanes. I might suggest a move to Montana, where heavy snows and occasional forest fires are all you have to worry about.

Profiling the People

For hundreds of years, people from all across the globe have made their way to North America. They brought with them their ethnic characteristics, languages, customs, and religions. Because America is indeed a melting pot, any generalizations about its culture must necessarily be broad. After years of "melting," North American culture is unique.

Where the People Are

Population distribution in Canada is easy to visualize: Think south. Almost all of Canada's relatively small population resides in the southernmost part of the country in a small east–west strip next to the United States. More than half of Canada's population lives in the area called Main Street, which runs from the north shore of Lake Erie northeast along the Saint Lawrence River.

In the United States, the largest population cluster is in the *northeastern megalopolis*. A second major population cluster that extends west from the megalopolis is in the manufacturing areas around the southern fringes of Lakes Ontario, Erie, and Michigan. Two more major population clusters are in Florida and southern California. Except for urban areas such as Denver and Salt Lake City, the interior of the United States remains relatively sparsely populated.

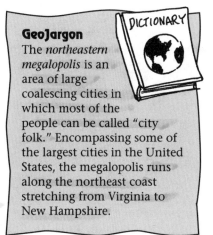

GeoJargon
The *northeastern megalopolis* is an area of large coalescing cities in which most of the people can be called "city folk." Encompassing some of the largest cities in the United States, the megalopolis runs along the northeast coast stretching from Virginia to New Hampshire.

Alabaster Cities Gleam

North America is a region of many large and well-known cities. Selecting important cities to include in this book was a treacherous business: Most people think that the most important city is their own. Nonetheless, this list shows the major cities in North America (in no particular order), as shown on the map.

North American population centers.

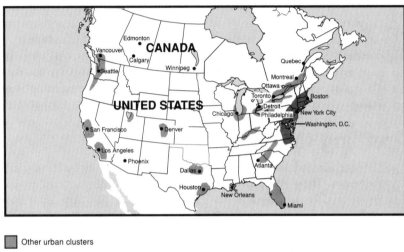

Other urban clusters

Megalopolis

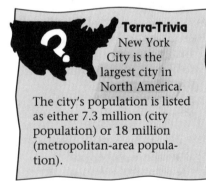

Terra-Trivia

New York City is the largest city in North America. The city's population is listed as either 7.3 million (city population) or 18 million (metropolitan-area population).

Canada's most important cities are all located in the south, close to Canada's border with the United States:

➤ **Toronto:** Largest city, financial center

➤ **Montreal:** Center of French-Canadian culture

➤ **Ottawa:** Capital, seat of Parliamentary government

➤ **Vancouver:** Canada's link to trade with the Pacific Rim

➤ **Other important Canadian cities:** Quebec, Winnipeg, Calgary, Edmonton

The following list of important cities in the United States reflects a population shift to the west (although at one time all the major U.S. cities were eastern cities):

➤ **New York, New York:** Largest North American city, financial center

➤ **Washington, D.C.:** Capital of the most powerful nation on earth

➤ **Los Angeles, California:** Entertainment capital, pop culture center

➤ **Chicago, Illinois:** Home of some of the world's tallest buildings (the Sears Tower is North America's tallest building, at 1,454 feet)

➤ **San Francisco, California:** Tourist center, hub for high-level technology

➤ **Other important United States cities:** Boston, Massachusetts; Philadelphia, Pennsylvania; Atlanta, Georgia; St. Louis, Missouri; Houston, Texas; Dallas, Texas; Detroit, Michigan; Denver, Colorado; Seattle, Washington; New Orleans, Louisiana; Phoenix, Arizona; and Miami, Florida

Roots of the People

In both the United States and Canada, a large percentage of the population is of European ancestry. On the west coast, Canada has a substantial Asian minority, and in the north indigenous Indian and Eskimo ancestry is common. The United States has a large African American minority distributed primarily in urban areas and in the south. Substantial Hispanic communities are located in many cities, the states that border Mexico, and in south Florida.

Finding the Faith

Christianity is the dominant religion throughout much of North America. In southeast and central portions of the region, Protestant faiths are most prevalent. Roman Catholicism is the dominant faith in northeastern and southwestern North America. Mormonism is common in the west central part of the United States (in and around Utah).

Judaism is practiced in many urban areas across North America. On reservation lands and in the far north, traditional Native American religions are being practiced with renewed vigor. A heightened interest in Eastern faiths and Islam has also marked North American faith in recent years.

Speaking of Language

For generations, immigrants have come to the United States speaking little or no English. Although the United States does not have an official language, English is the de facto official language and is spoken by the vast majority of its citizens. Some people have acquired the ability to speak English by inspiring means: I know of an Italian family who learned English by watching taped reruns of *The Beverly Hillbillies*. (I don't know, however, whether they still refer to a swimming pool as a "cement pond.") Although Spanish is spoken in many areas of the south, English is the norm for most of the United States.

GeoJargon
A *Francophone* is person who speaks French.

Canada's language story is a little more complicated because it has two official languages: English and French. *Francophones* (French speakers) are concentrated primarily in the province of Quebec, where more than 85 percent of the population speaks French. Language, a primary factor in people's culture, has led to a strong nationalist sentiment in Quebec. Several times in recent years, it has deliberated about splitting from Canada and forming a separate nation.

In the News: Indigenous People

The impoverished and increasingly threatened lifestyles of North America's remaining indigenous peoples have recently received a great deal of media attention. Off Hudson Bay, near a small offshoot called James Bay, the Cree people have fought an ongoing battle with the huge Hydro-Quebec utility over massive dams flooding enormous tracts of land and threatening the homelands of the Cree and their traditional way of life. Canada has begun to take dramatic steps toward reconciliation with its indigenous people through the creation of the new territory of Nunavut.

Terra-Trivia

In an ironic twist, the western world refers to the indigenous people of the northern plains as the *Sioux,* a French distortion of a Chippewa word meaning "snake" or "enemy." The Sioux have always called themselves Lakota (Dakota or Lakota, depending on dialect), which means "ally" or "friend."

In the United States, the general populace has become more aware and concerned about the squalid conditions that prevail on many of the nation's reservation lands. Although gaming facilities have provided an economic stimulus for some groups, the facilities have also threatened traditional culture. Many native groups seek the resolution of broken treaties with the United States government. The struggle of the Lakota (Sioux) people for the return of the Black Hills area in South Dakota is one such issue in the news.

The Least You Need to Know

➤ The region of North America contains only two countries: Canada and the United States.

➤ Canada is politically divided into ten provinces and two territories; the United States has 50 states.

➤ Mountain chains in North America run north–south. The largest are the Appalachians and the Rockies.

➤ The Mississippi River is the largest and most important river in North America. The major water bodies surrounding North America are the Pacific, Atlantic, and Arctic oceans and the Gulf of Mexico.

➤ Important North American cities include Toronto, Montreal, Ottawa, Vancouver, New York, Washington, D.C., Los Angeles, Chicago, and San Francisco.

➤ Canada is officially bilingual; the United States has no official language.

Chapter 7

The British Isles

In This Chapter

➤ How the history of the British Isles affects the region today

➤ How to differentiate between Great Britain, the United Kingdom, and England

➤ Why Ireland is known as the "green land"

➤ Why an umbrella is useful in this region

➤ Why everything revolves around London

➤ Why all Celtics are not basketball players

➤ How the Chunnel provides a link to the continent

The British Isles are located on the western fringe of Europe. Although their story parallels and meshes with much of Europe's history, their island status caused them to develop quite distinctly. Protected by water from every angle, the United Kingdom has withstood invaders for nine centuries. The islands have also dominated the surrounding seas for hundreds of years and have achieved a power and greatness well out of proportion to their size.

Once upon a Time

The British Isles have left a legacy of literary distinction, a philosophy of individual liberties, and the legal system to ensure them. Until recently, this small grouping of islands ruled more than a quarter of the earth. Although the empire is gone, the region remains an industrial power. At the dawn of the twenty-first century, the British Isles are still a major global player.

The Early Years: Celts, Romans, Angles, Saxons, Vikings, and Normans

Centuries earlier, the Romans had left fortified towns (London had been surrounded by two miles of eight-foot walls), constructed impressive road systems, and left behind not only the concept of republics and empires but also a system of laws. Germanic tribes of Angles, Saxons, and Jutes poured in from the south and southeast using the rivers and good roads. When William of Normandy (William the Conqueror) invaded the British Isles in 1066, he found an England fairly united.

The new immigrant groups pushed the existing Celtic residents back to the fringes of the islands. In Wales and Cornwall and in the south of England and western Scotland, pockets of these Celtic groups remain. Waves of Vikings brought a Danish and Norwegian influence to the British Isles. William, the last successful invader, combined his French with Anglo–Saxon to create the basis for the English language we speak today.

Claims for territory in France occupied the British Isles for many years, followed by the War of the Roses in the fifteenth century, a civil war that established the rule of the Tudor monarchy and ensuing prosperity. Trade expanded and spread London-style English. In the early sixteenth century, King Henry VIII (and his ill-fated wives) broke with the Roman Catholic church, and the king became head of the Church of England (Anglican). By the end of the sixteenth century, "Good Queen Bess," Elizabeth I, ushered in the age of naval power and expanded British colonies abroad. After Sir Francis Drake's defeat of the Spanish Armada in 1588, England became the firm ruler of the seas.

The Empire

Despite losing the American colonies in the eighteenth century, Britain acquired vast Canada to add to the "lifeline of the Empire." Thanks to the East India Company's spice trade, India was added to the realm. At home, parliamentary power grew, and technology and the Industrial Revolution continued to alter life on the islands. A processing economy developed, and commerce increased as imported raw materials flowed in from the colonies.

Terra-Trivia

On Salisbury Plains in southwestern England stand the ancient stones of Stonehenge. The huge, 21-foot-high stones date back 3,000 to 5,000 years. Thought to be an ancient astronomical device, the question remains: How were these huge stones moved? Today, this area is frequented by mysterious crop circles that bewilder experts.

Terra-Trivia
Wars over church and power resulted in the Magna Carta of 1215 A.D., which provided for representative government and a limit to the king's power. From these origins evolved the British concept of Parliament. (The word *parliament* comes from the French word *parler*, which means "to speak.")

By the late nineteenth century, British influence reached into Africa from Egypt to Capetown to Australia and to the South Atlantic and the Caribbean. The phrase "the sun never sets on the British Empire" was literally true. The empire's power was at its zenith as the British crown ruled over a fourth of the world's land and one-fifth of its people.

The empire dwindled after World War II. Using Churchill's philosophy of "blood, sweat, and tears," British tenacity and pride, and the Royal Air Force (RAF), England had staved off Hitler and survived the ravages of the war. During the following years, however, independence swept through the former colonies, and the once vast British Empire faded.

The Recent Past

After World War II, British industry grew more obsolete. With the conflicts in Northern Ireland plaguing it, England turned its attention to fine-tuning the "welfare state." The United Kingdom is at the forefront in wiping out economic inequality and poverty (although the "blitz," the German bombing raids, took out the East End slums of London). The country socialized medicine in the form of national health service care for all. At the same time, nationalization was introduced in services such as providing electricity and gas.

Terra-Trivia
Despite the United Kingdom's royalty having little more than ceremonial status, the British people love the royal family. Since 1952, Queen Elizabeth II has reigned as a model of royal decorum.

Concentrating on rebuilding at home and on its new social programs, the United Kingdom was slow to warm to the European Union, or E.U. (formerly the Common Market, a cooperative economical union of European countries). Island status kept the British Isles somewhat distant from the affairs of continental Europe. The islands maintained closer ties to the United States, in fact, than to their nearby European neighbors. Finally, in 1973, the United Kingdom and Ireland joined the Common Market.

The United Kingdom—It Doesn't Equal Great Britain

The region of the British Isles is composed of two large islands and a collection of smaller, peripheral islands. Although the entire region is known as the British Isles, only one island is named Britain. (The map of the United Kingdom and Ireland at the beginning of this chapter can help you sort out all this mishmash.) The two large islands are described in this list:

➤ **Great Britain:** The larger, more easterly island is sometimes called just *Britain.* People often get confused because almost any island, administrative division, and country in the entire area is referred to as British. (Calling the region the British Isles is an example.) Notice that Great Britain is an island and not a country.

➤ **Ireland:** Although Ireland is the name of the smaller, more westerly island, it's also the name of a country that occupies part of the island—so don't get confused.

Skip the smaller islands for now, and head for the two large islands—the countries of the British Isles:

➤ **United Kingdom:** Also called the United Kingdom of Great Britain and Northern Ireland, it's the larger of the two countries occupying the islands in both land area and population. The United Kingdom occupies the entire island of Great Britain and the northeast portion of the island of Ireland.

➤ **Ireland:** Also called the Republic of Ireland, or Eire, it's the smaller of the two countries in the region. Ireland, independent since 1921, is located on the island of Ireland. It occupies all except the northeast corner of the island (which, as noted, is part of the United Kingdom).

Although these terms cause some confusion, true bewilderment begins with the administrative divisions. The United Kingdom has four that were at various times in history separate political entities (now they all form the United Kingdom, but no one seems to be able to let go of the past):

➤ **England:** The largest of the four administrative divisions that form the United Kingdom and located on the island of Great Britain, England occupies the bulk of the southern and eastern portions of the island. Note that England is not a country.

➤ **Scotland:** Like England, Scotland is an administrative division of the United Kingdom and is not a separate country. It occupies the northernmost third of the island of Great Britain.

➤ **Wales:** Situated on the southwestern portion of the island of Great Britain, this bulge forms the administrative division of the United Kingdom called Wales—also not a separate country.

➤ **Northern Ireland:** This division is the source of all sorts of puzzlement. Perhaps I can help clarify by explaining what Northern Ireland is *not:* It is not located on the island of Great Britain, it is not part of the country of Ireland (Eire), and it is not a separate country. Northern Ireland is an administrative division of the United Kingdom, even though it's on the island of Ireland.

Although this chapter focuses on the four primary administrative divisions of the United Kingdom, remnants of the empire abound around the globe, such as Bermuda, the British Virgin Islands, the Cayman Islands, the Falkland Islands, Gibraltar, Pitcairn Island (settled by mutineers from the ship *Bounty*), and St. Helena.

Terra-Trivia
Each of these countries or administrative divisions contains smaller political subdivisions. England, Wales, and Ireland have counties; Scotland has regions; and Northern Ireland has districts.

Jolly Olde England?

The coal mines, steel mills, and factories that fueled the industrial revolution and the British Empire have played out and rusted, leaving some of England scarred and somewhat battered. In the industrial center of England are the Midlands, resting amid smokestacks and soot. The resulting "black country" that once brimmed with wool centers in Leeds, textile plants in Manchester, and coal in Birmingham now lies fallow.

England's economy has recently struggled with a lack of diversification and a need to adjust to modern demands. Manufacturing plants near London are now bright spots. England has moved from heavy industry to high-tech and design-based enterprises. The pound is prized as one of Europe's strongest currencies.

The Bonnie Land of Scots and Lochs

Though sharing a country with the English, the Scots have remained distinct. Helped by their highland location and tight clan system (which is the basis of family life), they have maintained a rugged character and an independent approach to life. Teased about their presumed frugality and production of haggis (a concoction of oatmeal and sheep's liver, heart, and lungs), the Scots maintain a strong sense of nationalism.

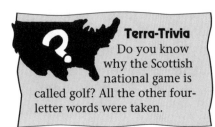

Terra-Trivia
Do you know why the Scottish national game is called golf? All the other four-letter words were taken.

The coal and iron once mined in Scotland have been depleted, and the manufacturing at textile mills and shipyards has waned. Scotland has rebounded, however, with research and development, electronics, and North Sea oil and gas. Tourism is also a mainstay, and playing golf at St. Andrews has become sort of a "Holey Grail" for golf players. Despite the draw of romantic beauty and the lure of the Highlands, life in Scotland can be difficult. One in six children younger than eight years old has asthma, and the second leading cause of death in men 20 to 40 is suicide.

The Slate and Coal of Wales

Wales, from the Anglo-Saxon word that means "foreigner," is the home of a tough people pushed into the infertile, wild, and mountainous western edge of Great Britain. Subsisting on laver bread (fried cakes made of oats and boiled seaweed to go with salted tea), the Welsh have survived a harsh life among the Cambrian Mountains.

The Welsh traditionally have eked out a living by mining coal and slate or working the machinery of the industrial revolution. In addition to the strip-mine scars that recall that era, the countryside is scattered with sheep, farms, and coastal fishing villages. The expanding businesses of refining tin, aluminum, and copper tubing now add a touch of optimism.

Both Wales and Scotland have seen a resurgence in nationalism in recent years. Road signs bear both English and Celtic names, and the traditional languages are increasingly being taught in school. These nationalist urges may portend a time when the United Kingdom will not be so united.

> **!** **Geographically Speaking**
>
> If you're a castle lover, Wales is the place for you. The most outstanding are the famous thirteenth century castles built by Edward I of England. Although most people think of castles as defensive structures, Edward constructed his string of castles for the offensive purpose of subduing Wales.
>
> Edward's castles were built along the sea and placed within a day's march to facilitate reinforcement. These impenetrable fortresses—complete with moats, gatehouses, drawbridges, and towers—housed the English troops that finally reined in the Welsh.
>
> The castle names are legendary: Aberystwyth, Builth, Flint, Rhuddlan, Harlech, Caernarfon, Conwy, and Beaumaris. No trip to Wales is complete without gazing at the sea from these storied parapets.

The Troubles of Northern Ireland

Northern Ireland was settled primarily by Scottish and English immigrants who crossed the Irish Sea to occupy one-sixth of the territory of the island of Ireland. Unlike the almost totally Catholic Republic of Ireland (Eire), only a third of Northern Ireland is Catholic (the rest are primarily Protestant). Protestants and Catholics frequently skirmish, and civil war constantly threatens.

Protestants suspect their Catholic compatriots of seeking consolidation with the Republic of Ireland, and Catholics decry discrimination from the Protestant-controlled government. Violence and terrorism involving police, parts of the Irish Republican Army (IRA), which is outlawed in the Republic of Ireland, and Protestant factions persist. Hoping to diffuse the tension, the United Kingdom eliminated Ulster's parliament in 1972 and began ruling directly from London—a nice try, but an abject failure. Bloodshed and sectarian fighting continue today.

In this uneasy atmosphere, efforts to spark new business (the Belfast DeLorean plant, for example) have had—not surprisingly—poor success. Traditional flax and linen making were made obsolete by the advent of synthetics. Unemployment, though less extreme than the mass departures following the devastating Great Potato Famine of the mid-nineteenth century, has forced many people to seek greener pastures.

Ireland: The Green Land

After resisting English domination for centuries, Ireland finally fell during the seventeenth century. Not until 1921 did the Republic of Ireland (Eire) again emerge as a separate country. Ireland never became as industrialized as the United Kingdom. The term "the green land" (or "emerald isle") is appropriate because, even now, its economy is based on agriculture.

In addition to agricultural exports, the country relies on tourism and a recent upsurge in manufacturing to complete its economic picture. Despite recent advances, the population is much less urbanized than most countries in Europe, and income levels lag well behind those of the United Kingdom.

During the potato famine of the preceding century, more than a million Irish people died, and many more were forced to leave the island. Even now, the population remains much smaller than it was before the famine. Ireland's green hills, lush countryside, and pastoral nature draw travelers from around the world. In a small village pub, you can still sample the famous Irish gift of gab and an equally famous pint of stout.

Natural Diversity

The British Isles feature much in the way of natural diversity. During the ice age, much of the region was enveloped by the massive ice sheets, as evidenced by the effects on the northern landscape. Because islands large and small comprise the region, the surrounding waters contribute significantly to its climate.

Landforms: From Scottish Highlands to Peat Bogs

As shown in the following map, the Northwestern Uplands, or Highlands, in Scotland are as high and spectacular as those in Norway, but are too rocky and wet for all except the hardiest settlers. The Highlands continue south from Scotland, turning into the Pennines Mountains in north central England and the Cambrian Mountains in Wales. The north country is also sprinkled with lakes that form the celebrated Lake District, highlighted by the romantic Lake poets William Wordsworth and Samuel Coleridge. The crags of the mountains are easily differentiated from stone-walled, sheep-filled dales underneath them.

Terra-Trivia
Separating Scotland from England is the famous 75-mile-long Hadrian's Wall. Built in England by the Romans in 200 A.D., it was designed to keep northern barbarians at bay. It symbolically separates the Scots and English even now.

Glacial remnants endure in the jagged, fjordlike terrain of northern Scotland that some say separates savage people and scenery from civilized Scotland. Bogs and moors abound as glaciers from the Grampian Mountains poke their way past lochs and through heather coastward, where the fishing is rich and the people more numerous. There, you can find Loch Ness and maybe glimpse its fabled monster, Nessie.

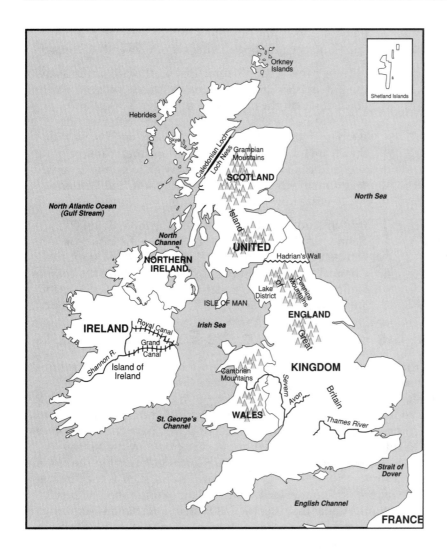

Physical features of the British Isles.

In addition to the Highlands of Scotland and Wales, most of the rest of the island of Great Britain is lowland, aptly called "downs." The lowlands are low (not surprisingly) and arable (easily cultivated) and have adequate rainfall. Southern England's lowlands form rolling hedgerowed farm country. The thick hedges were used rather than fences to separate farmlands. The poet William Blake called this area "England's green and pleasant land." In this agricultural land, every inch is cultivated to produce such delicacies as roast beef and Yorkshire pudding—and clotted Devonshire cream to go with its tangy strawberries. Even so, two-thirds of the food consumed on the British Isles must be imported.

Marking the westernmost edge of Europe, the Emerald Isle of Ireland is lush and green and rimmed with low hills spaced around the fertile central plain. Half of Ireland is made up of meadows; the other half is filled with lakes, mountains, and the peat bogs that cover one-seventh of Ireland.

Although this section has focused on the two larger islands, the British Isles have many smaller islands, most of which you can see on the maps in this chapter. To Scotland's northeast lie the Orkney Islands, replete with race tides, skerries (small, rocky islands), and rocky reefs; only 20 of the 70 are inhabited. Farther to the northeast are the Shetland Islands, the home of every child's pony dreams. Additionally, the sheep are especially soft, and Shetland sweaters are now the rage.

On Scotland's northwest coast are the more than 500 Hebrides Islands and the Isle of Skye. In the Irish Sea between Great Britain and Ireland is the Isle of Man. The Channel Islands are in the English Channel off the coast of France, and off Land's End, in extreme southwest England, are the tiny Isles of Scilly.

Waterways: From the Gulf Stream to the English Channel

As you can see from the preceding map, an important part of the geography of the British Isles is the water that surrounds them. The primary bodies of water surrounding the islands are the Atlantic Ocean to the west and the North Sea in the northeast. Separating the British Isles from continental Europe are the English Channel and the Strait of Dover. Southeast England's cornerstone, the famous white cliffs of Dover, have challenged every invader from Caesar to Hitler. Ireland is separated from Great Britain by the North Channel in the north, the Irish Sea in the middle, and St. Georges Channel in the south. Great Britain and Ireland are also home to many well-known rivers, such as the Severn, Thames, and Avon in Great Britain and the Shannon in Ireland. Canals abound, including the 60-mile-long Caledonian Canal that slices across Scotland. England is laced with hundreds of miles of canals that connect virtually every corner. Ireland features two famous canals, the Royal Canal and the Grand Canal.

A Balmy Climate at 55 Degrees North?

England is known for its fog and rainy skies, and Ireland is even rainier. What's surprising is that the region is not known for its heavy snows. Great Britain stretches from about 50 degrees to 59 degrees north latitude. Carried over to North America, the latitude would approximate that of Hudson Bay, with a climate that's ideal for polar bears. How is it that North America is adrift in ice and snow and the British Isles are relatively balmy?

The answer lies in the Gulf Stream. a vast, warming ocean current. Originating in the tropical waters of the Gulf of Mexico, it becomes the North Atlantic current that warms the British Isles. This huge flow of warm water tempers the region's climate and helps to make it milder and wetter than its latitude would otherwise dictate. Although umbrellas are requisite in the British Isles, snow shovels can stay largely in storage.

The Cities: Is London All There Is?

In the British Isles, everything revolves around London, the capital of the United Kingdom and one of the world's largest cities. Squatting on the Thames River and with access to the North Sea, London is also a large and bustling port. The London of Big Ben and Trafalgar Square is also the cultural focal point of the region. Pigeons and bobbies (London's unarmed street cops) mix with punksters sporting pink Mohawks. All stroll past the architectural wonders of St. Paul's Cathedral, the houses of Parliament, the Tower of London, and Buckingham Palace.

"The lungs of London" are its parks and gardens. The extensive network provides solitude, riding, and cricket to counter the unbelievable traffic jams of lorries (trucks), double-decker buses, and cars. You can people-watch at Piccadilly Circus, drop by Harrod's for high tea, or patronize "the local" (a pub) for bangers (sausages) and mash (turnips). A good pub makes everyone welcome to "stand a round" (buy a round of drinks) or challenge a patron to a heated match of darts.

Terra-Trivia
Those picturesque reed roofs you see all over the British Isles are made of *thatch,* which provides good insulation and can last 20 to 80 years. A *legget* is used to make the bundles of reeds fit over one another on the roof. Wire mesh is then laid on top to protect the thatch from birds, mice, and the weather.

The region has more cities than just London, however. Here are some of the more notable:

England

➤ **Bristol:** It's the center of shipbuilding, aircraft construction, and milling.

➤ **Bath:** Hot mineral springs in the first century brought the Romans and their baths. It's Jane Austen territory, too.

➤ **Oxford:** "That sweet city with her dreaming spires" (Matthew Arnold) has shaped the English elite for 800 years with its world-class university. Offering quiet punting (poling a flat boat along in shallow water) on the Cam River and intellectual training in the same milieu, the city was enjoyed by both Margaret Thatcher and Bill Clinton.

➤ **Leeds:** This industrial center for clothes and woolen milling is near Haworth, where the Brontë sisters' romantic novels were hatched.

➤ **York:** This walled city on the Ouse River is a manufacturing town with an eleventh century cathedral.

➤ **Liverpool–Manchester:** This conurbation (network of cities) was a leader during the industrial revolution. Manchester was known for its textiles, and Liverpool served as a major port and, more importantly, the hometown of the Beatles. This area is now largely economically depressed.

➤ **Birmingham:** A center of heavy industry, manufacturing cars, chemicals, and machinery, Birmingham also has a good orchestra.

Scotland

➤ **Glasgow:** This old industrial city benefits from oil and gas exploration, industrial research, and electronics.

➤ **Edinburgh:** Presbyterianism began in this capital and cultural center of Scotland. Lovely parks and stores contrast with tall, narrow houses made of granite. Nearby is Leith, where Scotch whisky, its most famous export, is blended.

Northern Ireland

➤ **Belfast:** This area alone sent 14 million Ulster Scots (Scots who moved to Ulster) to eighteenth century America. This capital and manufacturing center is now torn by strife and violence.

Wales

➤ **Cardiff:** Totally rebuilt after World War II bombing, this coal and ore port on the Severn River is a city of Arthurian tradition. It has more choirs, debating, and drama groups than almost anywhere.

Republic of Ireland

➤ **Dublin:** This capital city on the Irish Sea houses Trinity College (Dublin University). The huge Phoenix Park racecourse and zoo make it world-famous.

➤ **Cork:** Ireland's second largest city is marked by palms and azaleas and the presence of Blarney Castle, with its noted stone.

The People and Their Languages

The mix of ancient peoples that now comprises the population of the United Kingdom is primarily Protestant (the Church of England), except for a Roman Catholic minority in Northern Ireland. The Republic of Ireland is overwhelmingly Roman Catholic. Both islands are densely populated, except for the northern reaches of Great Britain.

Most people in the United Kingdom and Ireland speak English, a Germanic-based language. When invaders marched inland, the Celtic people inhabiting the islands were pushed into the backcountry perimeters of the islands. Sheltered in these refuge areas, the people of Celtic origin never completely adopted the invaders' speech.

Three remnant Celtic tongues are still spoken on the islands. Wales has the complex Welsh language, with long words and lots of tongue-twisting *yy*s and *dd*s. In northwestern Scotland, highlanders still speak the Gaelic language. Irish is spoken throughout Ireland, especially in remote clusters on the western coast.

Terra-Trivia
As you trek the Welsh countryside, you might encounter a sign announcing a town named Llanfairpwllgwyngyllgogerychwyrndrobwllantysilogogogoh. Known also as Llanfair P.G. and located on the Isle of Anglesey, this town has the longest name in the world.

Linking to the Continent—The Chunnel

The United Kingdom is no longer separate from continental Europe. The Chunnel, a tunnel underneath the English Channel, has forced Great Britain to reconsider its island mentality somewhat. The Chunnel now connects the British Isles to the rest of Europe for the first time since the Ice Age.

The Chunnel was drilled from Folkstone, England, to Coquelles, France. Measuring 31 miles, three lanes, and 150 feet below the sea through chalky marl, it's called an engineer's delight because water won't leak through it and it's no harder than soap. Some 220 banks from 26 countries signed on for the building project, and 15,000 workers were hired. In May 1994, 13.5 billion dollars later, the Chunnel became a reality. France is just a 35-minute ride away on the high-speed rail system that now links the regions.

The Least You Need to Know

➤ The people of the British Isles trace their origins to Celts, Romans, Vikings, Angles, Saxons, and Normans.

➤ The British Isles are made up of many small islands and two large islands, Great Britain and Ireland. The region consists of two countries, the United Kingdom and the Republic of Ireland.

➤ Despite the high latitude of the British Isles, the influence of the Gulf Stream makes its climate fairly mild.

➤ London is by far the largest city in the British Isles and is the focus of its culture.

➤ The people of the United Kingdom are largely Protestant; Ireland's population is decidedly Roman Catholic.

➤ English is the dominant language in the region, along with remnants of ancient Celtic tongues.

Western Europe: The Heart of the Continent

Western Europe has been the mother of great philosophers, scholars, writers, traders, and explorers. Though it's home to only a fraction of the world's population, it's a dominant player in trade, manufacturing, finance, art, fashion, and technology.

Small but complex, Western Europe is a collage of separate nations characterized by different economies, politics, and physical attributes. Divided by old animosities, religious differences, and language barriers, Western Europe has managed to perform as a heavy hitter in today's world.

A Legacy of War

Western Europe's location on the continent has proved to be both an asset and a curse. Certainly, Western Europe has served as a crossroads and has thrived on the resulting mixture of trade and culture that this strategic position encourages. The downside is that, as a crossroads, Western Europe has for centuries also served as the western world's battleground.

Apparently, if you're going to throw a war, this is the place to do it. Julius Caesar marched his Roman legions through the area in 55 B.C., crushing the Germanic tribes he encountered. Franks and Visigoths were the next to ravage the European countryside. Centuries of warfare later, Charlemagne established the Holy Roman Empire in Western Europe. The Thirty Years' War continued the warfare trend in the seventeenth century, and the early nineteenth century brought more turmoil and another empire builder in the person of Napoleon Bonaparte.

Life didn't improve in the twentieth century; the warring reached new heights, in fact. Western Europe became the primary battlefield for both World Wars I and II. Flanders Fields (in the Belgian countryside) and the Normandy beaches on the western coast of

France recall scenes of warfare at levels unheard of in human history. The loss of life during these conflicts staggers the imagination. Between Germany and France, more than three million people died in the First World War. In the Second World War, more than three million German men died in battle.

The effect of this hideous loss of life and economic devastation on Western Europe cannot be minimized. The heart of European culture reeled as priceless art and centuries-old architecture burned in these conflicts. Although Western Europe holds a strategic geographic position, it has paid a high price.

Countries in the Crossroads of Europe

Western Europe can be grouped into three primary geographic subregions composed of these countries:

➤ **France and Germany:** Core countries that form the largest subregion, in terms of land area and population

➤ **Switzerland and Austria:** Landlocked Alpine countries

➤ **Netherlands, Belgium, and tiny Luxembourg:** Low Countries that comprise the smallest subregion

The Biggies: France and Germany

The largest of Western Europe's countries (it's not quite as large as Texas), France covers more than 200,000 square miles. With nearly 60 million people, France is the second most populous country in Western Europe. The westernmost country in the region, the inhabitants of France enjoy both the Alps and the sea. The French countryside is laced with vast networks of rivers and canals that unite the region and connect it with the rest of the continent.

France is an agricultural leader in Western Europe. Grains from the north (France is the only European country able to export surplus wheat) are the dominant crop. Farms in the south benefit from long growing seasons for olives and oranges. Wines, of course, put France on the international map.

Terra-Trivia
France has more than 3,000 miles of canals. They connect the country's rivers in a way that grain grown in northern Normandy can float to Mediterranean Marseilles in the south. Because canal barging has become a popular tourist activity, travelers can view the vineyards, chateaus, and pristine scenery as they sample the wine and work the *ecluse* (manual canal locks).

From the casked brandies of Cognac to the red wines of Bordeaux and Burgundy to the sparkling "bubbly" of Champagne, each district of France has its vineyard. The white wines of the Loire Valley have particular appeal. For France, wine is serious business and obviously profitable.

Also glamorous and profitable is the haute couture business accompanying the textile industries in Lyon and the Rhone Valley. In France, fashion cannot be ignored, as evidenced by the distinctly "French" brands Dior, Chanel, and Givenchy. Paris reigns supreme as the world's fashion center.

Germany, the home of bratwurst and blood sausage, Volkswagen automobiles, and the speed-limit-less autobahn, is called the "locomotive" of Europe. Germany, with the largest population in Western Europe, lays claim to the world's greatest concentration of industrial centers and factories; it also dominates the coal, iron, and steel industries that are fundamental to the European economy. Every type of machine—military equipment, oil pipelines, and even roller coasters—is made by German industry, which in turn is dependent on coal and oil.

In northwestern Germany is the center of this industrial might, the cities of the Ruhr valley. Essen, Duisburg, and Dortmund are also in this enormous industrial complex. From this area convenient to the Rhine, goods are easily moved through the Rhine waterways to Rotterdam in the Netherlands and to the sea.

Lignite, or "brown coal," abounds in Germany, and despite its moderate-to-low energy potential, it is used to generate electricity. However, it and other factors are responsible for serious air and water pollution in southeast Germany. This part of Europe in general suffers from *waldsterben,* or forest death, from acid rain and other industrial pollutants.

The Alpine States

In the southeast of Western Europe lie the Alpine states. This subregion consists of just two countries: Switzerland and Austria. (Chapter 10 describes the microstate of Liechtenstein.) As the term *Alpine states* implies, these countries are dominated by Western Europe's vast mountain chain, the Alps.

Switzerland, like its Austrian neighbor, revels in Alpine beauty and picturesque qualities. With a heavy emphasis on dairy products (the basis for the famous Swiss chocolate), Switzerland stretches its meadow resources by using *transhumance,* whereby yodeling herders move cows or goats, *Heidi*-style, rather than buy imported grain to supplement limited grass on the farm proper.

Having few other resources, the Swiss have harnessed hydropower to fuel the factories and mills that create the renowned Swiss precision industry. Machine instruments, watches, and diesel engines make up the bulk of Swiss exports.

> **GeoJargon**
> *Transhumance* is the practice of transferring stock from mountain pastures in the summer to the valleys (where they eat hay and root crops) for the winter.

Austria uses "white coal" (hydropower) too, exporting power to Germany and Italy as well as firing the steel and chemical plants spaced along the mighty Danube River. Austria has been careful in using its Alpine forests and has long been a leader in forest management. Because the forests protect against erosion and avalanches, they have been harvested selectively to provide fuel and lumber—no clear-cutting here.

The Low Countries

A third subregion of Western Europe is composed of three countries often referred to as the Low Countries: the Netherlands, Belgium, and Luxembourg. These smallish countries are located in north central Western Europe.

> **Geographically Speaking**
>
> The Low Countries are called "low" because they're elevationally challenged (political correctness rears its head). Low is in the eye of the beholder, however. The Netherlands is low by any standard: Its average elevation is less than 40 feet above sea level, and much of the country is below sea level.
>
> Although Belgium isn't as low as the Netherlands, most of this flat country can still slide in under that label. Luxembourg, however, is mostly a hilly plateau and hardly belongs in the low club. Because Luxembourg is not big like France or Germany nor Alpine like Switzerland or Austria *and* because it's close to Belgium and the Netherlands, it is forever known as low.

Windmills, tulips, and wooden shoes are associated with the Netherlands (which means "low lands"), more familiarly known as Holland. With two-fifths of its land below sea level, the Netherlands really is the lowest country in the world. The country is buttressed against surrounding waters by a vast system of dikes. The Dutch have enlarged their small country with numerous *polders,* the Dutch word for reclaimed land below sea level. (For more information about polders, see the section "Holland's Fight Against the Sea," near the end of this chapter.)

Lately, the Dutch (with others) have learned to value the ecological benefits of wetlands and have called a halt to additional "poldering." Also, the heavy use of fertilizer and the proximity of the fields to Holland's maze of canals has led to water pollution, which is now being corrected. The country produces one-half of Europe's greenhouse produce and lots of beef, and mountains of colorful Edam and Gouda cheeses feed its neighbors at home and abroad.

Belgium is situated below the Netherlands and is in the middle of the Low Countries in terms of land area and population. In conflicts between the French-speaking Walloons of the south and the Flemish of the north, however, no middle ground exists. As a matter of course, students in the north study Flemish while students in the south are instructed in French Walloon. The Belgium government is often at a standstill as Flemish and Walloon factions collide. It will be interesting to see whether Belgium can remain intact over time.

Despite their differences, both Belgian cultures have introduced the delights of 300 types of beer, crepe, Ardennes ham, fresh mussels, and, of course, the concoctions of sausages and potatoes for which Belgium is famous. Tourists often skip Belgium, which is a pity: It has much to offer, including two cultures for the price of one.

The last of the Low Countries in this list is the pseudo-low country of Luxembourg. Although it's a notch above microstate, it's only the size of Rhode Island. Despite the country's woodsy reputation as an outdoor paradise filled with lush forests and rushing streams that beckon kayakers, it's very modern and has such a large steel industry that immigrant workers are hired from Portugal and Italy.

Along with Belgium and the Netherlands (also steel producers), Luxembourg completes the trio of *Benelux* (from the combination of *Bel*gium, the *Ne*therlands, and *Lux*embourg) countries that started the idea of advancing trade by lowering tariffs. Multilingualism and stability make Luxembourg attractive for foreign business, and its bucolic aura, pâtés, wild game, and Moselle wines don't hurt either.

Western Europe's Natural Side

Pushing into the Atlantic Ocean on the west and the North Sea on the north, much of Western Europe has a maritime feel. The remnants of the Gulf Stream warm the Atlantic, and this moderating influence tempers the plains of the northwest. Interior areas and the mountains face colder climes. Protected by mountains from the cold of the north and the Mediterranean shore of France, the French Riviera enjoys sunny temperatures with only a little rain.

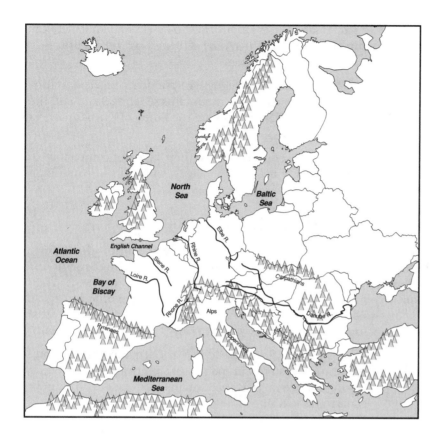

Physical features of Western Europe.

Landforms: Polders and Alps

Contrasting with the below-sea-level polders of the Netherlands is the most striking of Europe's physical features: the mighty Alps. As shown in the preceding map, this chain is a 680-mile-long collection of snowy peaks and valleys filled with glaciers and dotted with lakes. The Alps hook together to form an arc that stretches from France's Mediterranean coast through Switzerland and Germany to Austria's Danube, where the Carpathian range begins.

The French Mont Blanc tops the Alps with an elevation of 15,781 feet. In Switzerland, romantic castles and beflowered chalets nestle along the lakes and in the chinks of the famous mountains Jungfrau and Matterhorn (on the Switzerland–Italy border). The

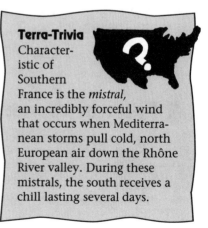

Terra-Trivia
Characteristic of Southern France is the *mistral*, an incredibly forceful wind that occurs when Mediterranean storms pull cold, north European air down the Rhône River valley. During these mistrals, the south receives a chill lasting several days.

Bavarian Alps in southern Germany feature the famous cuckoo clocks and chocolate cake of the Black Forest and Oberammergau's Passion play, the centuries-old celebration marking the area's deliverance from medieval plagues.

Western Europe is divided from Southern Europe by a natural boundary between southern France and Spain, the Pyrenees Mountains. These mountains separate Spain and the Iberian Peninsula from the rest of Europe. The Pyrenees are the home of the Basques, who periodically demand separate national status.

Despite the size of these huge and magnificent mountains, they have not completely barricaded nor isolated areas. Because of natural passes, valleys, and rivers, the mountains have not fostered complete isolation.

Waterways: From the Danube to the Rhine

Snowmelt from the Alps gives rise to three of Europe's key rivers: the Rhine, the Rhône, and the Danube, as shown on the preceding map. Other notable rivers are the Seine and the Loire. These rivers form great east–west corridors through the plains of France. The flat, level plains of Northern France have made it easy to join the all-important rivers. The Rhine, Meuse, Seine, Marne, and Rhône are interconnected in a complicated network of bridges, rail links, and canals.

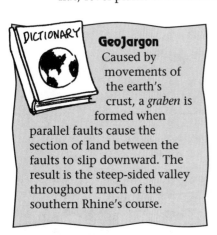

GeoJargon
Caused by movements of the earth's crust, a *graben* is formed when parallel faults cause the section of land between the faults to slip downward. The result is the steep-sided valley throughout much of the southern Rhine's course.

Dotted with castles and situated in Germany's midsection is the Rhine, Western Europe's most significant river. Set among the scenic Hartz Mountains (complete with singing canaries), the Rhine and nearby spas in the Thuringian Forest create a major tourist attraction. For much of its course, the Rhine flows through a gorge called a *graben,* or rift.

On a romantic note, the Lorelei of German legend lurk near the Rhine's dangerous rapids and lure sailors to their death. Various white wines are produced along the banks of this magical river. As the aorta of central Europe, the beautiful Rhine gorge is the lifeline for products flowing from the heartland to the sea.

A Place of Wondrous Cities

Western Europe is the home of many of the world's most spectacular and storied cities. Despite the natural beauty of the European countryside, many travelers to the region do no more than visit the cities—and they could easily spend a lifetime at it. Because this region has too many cities to cover in-depth in this book, this section presents just a few highlights (all labeled on the following map).

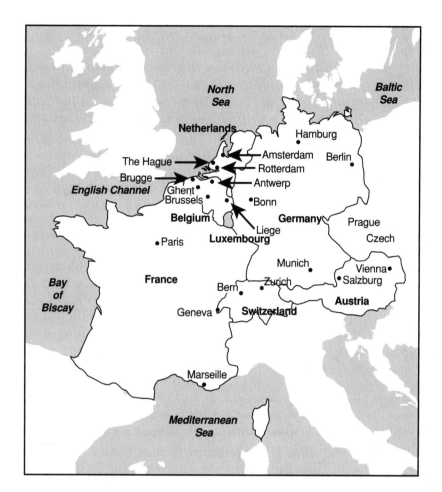

Cities of Western Europe.

France

➤ **Paris:** Most roads and railroads lead to Paris. The center of just about everything in France, Paris is the seat of government and the expression of national pride and culture. Unquestionably one of the world's most beautiful cities, it graces the Seine River. In midriver is the Ile de la Cité, the ancient heart of Paris. You can also visit Notre-Dame de Paris, the Gothic cathedral replete with gargoyles, and the smaller but exquisite Sainte-Chapelle, with its beautiful stained-glass windows.

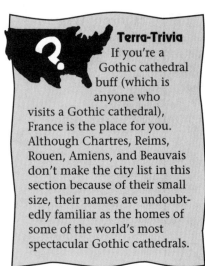

In Paris are the Tuileries gardens, Place de la Concorde, the magnificent Louvre (home of da Vinci's Mona Lisa and countless other art treasures), the Arc de Triomphe. For a comprehensive view of "the city of light," climb the Eiffel Tower and see the world's greatest ornament spread before you.

➤ **Marseilles:** The principal port on the Mediterranean and the gateway to the east, France's oldest city offers the finest bouillabaisse (fish stew) to start off your vacation in nearby Nice or Cannes or any of the terraced and windy hill towns of Provence.

Germany

➤ **Berlin:** This capital city of three million people is the site of the remnants of the notorious Berlin Wall. It's also the home of Checkpoint Charlie of Cold War fame and the Bradenburg Gate, the symbol of freedom and the 1990 reunification of Germany.

➤ **Hamburg:** A port at the mouth of Elbe River, this cosmopolitan city is the place to go for sole and eel soup.

➤ **Bonn:** This former West German capital and university town is also notable as Beethoven's birthplace.

➤ **Munich:** The primary city of southern Germany and Bavaria, Munich is the site of the annual September revelry Oktoberfest, Germany's raucous beer festival.

Switzerland

➤ **Zurich:** Only in Switzerland's largest city and center for banking and business can you celebrate spring's arrival by watching "the burning of the Boog" (burning snowmen stuffed with firecrackers).

➤ **Bern:** The capital and seat of Swiss government, Bern is located on a distant tributary of the Rhine. Its town mascot is the bear; live specimens are found in its famous bear pits.

➤ **Geneva:** The home of the Red Cross and the League of Nations, Geneva also boasts the towering Reformation monument to remember Calvin, Knox, and others who made the city a center of Protestantism.

Austria

➤ **Vienna:** This capital and the heart of Austria is adjacent to the Vienna Woods made famous by the Strauss Waltz. Don't miss the state opera house or the whipped cream and sacher tortes.

➤ **Salzburg:** This city, the home of Mozart, is also the city of music and *Sound of Music* beauty. Spectacular mountain scenery provides the backdrop for Salzburg's famous music festival.

The Netherlands

➤ **Amsterdam:** The capital of the Netherlands is linked to the North Sea by canal. *Canal* is the key word to use in describing this beautiful city because it's impossible to stray more than a few feet from one. Amsterdam's museums house the works of Rembrandt and the other Dutch masters as well an extensive Vincent van Gogh collection.

➤ **Rotterdam:** Almost destroyed during World War II, this city was rebuilt and extended its former self as the world's busiest port. Rotterdam is known as the "gateway to Europe" because of its spot on the entrance to the Rhine. Its huge harbor complex includes oil refineries and other support for Dutch industry and shipping.

➤ **The Hague:** This city is the Dutch seat of government, the location of the World Court, and the site of foreign embassies.

Belgium

➤ **Brussels:** This lovely but linguistically divided city struggles to be Belgium's capital; it also serves as the home of NATO's European headquarters and the European Union.

➤ **Antwerp:** This wealthy inland port on the Scheldt River is famous for its Rubens museums.

➤ Other important cities are Ghent, with its Flemish textiles, and Liège, the cultural focus of French-speaking Belgium.

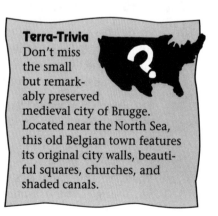

Terra-Trivia
Don't miss the small but remarkably preserved medieval city of Brugge. Located near the North Sea, this old Belgian town features its original city walls, beautiful squares, churches, and shaded canals.

The People

Although it's not a large region in area, Western Europe does have a very large population. Especially heavy concentrations of people live in northern France, the Low Countries, the Rhine River valley, and along the Rhône in southern France. Western Europe is highly developed and industrialized, and most of its people live in cities.

The Netherlands, with 1,100 Dutch squashed into each mile, is the densest of Europe's population bases; Belgium is a close second. Because Europe's birthrate is relatively low and life expectancy is high, however, you might hear mutterings about a "demographic winter" whose projections suggest a high old-to-young ratio. As this ratio becomes reality, Western Europe faces a situation of having fewer workers and higher costs for social programs.

The exception to the trend of declining populations lies in the presence of the *gastarbeiter*, or guest worker, in Germany. As trade barriers are lowered, immigrants arrive to do the low-level jobs others reject. Two million migrant workers from Turkey, Greece, and Spain have gone to work in Germany.

With the end of the empires, France has recently had an influx of North Africans. These new immigrants have settled amid the Europeans, whose prejudice toward the immigrants has resulted in profound cultural conflicts. Questions of citizenship, education, and health-care costs abound as the birthrate of these ethnic minorities rises.

Protestants and Catholics

Religion throughout Western Europe is primarily Christian. In the north, the Protestant faiths are most prevalent; in the south, most Western Europeans are Roman Catholic. The Dutch, Swiss, and Germans are fairly equally divided between Protestant and Catholic. The French, Austrians, Belgians, and "Luxembourgians" are primarily Catholic.

Languages: Germanic and Romance

Western Europe has two primary language subfamilies: Germanic and Romance. Both subfamilies belong to the Indo-European language family. French, the primary representative of the Romance subfamily, is spoken primarily in—France. French is also spoken in southern Belgium and western Switzerland.

The Breton Peninsula in northwestern France is considered a *refuge area*. It has an isolated pocket of people who speak Breton, a language in the Welsh group of the Celtic subfamily. Celtic languages are considered *relic languages*. Once spoken over a wide area, they were overwhelmed by other languages and now remain only in refuge areas.

The Germanic languages (including Dutch, Flemish, and German) are dominant in the remaining countries in the north and east. Almost all Europeans are at least bilingual, with much of the population proficient in three or more languages. Switzerland features four official languages: Italian, French, German, and Romansh (an obscure group of the Romance subfamily). By the way, many Western Europeans also have mastered English.

Holland's Fight Against the Sea

You already know that the Netherlands holds the title of the "world's lowest country." Because half the land area is below sea level, the title is richly deserved. How did those clever Dutch do it, and what problems does the future hold for this very low country?

The Netherlands is cradled around what was once a shallow inland sea called the Zuider Zee. At the mouth of this sea, the Dutch built a huge earthen dam, or *dike*, that cut off the Zuider Zee from the North Sea. The dike created a lake from the former Zuider Zee that the Dutch renamed the IJsselmeer. Over the years, the Dutch have drained portions of this shallow lake bed to form reclaimed lands called *polders*.

The poldering process involves the construction of massive dikes around areas to be reclaimed. After the dikes have been constructed, the brackish water is pumped out. The area's famous canals are then used to drain the land. (The windmills had the job of constantly pumping out the ever-seeping water, although now diesel pumps are used and the windmills are primarily scenic elements.)

If you live on land that's well below sea level, ocean storms and rising sea levels are the sorts of things that put you on edge. So it is with the Dutch. North Sea storms have breached the dikes in the past, and the loss of life and property has been catastrophic. The Dutch take great care in the construction and maintenance of the dikes. They also view global warming and potentially rising sea levels with great concern.

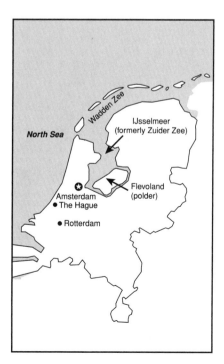

The Netherlands' polders.

The Least You Need to Know

➤ Western Europe has three subregions: the large countries of France and Germany; the Alpine states of Switzerland and Austria; and the Low Countries of Belgium, the Netherlands, and Luxembourg.

➤ Western Europe has historically been the site of many wars.

➤ The Alps are the dominant physical feature of the region.

➤ Western Europe is home to many of the world's great cities.

➤ The region is relatively small but densely populated.

➤ The people of Western Europe are primarily urban, affluent, and Christian, and they speak Indo-European languages.

Chapter 9

Northern Europe: Land of Ice

In This Chapter

➤ Weathering the wrath of the Norsemen

➤ Chilling in the cold climes of Scandinavia

➤ Heralding Viking heritage

➤ Highlighting important Northern European cities

➤ Preserving the future

The region of Northern Europe stretches northward from the continental core of Europe through the Baltic Sea and the North Atlantic Ocean to the frozen Arctic Circle. Northern Europe is appropriately named because all of this region is far to the north (it's the northernmost group of countries on earth). The sea has provided a livelihood, albeit a difficult one, but has distanced people of the North from the rest of Europe. The result is a sense of neutrality and unity—and no small degree of isolation.

Democratic and literate, sturdy and industrious, Northern Europe's sparse population is homogeneous and mostly Protestant and enjoys a high standard of living. An egalitarian outlook on life is typical, and the "cradle to grave" social support system is an accepted part of life. Sports (especially winter sports) pervade the national scene, adding color and glamour to the incredible natural beauty of the northern lakes, vast wilderness areas, and rugged mountains. The romance of Viking lore combined with the influence of Hans Christian Andersen's forest creatures and Christmas gnomes, the modern-day Legoland, and a theme park in Denmark paint a quaint fairy-tale setting for life in "the land of the midnight sun."

Northern Europe.

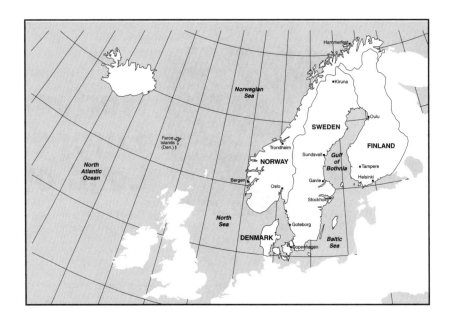

Northern Raiders and Resisters

Beginning in the eighth century, Viking warriors launched invasions from Northern Europe. As they sailed their open longboats (adorned with dragon heads), they raided, plundered, and generally terrorized coastal inhabitants, leaving a path of destruction in their wake. Despite their "wrath of the Norsemen" reputation, the Vikings were more than raiders: They explored and discovered as they traveled, and Viking settlements existed far beyond Scandinavia for centuries.

The seafaring Norsemen also made forays across the Atlantic. Settlements were established in Iceland and even in Greenland. Around the year A.D. 1,000, Leif Eriksson and his comrades abandoned Greenland and sailed west to become the first Europeans to land in North America. Remains of a Viking village from Eriksson's era were unearthed in Newfoundland, Canada, in the 1960s. The arrival of the Black Death (bubonic plague) in the fourteenth century weakened the Norse (and just about everyone else in Europe). Germany gained control of trade, and the shipping lines to Greenland were broken. Although the colony in Greenland died out as a consequence, Iceland managed to survive.

During the Second World War, this region again distinguished itself. For the five years of occupation, Norway's government was taken over by the Nazis and run by Vidkun Quisling, whose surname has become synonymous with "traitor." The resistance movement was universally supported, especially in Denmark, where the Danes defied the Nazis with varying degrees of subtlety. When rumors of plans to arrest all Jews surfaced, the Danes acted immediately and managed to spirit away, conceal, and rescue most of Denmark's 8,000 Jews in one day.

The Land of the North

Europe's northern tier includes five countries, which as a group are referred to as Norden. Scandinavia refers to the Scandinavian peninsula containing Norway and Sweden. Finland, Denmark, and far-off Iceland (900 miles away) complete the Nordic unit. Influenced by a shared heritage and culture, a harsh wilderness, and a lengthy and close association with each other, these countries have achieved autonomy for themselves to go along with their strong and unique economies.

Geographically Speaking

Northern Europe has long been called the "land of the midnight sun." During the Arctic summer, the sun never dips below the horizon. The $23^1/_2$-degree tilt of the earth's axis causes the polar areas never to leave the illuminated half of the earth around the summer solstice. Summertime brings 24-hour sunshine, midsummer's Arctic bramble bush wine called "mesamarja," and long evening celebrations.

This phenomenon has a downside. Northern Europe can also be called land of the noon moon (it's not astronomically correct, but I like the rhyme). During the dead of winter, the sun never rises above the horizon. Even in southern Northern Europe, winter days are depressingly short. The region suffers from some of the world's highest rates of alcoholism and suicide.

Norway, Sweden, and Finland all lie in the Northwestern Uplands, Europe's oldest mountain range, which runs all the way through central Europe to England. The countries of Northern Europe benefit from the moderating temperatures generated by the Gulf Stream (a warm current) as it pours into the North Atlantic. Southern coastal Norway does not have a frigid climate; harbors remain open and free of ice year-round. The same cannot be said, of course, of the interior or mountain regions, where the weather, like the terrain, is much more severe.

Despite some relief from the Gulf Stream, the region still encroaches on the Arctic Circle. Moisture and wind from the Atlantic storms are—pardon the expression—nothing to sneeze at. Winter days in this region are short, dark, and cold. The poor, rocky soil clings to the steep terrain, and the growing season lasts about 15 minutes.

The far north is wild and mountainous, with much of it above the Arctic Circle. Most of this area is blanketed with vast stands of coniferous forests, called *taiga,* and with extensive tracts of birches (schoolchildren plant 100 million saplings a year). The realities of this land explain the importance of wood to life in the North.

Physical features of Northern Europe.

The Fjords of Norway

Norway features one of the most famous coastlines on earth. Jagged and indented, the coast features the magically beautiful *fjords,* shown on the preceding map. These glacially formed inlets jut inland from the Norwegian Sea, some as long as 100 miles. Running inward along green valleys, the fjords are fed by river waterfalls that cascade from the mountains to join the sea. As the fjords reach the sea, *skerries* (small, rocky islands) form a buffer between the Atlantic and the shore. Visit this must-see tourist destination in the summer.

In addition to the tourist bonanza the fjords attract, thousands of rivers and streams from Norway's high mountain ranges produce more kilowatt hours of energy per person than anywhere else in the world. In addition to hydro-electricity, the country enjoys other resources in the form of oil and natural gas yielded by the North Sea. Half of Norway's exports are petroleum products, and much industry has developed around power sources, both hydro and petroleum.

Terra-Trivia
Measuring 125,000 square miles and reaching 1,000 miles from tip to toe, Norway is huge compared to most European countries. Despite Norway's size, much of its territory is inhospitable. Only four million people live here, most of them on the coast. Of the Northern European countries, only Iceland has fewer residents.

The Swedish Core

On Sweden's west and north sides, rugged highlands mark the boundary with Norway. Coniferous forests mark the northeast, and large lakes and low agricultural land are found in the south. The country is laced with rivers that originate in the highlands of the west and drain to the Gulf of Bothnia in the east. Although lacking Norway's oil and gas deposits, Sweden does possess substantial mineral wealth, primarily in the form of metal ores.

The Forests of Finland

Although Finland's terrain is flatter than its neighbors' to the west, poor soils and cold temperatures make agriculture nearly impossible. Mineral resources are few. Although this situation sounds dismal, Finland does have trees; they have led to prosperity for the country in the form of paper, pulp, and lumber products.

Although nearly a third of the country lies above the Arctic Circle (it's neighbor to the east is Russia), Finland, like its neighbors, benefits from the warming trends of the Gulf current. Despite its northern location, Finland never has permanent snow cover or ground frost. The Baltic Sea and the Gulf of Bothnia help make for the maritime climate present elsewhere. Some 30,000 islands form micro-*archipelagos* (island clusters) along the Finnish coast.

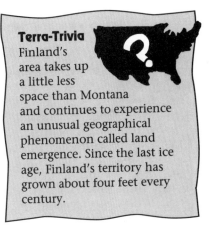

Terra-Trivia
Finland's area takes up a little less space than Montana and continues to experience an unusual geographical phenomenon called land emergence. Since the last ice age, Finland's territory has grown about four feet every century.

Denmark's Peninsulas and Islands

As the southernmost Nordic country, Denmark occupies yet another peninsula, Jutland. In the south, the Jutland peninsula shares a mere 42-mile border with Germany. The low flatlands of the Jutland peninsula strike a sharp contrast to the rugged highlands of the Scandinavian peninsula to the north.

Also notable are Denmark's 482 islands. They range from mite-size to medium-size (such as Zealand, which is home to Copenhagen, the capital) to Greenland, the world's largest island. Denmark also has the 21 Faeroe Islands, off Scotland's north shore. Some of the Faeroes, meaning "sheep islands," have a language and flair of their own and even a separate set of laws. All make up the Kingdom of Denmark.

The country is more agriculturally blessed than other Nordic countries. With fair soils, milder temperatures, and level lands, Danish farmers are successful at growing food and producing dairy products. Rolling moors have been built up and turned into farmland that covers three-quarters of the countryside. Compared to its northern neighbors, Denmark is indeed a Nordic breadbasket.

109

Iceland's Heat from Below

Perhaps Iceland has it the worst of any country in the region. Three-quarters of the interior is filled with volcanoes, glaciers, lava formations, frozen lakes, and varied wastelands—most of Iceland is uninhabitable. All the island's settlements, and the few farms that make hay for the famous Icelandic pony and hardy sheep, are coastal. Nonetheless, Icelanders have learned to make the best of their situation.

Perched on the spreading mid-Atlantic ridge, a volcanic hot spot, Iceland treats its 100 volcanoes as an additional resource. Using *geothermal* heat in combination with other energy forms (hot water and steam produced by geysers and waterfall- and river-generated hydroelectricity), the country can heat apartment buildings for habitation and greenhouses for vegetable production. Not bad for a barren land with no fossil fuels.

Terra-Trivia
With fewer than 265,000 residents, Iceland is one of the world's least populated countries (the population of the entire country is less than Birmingham, Alabama). Population density in Iceland is a paltry 7 people per square mile. To put this number in perspective, Iceland has about 14,565 fewer people per square mile than Hong Kong and only 7 more people per square mile than the moon.

The Descendants of the Vikings

Considering Norden's location at the top of the world, you probably aren't surprised to hear that few people live there. The hardy souls who do reside there live on the coasts and as far south as possible. With just fewer than nine million inhabitants, Sweden boasts the largest population in the region. Population densities are fairly low throughout Norden (Denmark has the highest), and the bulk of the people are urban dwellers.

The typical Nordic person has blond hair, a fair complexion, and blue eyes. If you travel to the region in the summer months, you may see more Nordic features than you bargained for. During these months, people stream from their offices for lunch in the parks, peeling off every stitch of clothing and reveling in the few scant hours of sunlight that the region has to offer.

The Lapps: Life on the Top of the World

Terra-Trivia
Language is often an excellent way to gauge the importance of something to a society. In Lapland, the Lapps have a special word for the male reindeer in each of the first six years of his life.

One deviation from typical Nordics live along the Arctic Circle across the top of Norway, Sweden, Finland, and part of Russia. The Lapps, a nomadic and hardy people, inhabit these harsh mountainous regions. Living well above the tree line, the Lapps raise reindeer on the low-lying tundra. Utilizing flat fardmoss and lichen-covered areas over frozen subsoil, they work the wild dogs they have trained to sled. The Lapps' invention of skis has helped them in their daily life built around the reindeer herd.

From the reindeer come food, transportation, clothes, and shelter. Tools are made from reindeer bones. Much like the buffalo in the old American West, reindeer have been the basis for Lapp life and remain their main source of income. In recent years, northern mineral deposits have introduced more outside people and their modern trappings to the area of Lapland.

Almost Intelligible

Most of Norden shares a common linguistic heritage. Danish, Icelandic, Norwegian, and Swedish are all Germanic languages in the Indo-European language family. Although these closely related languages are more or less mutually intelligible, Finland is the oddball. Most people in Finland speak Finnish (no surprise here), but Finnish belongs to a separate language family. The Finnish language originated in central Asia and is part of the Uralic language family (as is the Lapp language Lapponian). Although Swedish is the second official language in Finland, Finnish remains an anomaly in the Norden group.

Norse Food and Leisure

Denmark is sometimes called the "land of balance," which alludes to the hard-working, practical side of the Nordic people that has become a stereotypical image to much of the world. Another side to the people of the north, however, is reflected in Denmark's other nickname, "a toyland run by adults." Yes, the people of the north have their playful side.

A national obsession for many in the north is the sauna. After a thorough roasting in an extraordinarily hot sauna, tradition dictates a plunge into an icy lake or snowdrift and a few good lashes with a birch switch. Is this invigorating or intimidating? Either way, it's a great way to work up a hearty appetite.

Start with a delicious, cholesterol-laden breakfast of ham, bacon and eggs, cheese, and pastries. You might even choose a plate of pickled herring for lunch. After an afternoon on the cross-country ski tracks, you're ready for a real Nordic feast: the *smorgasbord*. Wash it all down with a Carlsberg or Tuborg beer, and you're ready for another sauna.

> **Terra-Trivia**
> Profits from the sale of beer, the national drink in Denmark, are poured into education, museums, and the arts by the way of the Carlsberg Foundation. Similarly, Tuborg sales finance foreign-study grants for young people. Beware, however: Because alcohol is heavily taxed, buying a couple of beers is likely to set you back a few kroner.

The Many Religions of Northern Europe?

Think Lutheran. Lutheran is the main religion in all the countries in this region, and about 90 percent of the population is Lutheran.

111

Making Ends Meet on the Tundra

Despite its location on the edge of the world with freezing temperatures, poor soils, and few minerals, Northern Europe has emerged as a prosperous region. The economies in the north have thrived on fishing (especially Iceland), wood products (especially Finland), industry and manufacturing (especially Sweden), dairy products (especially Denmark), and petroleum revenues (especially in Norway). The people of Northern Europe generally have high incomes and are well educated.

> ## ! Geographically Speaking
>
> Known as "the land of fire and ice," Iceland shares a maritime economy with its Nordic neighbors. It sits amid the richest fishing grounds in existence and annually delivers more than 1.5 million tons of fish to the rest of the world. Because fish makes up 83 percent of Iceland's exports, it has no interest in having foreign fleets encroaching on its domain and takes great pains to make that information public knowledge.
>
> Cod and capelin that are air-dried on open racks (a tried-and-true Viking method) are called *stockfish.* Canning and salting are also common preserving practices used to prepare fish before shipping to Southern Europe and Africa. Think of Iceland, and think of fish.

The Cities of the North

Norway

➤ **Oslo:** The capital and the center of commerce and culture in Norway. Contemporary furniture, crystal, and furs are all available here. Delicious seafood and reindeer steak are found near Frogner Park, which houses Gustav Vigeland's bronze and granite sculptures showing man's life cycle. You can also check out Thor Heyerdahl's *Kon Tiki,* the raft the famous explorer used to cross the Pacific.

➤ **Bergen:** A favorite starting point for many tours of the fjords and the host of an international festival of arts. Edvard Grieg is highlighted here, as is playwright Henrik Ibsen.

➤ **Hammerfest:** Deserves mention as the northernmost city in the world.

Sweden

➤ **Stockholm:** The capital of the most industrialized of the Nordic nations and known as "the queen of the waters." Stockholm is built on islands where the Baltic meets

Lake Malar. The modern city grew from the island that became the center of business as well as the government seat.

➤ **Göteborg:** The country's greatest port, with a modern shipyard and an active automobile-manufacturing center.

➤ **Malmö:** The third largest city in Sweden. Resting opposite Copenhagen, it's a major port and shipbuilding center.

Finland

➤ **Helsinki:** By far the largest and most important city in Finland and also the Finnish capital. The "white city of the north" is the home of the National Museum and the Helsinki railroad station, designed by the famous architect Eliel Saarinen and his son, Eero.

Denmark

➤ **Copenhagen:** The capital of Denmark and called the most cosmopolitan of Nordic cities because it's closest to Europe. Home to the world-famous amusement park Tivoli Gardens, which contains symphonies and ballets, glitzy rides, and tranquil gardens, all lit up by fireworks at midnight according to City Hall's clock. The Royal Ballet exemplifies the Danes' love of dance. A maze of bridges enables trains and cars to go to and fro while a bronze statue of Hans Christian Andersen's *Little Mermaid* observes the hundreds of ferries scuttling around the harbor.

Iceland

➤ **Reykjavik:** The capital of Iceland's republic and home to more than one in three Icelanders. It's not a large city by world standards (only about 100,000 people live there) but is by far the most important in Iceland. Reykjavik hosted the summit between the U.S.S.R. and the United States that led to the 1987 ban on medium-range missiles. (The city is a geographic midway point between Moscow and Washington, D.C.)

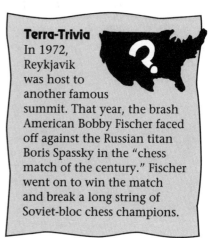

Terra-Trivia
In 1972, Reykjavik was host to another famous summit. That year, the brash American Bobby Fischer faced off against the Russian titan Boris Spassky in the "chess match of the century." Fischer went on to win the match and break a long string of Soviet-bloc chess champions.

Recent Developments in Norden

The people of the north share an appreciation of and commitment to preserving and protecting the environment. Despite their efforts, the people of this region have been victims of environmental woes brought on by their location. Air pollution from the industries of Europe drift north with the prevailing winds to collect in Norden. Falling as acid rain, the toxic brew is killing forests and rendering Nordic lakes lifeless. (Pollution knows no human boundaries.)

Fears are real in this part of the world, with the specter of glowing reindeer noses resulting from wind-borne radiation from the Russian reactor accident at Chernobyl. Sweden and Finland are the only countries with nuclear power, and Sweden has committed to shutting down its facilities by the year 2010. Opposition to the construction of nuclear weapons is universally strong, and all five countries have signed the Nuclear Test Ban Treaty. Denmark and Norway do not permit nuclear weapons on their soil, and the Swedes, despite their strong feeling for defense, have unilaterally renounced all nuclear, biological, and chemical weapons.

The Least You Need to Know

> ➤ Five countries are in the Northern European region: Norway, Sweden, Finland, Denmark, and Iceland.

> ➤ The Northern European countries are the most northern countries in the world.

> ➤ The region is relatively poor in resources but highly developed and affluent.

> ➤ The region's relatively small population is located in the south and along the coasts.

> ➤ The most important cities are the capitals: Oslo, Norway; Stockholm, Sweden; Helsinki, Finland; Copenhagen, Denmark; and Reykjavik, Iceland.

Chapter 10

Southern Europe: Embracing the Mediterranean

In This Chapter

➤ The domination of the legendary Roman Empire

➤ The diversity of the region's major cities and microstates

➤ The historical significance of Southern Europe's major cities

➤ The coastal constitution of Southern Europe's people

➤ An eye toward independence?

The four countries of Southern Europe have many similarities in addition to their occasional differences. (Turkey is not considered part of this region in this book because only a small fraction of the country is within continental Europe.) All have their cultural roots in ancient Greek and Roman political and social foundations. The entire region shares a Mediterranean climate and vegetation, mountainous landscape, and poor soil. The countries are on peninsulas, and their people are located primarily along the coasts.

The Mediterranean is also a region of exceptions. All its countries border on the Mediterranean Sea, except Portugal. All are Roman Catholic, except Greece. None of them touches another, except Spain and Portugal. All share Romance languages, except Greece. All lack major industry, except Italy. The list could go on, but you get the idea. As interesting as they are varied, the southern countries share a Mediterranean way of life.

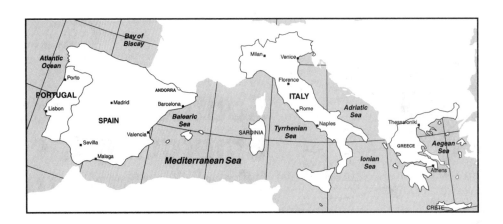

The Legacy of the Greeks and Romans

The ancient Greeks established an extensive trading network that spanned the Mediterranean. By the fifth century B.C., classical Greece had established crowning achievements in the arts, science, architecture, mathematics, philosophy, literature, and democratic principles. The accomplishments of the Greeks were absorbed and later expanded on by the Romans.

The Roman Empire that followed Greece eventually controlled all the Mediterranean and beyond. The territory under the empire shared one law, one language, and, later, the Roman Catholic Church. Rome dominated the Mediterranean for almost 1,000 years, until the Germanic tribes swept south in the early fifth century (420) A.D. The Lombards occupied northern Italy, the Visigoths occupied Spain, and the Vandals dominated Africa and (later) Spain. In the eighth century, Muslim invaders pushed north into Spain from Africa and established a Moorish dominion that lasted until the late fifteenth century. These varied peoples formed the seed of the Mediterranean lineage of today.

Southern Europe became a wellspring for great explorations. The Venetian Marco Polo traveled to China in the thirteenth century and wrote of his exploits. Portugal's great geographer and ruler Prince Henry the Navigator sent forth expeditions to the southern coasts of Africa. Other great explorers and adventurers followed, including Christopoher Columbus, Vasco da Gama, Hernando Cortés, Fernando Magellan, and Francisco Pizarro. Through these explorations and conquests, Mediterranean culture was spread (often brutally) across the entire globe.

Later centuries saw great advances in art, philosophy, literature, and architecture. Botticelli, Dante, Giotto, and Petrarch paved the way for the Renaissance (rebirth). Beginning in the fifteenth century, it established Italy as a cultural center for centuries to come, with Raphael, Michelangelo, Bramante, da Vinci, and Bernini creating masterpieces that would last for posterity.

Countries in the South: Major Cities and Microstates

As shown on the map, Southern Europe contains four countries (and a host of microstates, mentioned later in this chapter). In the west are Portugal and Spain, the only countries in the region to share a border; in the center lies the famous boot of Italy; and to the east is Greece, with its many islands. This section looks more closely at each country's particular Mediterranean character.

Portugal and Wildflowers

Portugal is surrounded by Spain and the Atlantic Ocean. Filling 35,550 square miles, Portugal is Southern Europe's smallest and westernmost country. Though Portugal's colonies were extensive and its empire lasted into the twentieth century, only Macau remains from an empire that was once 23 times the size of Portugal. (Angola, Mozambique, Timor, and others all are gone.) Even Macau will soon be gone: Portugal is scheduled to return it to China in 1999.

Income levels in Portugal are lower than in most European countries. Agriculture, a mainstay of Portugal's economy, includes sheep raising and the cultivation of typical Mediterranean crops (olives, fruit, grapes, and grain). Don't miss spring in Portugal—the wildflowers are unforgettable. Tourism in the southern Algarve area is a mainstay of the economy.

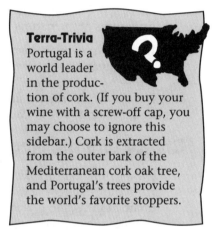

Terra-Trivia
Portugal is a world leader in the production of cork. (If you buy your wine with a screw-off cap, you may choose to ignore this sidebar.) Cork is extracted from the outer bark of the Mediterranean cork oak tree, and Portugal's trees provide the world's favorite stoppers.

Sunny Spain

Spain is Southern Europe's largest country in land area and has the second largest population (after Italy). Almost 80 percent of Spaniards live in the cities, which makes Spain more than twice as urbanized as its neighbor Portugal. Spain's income levels are also higher than those of Portugal's, and its economy is more industrialized. As with the other countries in the region, agriculture figures prominently in the Spanish economy.

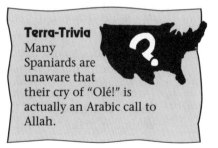

Terra-Trivia
Many Spaniards are unaware that their cry of "Olé!" is actually an Arabic call to Allah.

Tourists from Northern Europe and beyond flock to sun-drenched Spain, replete with bubbling fountains and the smell of roses. The local café is a perfect place to enjoy *churros con chocolate* (deep-fried dough eaten with dark chocolate) or a refreshing sangria made of red wine and fruit. You can top off an afternoon with a *corrida de toros* (bullfight) in the bull ring and spend your evening enjoying the flamenco (an Andulacian gypsy-based dance) and nibbling curls of *jamon serrano* (thinly sliced spirals of aged ham).

Italy: Hill Towns and Piazzas

Italy, the giant of the region, has the largest population and the second largest land area of any country in Southern Europe. Its population density far exceeds other countries' elsewhere in Southern Europe. Italy is also the Mediterannean's economic leader, with the bulk of the region's industry and its highest income levels.

Although many visitors are drawn to Italy's famous and exquisite cities (described in more detail later in this chapter), if you visit, you should take time to stray from the large cities to visit the tiny hill towns typically perched atop the rugged hills that dot Italy's landscape. Built from the same stone on which they stand, these towns seem to be simply extensions of the natural terrain.

In the heart of each town is a central *piazza* (plaza) and the local church, almost always an architectural gem. If the town has a *campanile* (bell tower), be sure to climb it if you visit—the view never fails to be spectacular. Some favorite hill towns are Orvieto, Assisi, Gubbio, Urbino, Siena, and San Gimignano. Searching out the many bell towers in the Italian countryside is loads of fun.

Greece and the Aegean

Although not as rural as Portugal, Greece is rural by European standards. Income levels in Greece are among the lowest in the region. Despite an emphasis on agriculture, only a quarter of the land is arable, plots are small, and rainfall is unreliable. Greeks to the north produce tobacco, cotton, and wheat in addition to the typical Mediterranean crops you might expect (olives, grapes, oranges, and vegetables).

Although Greece has been independent from its eastern neighbor, Turkey, since 1830, bad feelings still remain. Constant flare-ups between the two countries have continued to this day. The most recent conflicts have centered around the island of Cyprus. An unsuccessful Greek-sponsored coup in 1974 resulted in an invasion by Turkish troops and the formation of the Turkish-supported state of North Cyprus.

Directly and indirectly, many Greeks are tied to the sea. The population is clustered largely along the coasts, and even now many fish for their livelihood. Shipping is also an economic mainstay, and Greece boasts a substantial merchant marine fleet. The sea, fishing villages, and Greece's many islands provide a perfect backdrop for tourism. Dazzling white houses and lovely harbors are the order of the day on these islands; the chapels and windmills of Mykonos make it a favorite tourist destination.

The Miniature Microstates

In addition to the large countries of Southern Europe, this region has a handful of tiny (but full-fledged) countries called *microstates:*

➤ **Monaco:** Next to Southern Europe, on the French Riviera, this infinitesimally small constitutional monarchy is occupied almost entirely by the city of Monte Carlo. Monaco is known for Grace Kelly (Princess Grace), casinos, and its jet-set residents and visitors.

➤ **Liechtenstein:** This microscopic principality in central Europe is set high in the Alps astride the Rhine River and between Switzerland and Austria. Gorgeous mountains and fairy-tale castles make Liechtenstein a favorite tourist destination.

➤ **Andorra:** This miniscule microstate is sandwiched between Spain and France, where Catalan-speaking farmers tend to their goats deep in the Pyrenees.

➤ **Malta:** South of Sicily in the Mediterranean Sea, this tiny island did not become a republic until 1974, when the British departed after almost 200 years of military presence.

➤ **San Marino:** Perched atop Italy's Mount Titano, it was a haven for refugees during World War II.

➤ **Vatican City:** In the heart of Rome, this Lilliputian state, wrapped around Saint Peter's Basilica, is the home of the Pope and is the spiritual center for more than one billion Roman Catholics worldwide.

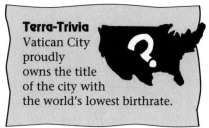

Terra-Trivia
Vatican City proudly owns the title of the city with the world's lowest birthrate.

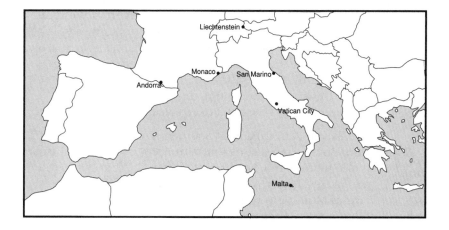

Microstates.

Mother Nature's Chosen One

For centuries, writers have waxed poetic about the natural beauty of this region, graced by sunshine and spectacular coasts. Although the Mediterranean was once cloaked in virgin forest, centuries of human occupation have left it largely deforested. Where forests once stood, vineyards and olive groves now grace the gentle slopes of the region.

The Lay of the Land: Peninsulas and Mountains

Physical features of Southern Europe.

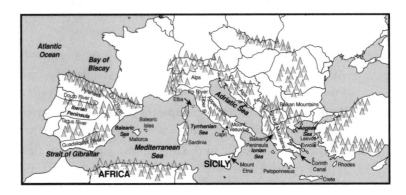

As you can see from the map, Southern Europe can be likened to a series of peninsular fingers poking into the Atlantic, the Mediterranean, and the region's assortment of seas. The largest of these peninsulas, the Iberian Peninsula, is located at the region's western edge. Occupied by Spain and Portugal, Iberia is bounded on the north by the Pyrenees Mountains, which separates the peninsula from France. To the south, only eight miles and the Strait of Gibraltar separate Iberia from Africa. The interior of the Iberian Peninsula consists largely of high, rugged plateaus.

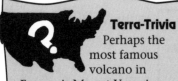

Terra-Trivia

Perhaps the most famous volcano in Europe is Mount Vesuvius, near Naples, Italy. In A.D. 79, Vesuvius erupted, spewing toxic gas, volcanic ash, and mud across the surrounding countryside. Although some 15,000 people were estimated to have perished in the cataclysm, the Roman cities of Pompeii and Herculaneum were buried in ash and preserved remarkably intact for archaeologists of our era to study.

To the east of Iberia is the boot-shaped peninsula of Italy. A mountain range also separates this country from Western Europe: This time, it's the Alps. Notable passes through the Alps include the Brenner from Austria, Maloja from Switzerland, and Little St. Bernard (yes, of canine fame) from France. Running the entire length of the "boot" is Italy's spine, the Appennine Mountains. To the south, on the island of Sicily, is the still-active volcano Mount Etna.

The Balkan Peninsula is on Southern Europe's eastern edge. Containing Greece at its southern tip, the Balkan Peninsula is also mountainous and rugged. From the snarling southern end of the Dinaric Alps and the Balkan Mountains, the Pindus Mountains run south through the peninsula and dominate the Grecian landscape.

South of the Balkan Peninsula and separated from it by the Corinth Canal is the Peloponnesus. It's difficult to know whether the Peloponnesus, made famous by the Peloponnesian Wars between Sparta and Athens, should be called a peninsula or an island.

Islands Everywhere

In addition to the peninsular nature of the region, islands abound. The most abundant are the more than 2,000 islands of the Greek archipelago (string of islands). The largest of the Greek islands are Rhodes, Lesvos, Evvoia, and Crete, shown on the preceding map.

Italy also has its islands, including the large islands of Sicily and Sardinia. Another Italian island of note is a favorite tourist spot, the tiny Isle of Capri. Located off Sorrento (near Naples), Capri is home to the famous Blue Grotto, where limestone formations are visible underneath the iridescent waters.

Spain controls the Balearic Islands (Mallorca is the best known) on its Mediterranean east coast and also owns the Canaries off the west coast of Morocco. Portugal's islands include the Azores and the Madeiras (as in the wine), located to the west and southwest in the Atlantic Ocean.

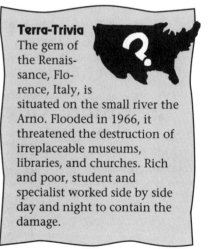

Terra-Trivia
After the French emperor Napoleon Bonaparte's horrible losses after the invasion of Russia, he was exiled to the small Italian island of Elba (between Italy and Corsica). He returned from exile only to watch his troops be defeated by British and Prussian forces in the Battle of Waterloo (1815).

Warm Waterways

As you can see from the preceding map, the Mediterranean Sea dominates Southern Europe. Only Portugal has no coastline on this sea (its coastline is exclusively on the Atlantic Ocean). A myriad of smaller seas subdivide the Mediterranean.

Spain shares coastlines with the Atlantic Ocean on the west and the Mediterranean's Balearic Sea on the east. To Spain's south is the Strait of Gibraltar, which connects the Atlantic and the Mediterranean. To the north of Spain lies the Bay of Biscay.

As Italy juts southward into the Mediterranean, it edges the Adriatic Sea on its northeast and the Tyrrhenian Sea on its southwest. Between the sole of the Italian "boot" and Greece is the Ionian Sea. The bulk of the Greek islands are located in the Aegean Sea between mainland Greece and Turkey.

Terra-Trivia
The gem of the Renaissance, Florence, Italy, is situated on the small river the Arno. Flooded in 1966, it threatened the destruction of irreplaceable museums, libraries, and churches. Rich and poor, student and specialist worked side by side day and night to contain the damage.

Southern Europe has several rivers of note, as shown on the preceding map. The most important, the vital Po River valley, cradles the bulk of Italy's agriculture, industry, and population. Historically important is Italy's Tiber River, which still flows through Rome.

Spain also has several significant rivers. The Ebro's waters cut deep gorges in the landscape, creating beautiful canyons in the headwaters. Because Northern Spain uses the Ebro (or Spanish Nile) for irrigation, a fertile, populous region developed around the Ebro River basin. The tiny district of Rioja boasts the best wine in Spain. Matured in special oak kegs, this light red wine is a famous representative of the Ebro River basin.

The Guadalquivir River in the southwest passes through Córdoba before finally flowing to the Atlantic. The fertile lowlands of the Guadalquivir valley are one of Spain's primary agricultural areas. Originating deep in the heart of Spain is the Tagus River, which passes through Toledo on its way to bisecting Portugal and emptying into the Atlantic Ocean at Lisbon.

Climate: The Mediterranean Sun

Almost all of Southern Europe enjoys a Mediterranean climate. Hot and dry summers prevail, and the brown land lies dormant until the winter brings cooler and wetter days. Northern mountainous areas throughout the region can be cold and snowy. On the Spanish interior plains, arid is the word of the day, and temperatures are extreme. Endlessly flat and treeless, this interior region called *secano* (unirrigated) is by no means barren. Fields of corn, rows of gnarled olive trees, and vineyards crawl over the prairies of La Mancha (which means "parched earth"), the home of the legendary Don Quixote, complete with windmills to grind the corn.

Maquis or *makis* (or what Californians call *chaparral*) is the common vegetation of the region. This scrubby vegetation can survive arid conditions: Its thorns and leaves are hard enough to hold water—if the goats don't eat them first. The maquis has replaced the native forests, which were long ago burned or cut for lumber. Remnants of these once prolific forests still remain in Italy's Po River valley.

The Historic Cities

Major Mediterranean cities.

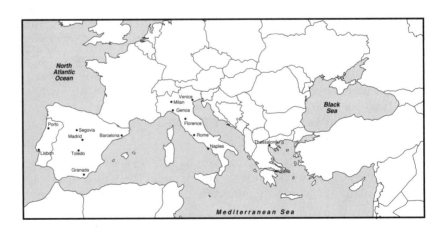

History abounds in the cities of Southern Europe. Although the capital cities of Lisbon, Madrid, Rome, and Athens each have storied pasts and were primary contributors to the course of Western history, many secondary cities in the region are also important:

Greece

➤ **Athens:** The port of Greece's principal city (its largest and most important), Piraeus handles most of the country's massive amount of shipping. Postwar urban sprawl marks this ancient city of the Parthenon and the Acropolis (the citadel of the city once dedicated to the Greek goddess Athena). The many remaining relics are threatened by the city's terrible, vehicle-caused smog—exacerbated by the typical Mediterranean double commute that results from the midday siesta. Athens was also the site of the first modern Olympics, in 1896.

> **GeoJargon**
> Athens is a classic example of what geographers call a *primate* city, a country's largest and the most important historically, culturally, and economically. A primate city is often the capital of its country.

➤ **Thessaloniki:** This city serves as a cultural and religious center for the northern part of Greece. Byzantine art graces its museums.

Italy

➤ **Rome:** The Eternal City offers a mix of the ancient, medieval, Renaissance, and, of course, modern eras. Built on the famous seven hills, Rome embraces its ancient lifeline, the Tiber River. The city is Italy's capital and is home to spectacular piazzas, churches, and concerts at the Baths of Caracalla, with its huge open-air opera stage.

The Colosseum, the Roman Forum, and the Circus Maximus hearken back to the heyday of the empire. Tourists throw coins into the Trevi Fountain and explore the shops of the Via Veneto. Local wine from the surrounding Alban Hills makes its way to the white linen-covered tables of Rome's many small trattorias.

➤ **Milan:** Unlike cities farther south, Milan suffers from twentieth century overload: This fast-paced city has both skyscrapers and subways. Textiles and fashion chic also mark this cosmopolitan city. The world-renowned La Scala opera house, Leonardo da Vinci's *Last Supper*, and the ornate Gothic Duomo (cathedral) are its historic treasures.

➤ **Turin:** Italy's industrial capital is the home of Fiat and more than 20,000 factories. Sophistication and elegance match the leather goods and vermouth. Turin is also home to the controversial Shroud of Turin, claimed by some to be the burial shroud of Jesus Christ.

➤ **Genoa:** In Italy's leading seaport and shipbuilding port, its local boy of note, Christopher Columbus, is suitably memorialized.

➤ **Venice:** A city built on 100 lagoons, this former maritime power traded in silk and spices from the East. The wealth this trade generated financed palaces on the canals that Venice uses as streets. A city of incredible beauty, it struggles with modern depredations: Motorboat vibrations, fumes from boats and on-shore industries, sea salt, and the draining of subterranean aquifers for industrial use have caused the city to slowly deteriorate and sink.

➤ **Florence:** The birthplace of the Renaissance and the modern age, Florence was the home of Galileo, Dante, Michelangelo, Luca della Robbia, Donatello, Ghiberti—the list goes on. The arts of Florence flourished under the ruling Medici family and are all exhibited for the world's art lovers to enjoy.

➤ **Naples:** This major port is the traditional center of culture, learning, and commerce.

Spain

➤ **Madrid:** Resting high on the central plateau, Spain's capital is a political hub, financial center, and industrial complex. Broad boulevards and the Puerta del Sol (Gate of the Sun) mark the central city. It has the Botanical Gardens, the Parque del Retiro, and one of the world's most distinguished museums, the Prado, in which priceless works, including those of Goya and El Greco, are proudly displayed.

➤ **Toledo:** Toledo was called "the imperial and crowned city" until Philip moved the court to Madrid. Beyond the city walls is the Fabrica de Armas, where the steel weapons Toledo is famous for are still made. A lovely cathedral incorporates several Spanish architectural styles.

➤ **Segovia:** This city is known as the royal Alcázar, the legendary El Cid's burial spot, and the ancient Roman aqueduct from the first century A.D.

➤ **Barcelona:** With a mild climate that features more plentiful rain than elsewhere in Spain, Barcelona is the hub of Catalonian Spain. This wealthy city, a main seaport and the center of commerce, hosted the 1992 Summer Olympics. The Picasso Museum and the buildings of Antonio Gaudi, who popularized the brash "modernisme" architectural style, draw tourists here.

➤ **Seville:** The home of Cervantes' *Don Quixote* and Bizet's *Carmen* is also a major seaport to the Atlantic by way of the Guadalquivir River.

➤ **Grenada:** The Alhambra Palace is a beautiful reminder of this city's Moorish past.

Portugal

➤ **Lisbon:** The country's capital and political center has Portugal's largest population. Overlooking the harbor at the mouth of the Tagus River, Lisbon is a city of black-and-white tiled streets and figs, carobs, and pomegranates. In this historic maritime center, seafarers prayed in the chapel before their expeditions.

➤ **Porto:** In northern Portugal at the mouth of Douro River is the port of Porto. From the terraced countryside, grapes flavored with brandy, making the delicious port wine, float in casks on flat-bottomed boats to Porto. Caves there act as cellars while the port awaits shipment.

People of the South

As might be expected in so mountainous a region, the people tend to be clustered in river valleys and along the coast, especially in Greece. More people derive their livelihoods from agriculture than is typical in Europe, and urbanization is lower than the European norm. Language and religion are generally straightforward, with a few twists, as described in this section.

> **Terra-Trivia**
> "Green Spain" refers to Galicia in northwestern Iberia. The area is known for its jagged granite shores, misty lochs, hillside apple orchards, and terraces for grapes. This beautiful corner produces 25 percent of Spain's timber and is now in the grip of overdevelopment.

Roman Catholic and Greek Orthodox

Although the Roman Catholic and Orthodox churches share common origins, the schism of A.D. 1054 finally separated the Eastern and Western churches. Southern Europe is overwhelmingly Roman Catholic, with the notable exception of Greece. The Greek people, who are almost entirely Greek Orthodox, represent the sole enclave for the Eastern Church in the region.

Romance Languages—and a Few Others for Good Measure

Romance languages aren't the sole tongues spoken in this region; not all the languages spoken there, in fact, even belong to the Indo–European language family. A few flies in the language ointment make sweeping language statements about the area a little sticky.

The Romance branch of the Indo–European family is dominant in three of the four countries in the region. Virtually everyone speaks Portuguese. In Spain, people speak either Castillian Spanish or Catalan Spanish (in the northeast). In the far northwest portion of Spain, people of Celtic stock speak Galician, which is related to Portuguese. The Basque language, spoken in north central Spain, is an odd bird discussed later in this chapter.

Italy is also in the Romance camp: Italian is the language of the land. Regional differences exist on Sardinia, where the Sardinian version of Italian predominates. In the far north, pockets of Romansh (another Romance language) exist, and even a little Upper German is spoken there.

In a break from the Romance branch, Greek is in the Hellenic branch of the Indo–European language family. The Greek language is spoken across almost all of Greece and its islands. It may not be a Romance language, but at least it's in the Indo–European family.

The same thing can't be said of Basque. Unique to Southern Europe, Basque is not a Romance language and is unlike any other language known today. Iberia's oldest native inhabitants, some 50,000 Basques live in a beautiful pocket of the Pyrenees in northern Spain. With their unique language and Asian nomadic stock, the Basque people tend to be taller and fairer than most Spaniards. These tough people enjoy stone-lifting contests and play *pelota,* or jai alai.

Though agricultural, the Basque region has led Spain's industrial growth in iron and steel, contributing much to the Spanish economy. Unemployment and dissatisfaction have spawned the E.T.A., a powerful terrorist organization, and has led to nationalist desires and violence. Remember that language differences mark cultural differences and often give rise to autonomy movements.

Geographically Speaking

Enclosed by land on all sides, the Mediterranean Sea's only exits are by way of the Strait of Gibraltar abut and the tight Suez Canal. Most of the rivers from Southern Europe empty into the Mediterranean, as do rivers from Western Europe, Eastern Europe, the Middle East, and Saharan Africa.

Pollution is especially acute in the Mediterranean Sea because with so narrow a link to the Atlantic Ocean, little flushing action occurs. Salinity (saltiness) and pollution prevail as pollutants from northern Italy's industry, rivers, and coastal cities and resorts pour into the Mediterranean. With so many people's livelihoods depending on these waters, it's no surprise that Southern Europeans are deeply concerned.

Looking Toward Independence?

Although the Basques are quite active in their nationalist demands, they're not the only rumblings in Spain. As mentioned, in northeastern Spain, Catalan is spoken rather than the official Castillian Spanish spoken elsewhere in the country. This area, called Catalonia, is centered around the principal city of Barcelona.

Catalonians feel more European than people in the rest of Spain, perhaps because of their progressive outlook and a proximity to France and Europe's core. Like the Basques, Catalonians believe that they do more for the country than do other areas. Economically, they're correct because textile and chemical production have put the area way ahead of the rest of Spain. Catalonia also benefits from the mild coastal climate and its strategic geographic position for trade and industry.

In Catalonia, people look more to Barcelona as their cultural center than to the Spanish capital, Madrid. Economic disparity, language differences, and different cultural affinities all tend to point toward separatism. Demonstrations and political maneuverings have punctuated the Catalonians' point. Will independence happen? Stay tuned.

The Least You Need to Know

➤ The Greeks and Romans left a lasting cultural imprint on the region of Southern Europe.

➤ All four countries in the region (Portugal, Spain, Italy, and Greece) are located on peninsulas and have many islands, especially the archipelago of Greece.

➤ All the countries except Portugal abut (border) the Mediterranean Sea, and all share a Mediterranean climate.

➤ The region is mountainous and tends to depend more on agriculture than does the rest of Europe.

➤ Most of the region shares Romance languages and the Roman Catholic faith, except for the Greeks, who speak a Hellenic language and practice the Greek Orthodox religion.

➤ Basque nationalist feelings threaten northern Spain with violence, and Catalonia has separatist tremors.

Unbloc-ing Eastern Europe

Eastern Europe is a collection of "formers": former eastern bloc countries that were, until recently, satellites of the former Soviet Union; former Soviet republics that, until recently, were portions of the huge country of the Soviet Union; former pieces of Yugoslavia that, until recently, were united under a single totalitarian government. In addition to the word *former*, the phrase *until recently* pops up frequently whenever someone talks about these countries.

To say that Eastern Europe has undergone some changes in the past decade is the ultimate geographic understatement. Even if you're a geography aficionado, the events of the past several years that have redrawn the lines on the map can be confusing. If you're a geographic novice, only a few names may be familiar—and only because they've flashed across the nightly news so often in the past few years.

History of the Turmoil

The latter part of the twentieth century has not been particularly kind to Eastern Europe. During World War II, more than 1.3 million Eastern Europeans were killed on the battle-field. Many more civilian casualties were suffered as first the Nazis and then the Soviet Red Army carved swathes of destruction through the lands of Eastern Europe.

In addition to battle deaths, the Nazi occupation during World War II brought unprecedented genocide to the region. In Nazi concentration camps, many sited in Eastern Europe, more than six million Jews (approximately two-thirds of the entire prewar European Jewish population) were exterminated. Jews, however, were not the only victims of the death camps. The Nazis also targeted Eastern European Gypsies and Slavs (Poles, Ukrainians, and Belarussians), and it's estimated that nine to ten million of these peoples were murdered.

The Soviet Bear

Following World War II, much of Eastern Europe fell under the sway of the Soviet Union. Economically and militarily, the Soviet Union dominated the original eastern bloc countries (Poland, Czechoslovakia, Hungary, Romania, and Bulgaria). Yugoslavia resisted Soviet incursions under its strong-armed leader Marshal Tito. Nonetheless, Yugoslavia followed the communist path, albeit its own. Albania also took a communist path, but aligned itself with Maoist Red China rather than with the Soviet Union. This odd alliance made Albania unique in Eastern Europe.

The reshuffling of Eastern Europe after the war meant the movement of more map boundary lines. The entire country of Poland was shifted west (gaining territory from Germany and losing territory to the Soviet Union). The Baltic states of Estonia, Latvia, and Lithuania were absorbed into the Soviet Union and ceased to exist as separate nations. Romania lost to the Soviet Union the territory known as Bessarabia, and Hungary's war gains were lost following the war. As a result, cartographers were busy in the years following World War II.

As the Cold War intensified in the 1950s and 1960s, several Eastern European countries attempted to gain greater autonomy from the Soviet Union. Uprisings and general strikes in Hungary (1956), Czechoslovakia (1968), and Poland (1960s to 1970s) were quelled by the Soviets, sometimes through brute force. Personal incomes lagged behind those in Western Europe as Eastern Europe foundered economically under the communist system.

Countries in Bloom

By 1991, the Soviet Union had finally crumbled, and, in the process, new nations began to appear on the Eastern European map. In the northern part of the region, the Baltic States (Estonia, Latvia, and Lithuania) were formed. Along the former Soviet Union's western edge that bordered its former satellites, Belarus (or Byelorussia or Byelarus), Ukraine, and Moldova were formed. Among the satellites, the former Czechoslovakia split peacefully to form the Czech Republic and Slovakia.

Peaceful is not a word that describes the dissolution of the former Yugoslavia. Tensions in that country had run high ever since the various ethnic groups (including Croats, Slovenes, Macedonians, Serbs, and Montenegrins) were patched together after World War I to form a country. Although the strong leader Marshal Tito had held the country together after World War II, autonomy movements spread throughout Eastern Europe in the early 1990s, and Yugoslavia split along ethnic lines.

Because of years of mistrust and ethnic tensions, the split was not cordial. "Ethnic cleansing" became a household word as Serbian troops massacred thousands of Bosnian Muslims. Croats, Serbs, and Muslims engaged in a vicious land grab as borders (along with people) were pushed back and forth across the former Yugoslavian landscape. Finally, Slovenia, Croatia, Bosnia and Herzegovina, Yugoslavia, and Macedonia emerged from the rubble.

What's left is a jumbled mess of 18 separate countries in Eastern Europe. Be thankful that the former Eastern and Western Germany united—that makes one less Eastern European country to keep track of. The situation can be simplified somewhat by subdividing the region into three subregions each (quite neatly) containing six countries. (This grouping works out well for now—as long as no more new countries spin off before this book is printed.)

! Geographically Speaking

The Cold War, the Iron Curtain, the eastern bloc, and the era of Soviet satellites were most dramatically symbolized by the Berlin Wall. Located in the once-divided city of Berlin in the former East Germany, the "wall" consisted of the wall itself, a minefield, barbed wire, spotlights, and guard towers. The Berlin Wall divided Berlin and prevented East Germans from emigrating to the West.

Like many countries in the region of Eastern Europe, East Germany was a Soviet satellite. For that reason, it would have been included in this chapter if the reunification of East Germany and West Germany had not occurred. It's a little strange to discuss in this chapter the bulk of the eastern bloc countries and not include the most tangible symbol of the Cold War era, the Berlin Wall.

Out on Their Own: The Former Soviet Republics

The Eastern European subregion of the former Soviet Republics can be split into two groups. As you can see on the map of Eastern Europe, earlier in this chapter, in the north are the three small former republics called the Baltic States: Estonia, Latvia, and Lithuania. The second grouping consists of three more former republics that now are the countries of Belarus, Ukraine, and Moldova.

GeoJargon
The *Baltic States* take their name from their position on the Baltic Sea, a long arm of the Atlantic Ocean via the North Sea. Poland shares with the Baltic States a coastline on the Baltic Sea.

Estonia

As the term Baltic State implies, Estonia is located on the Baltic Sea. The smallest of the Baltic States, Estonia is flat and dotted with lakes. Its capital is Tallinn; with almost half a million residents, it has about one-third of Estonia's people. More than one-third of Estonia's population is Russian and other non-ethnic Estonians. Russia still has a stake in Estonia because of Russians still living there.

Although Estonia is rich in oil shale, exploitation under the Soviets' rule resulted in extreme environmental degradation.

Only the Gulf of Finland separates Estonia from Finland; as you might guess, the two countries have much in common. Many Estonians speak Finnish, and the official language, Estonian, is a Finnic language of the distinctive Uralic family. As in Finland, Lutheranism is the dominant religion in Estonia.

Latvia

Another Baltic State, Latvia, is the home of the largest city in the Baltic States. Riga, its capital, is sometimes called "the Paris of the Baltic" and has almost 900,000 people. Latvia has the highest per capita wealth of the three Baltic States, with an established industrial base that can be built on as Latvia restructures its economy. It must also clean up the severe pollution that plagued the country under the Soviets.

Just more than half of Latvia's people are ethnic Latvians. As in Estonia, the country has a large Russian minority (30 percent) and also has substantial Ukrainian and Belarussian minorities. How Latvia deals with its ethnic mix and rebuilds its economy is a key to the future success of this small country.

Lithuania

Lithuania is the southernmost Baltic State (with a border on Poland) and the most populous (with a population of about 3.7 million). Through the centuries, its shared border with Poland has forged close cultural ties with that country. Lithuania differs from its fellow Baltic States in religion because, like Poland, it's heavily Roman Catholic.

This gently rolling, forested land has a strong agricultural base. In addition to exporting meat and dairy products, Lithuania is still a prime source of amber, for which it is internationally renowned. Its people speak Lithuanian, which is a Slavic tongue closely related to Latvian (Russian and Polish are also common). Lithuania is more ethnically homogeneous than the other Baltic States, and its capital, Vilnius, has 600,000 people.

Belarus

South of the Baltic States and bordering Latvia and Lithuania is the country of Belarus. This low-lying country is dominated by rivers, lakes, forests, and lots of swamps. The population center and industrial center of Belarus is its capital, Mensk (or Minsk), with 1.6 million residents. The people speak Byelo–Ruthenian (White Russian), which is closely related to Russian. The entire country, in fact, maintains close cultural and economic ties with Russia, and the two are discussing reunification.

Terra-Trivia

Before the Soviet Union crumbled, the huge country was composed of 15 Soviet Socialist Republics. Six (Estonia, Latvia, Lithuania, Belarus, Ukraine, and Moldova) are now countries in Eastern Europe. The other nine (Russia, Georgia, Armenia, Azerbaijan, Kazakstan, Uzbekistan, Turkmenistan, Kyrgyzstan, and Tajikistan) are described in Chapter 20.

Although Belarus was heavily industrialized under Soviet rule, its agricultural industry employs one-quarter of the country's workforce. The people of Belarus were hard hit in the aftermath of the 1986 Chernobyl nuclear power plant disaster that occurred in nearby Ukraine. More than 20 percent of the country was contaminated by radiation from the accident, and the response has since diverted much of Belarus' resources.

Ukraine

Although Ukraine is one of Europe's newest countries, it's now Europe's largest in terms of area, with more than 233,000 square miles. Because of its dark, fertile soils that cover most of its large expanse, it was once known as the "breadbasket of the Soviet Union." During Soviet rule, drawing on Ukraine's huge coal reserves, heavy industry (as well as heavily polluting) was stressed. The economy struggles to recover from aging industrial plants and the huge monetary drain of the Chernobyl clean-up.

The people of Ukraine trace their ancestry to the steppe Cossacks. Although almost three-quarters of them are ethnic Ukrainians, more than one-fifth of the population is Russian. The Slavic people of Ukraine speak Ukrainian, which is similar to Russian and Byelo–Ruthenian. The predominant religion in the country is Eastern Orthodox, with some Ukrainian Catholicism. The capital and largest city is Kyiv (or Kiev), with more than 2.6 million people. Including the capital, Ukraine has four cities with more than one million people, including the Black Sea port city of Odesa (Odessa).

Moldova

Another former Soviet Republic country is very much a transitional zone between eastern Slavic peoples and western Romanians. Part of Moldova is the area called Bessarabia, which was taken from Romania by the Soviets after World War II. As a result, much of the ethnically diverse populace speaks the Romance language Romanian rather than the Slavic tongues of the east. Because the land is an extension of the dark, rich soils found in Ukraine, it's heavily agricultural.

Although Moldova is a small country, its rich soils have nurtured a large population and produced the highest population densities in the region (approximately 345 people per square mile). The capital and largest city is Chisinau, with 750,000 inhabitants. As with its neighbor and cultural soulmate to the west, Moldova's people are largely Romanian Orthodox in their faith.

Out of Orbit: The Former Soviet Satellites

Shown on the map of Eastern Europe earlier in this chapter, the six countries that comprise this subregion were all considered Soviet satellites, or dependents, after occupation at the conclusion of World War II. Although not directly governed by the Soviet Union, these countries were almost completely controlled politically and economically by their big brother to the east. At various times, the Soviets even intervened militarily if one of their flock ventured too far from the fold. Since the disintegration of the Soviet Union, these countries have abandoned communism and are in transition as they forge their places in a market economy.

Poland

Poland is the largest of the six former Soviet satellites in land area (more than 120,000 square miles) and population (38.6 million). Much of Poland is a vast plain with its only highlands in the south where the Carpathian Mountains run around Poland's southern border. Although heavy industry spread during the years of Soviet domination, almost one-quarter of Poland's workforce is involved in agriculture. Coal, shipbuilding, and iron and steel are now all key industries.

GeoJargon
A *satellite*, as used in the political sense, is a country that's subordinate to or dominated politically or economically by another country.

The Slavic language Polish is spoken almost universally in Poland, and the people are overwhelmingly Roman Catholic in their faith. Warsaw, with a population of 1.6 million people, is the capital, the largest city, and the cultural heart of Poland. Smaller but well-known cities include Krakow (in Poland's industrial belt) and Gdansk (the site of Solidarity labor-union shipyard strikes).

The Czech Republic

After World War I and following the disintegration of the Austrian-Hungarian Empire, Czech territory was united with Slovakia to form the country of Czechoslovakia. After the breakup of the Soviet Union, the country's citizens voted to split along ethnic lines and form the two new countries of the Czech Republic and Slovakia. The Czech Republic includes the old Czech lands of Bohemia and Moravia.

Terra-Trivia
Running through much of Eastern Europe is the mighty Danube River. It runs through or touches more countries than any other river on earth, including Ukraine, Moldova, Romania, Bulgaria, Yugoslavia, Croatia, Hungary, Slovakia, Austria, and Germany (the Danube's source is in Germany's Black Forest). Called the beautiful Blue Danube (of waltz fame), the brown river is still beautiful but far from blue.

The Czech Republic contains the more urbanized, industrialized, and westernized portions of the former Czechoslovakia. Despite Czechoslovakia's industrial base, however, its industries are extremely outmoded and in need of modernization. The heart of the republic is the beautiful capital of Prague (population 1.2 million). The people speak Czech, a Slavic language close to Slovak and Polish. Roman Catholicism is the predominant religion in the republic.

Slovakia

The other half of the former Czechoslovakia is the present country of Slovakia. Not to be confused with Slovenia (a small country to the south resulting from the breakup of Yugoslavia), Slovakia is less developed and smaller in both size and population than the Czech Republic. Slovakia is a land of rough mountains, forests, and high pastures.

Slovakia has ties to its neighbor Hungary via a significant ethnic Hungarian minority population. Although Slovak is the official language, Hungarian is also spoken in many areas. Some tension exists in Hungary over the role these Hungarian minorities will play in the new country. The capital of Slovakia is Bratislava, with almost half a million inhabitants. Roman Catholicism is the principal religion, although significant minority religions include Protestantism, Eastern Orthodox, and Judaism.

Hungary

The fertile plains of Hungary have helped to make this agricultural country fairly prosperous compared to other Eastern European countries. Although initial Soviet occupation was harsh (Hungary had sided with the Nazis in World War II), eventually some privatization and free markets were allowed. Despite Hungary's agricultural nature, its capital, Budapest (population two million), is one of the largest cities in Eastern Europe. One in five Hungarians calls this beautiful city home.

Hungary, the ancient home of the Magyars (an Asian tribe), still preserves this cultural legacy through its language. Hungarian is a unique language, in the Ugrian branch of the Uralic family, that is unlike any other in Europe (its closest relatives are Lapp and Finnic). Almost 70 percent of Hungary's people are Roman Catholic, although 25 percent are Protestant.

GeoJargon
The Balkan Peninsula includes the land area below an approximate line drawn from the northern Adriatic Sea to the northern Black Sea and is home to the Balkan Mountains, where it gets its name. The term *balkanization* has come to mean breaking up into smaller, hostile units, which certainly describes the Balkan Peninsula, where its many ethnic groups have been fighting among themselves for centuries.

Romania

Romania's core is dominated by the rugged Carpathian Mountains running north–south and the Transylvanian Alps that run east–west. By the way, it's the same Transylvania of Count Dracula fame (who did exist and was no humanitarian, although the vampire stuff may be somewhat exaggerated). Although Romania has ample mineral resources, its oil and gas fields have begun to play out.

The capital of Bucharest is by far Romania's largest city, with more than 2.3 million people. Although the country's principal ethnic group is Romanian, a sizable Hungarian minority also exists. The language spoken is primarily Romanian, although Hungarian and German are spoken in some areas. The main religion in Romania is Romanian

Orthodox. For almost 25 years, Romania labored under an authoritative regime headed by Nicolae Ceausescu. It took a bloody coup in 1989 for Romania to finally free itself from Ceausescu's maniacal grip.

Bulgaria

Bulgaria, a mountainous country bisected by the Balkan Mountains, is one of three Eastern European countries (along with Romania and Ukraine) to have a Black Sea coastline. Before the late 1800s, this area was dominated by the Turks, who still have a notable ethnic minority in the country. Throughout years of Soviet domination, Bulgaria pretty much toed the party line. The country was heavily agricultural, but was somewhat industrialized under the Soviets.

Bulgarian is the language of the land, with a smattering of Turkish spoken in mountain enclaves. As with most Eastern European capital cities, Sofia (population 1.1 million) is several times larger than the next largest city. A shaky government, high crime rates, and a faltering economy form a dark cloud over Bulgaria's future.

Ethnic Powder Keg: The Former Yugoslavia and Albania

This subregion of Eastern Europe contains six countries that occupy the western side of the Balkan Peninsula, as shown on the map of Eastern Europe, earlier in this chapter. The fractured, mountainous landscape reflects the ethnic splintering that has long dominated the history of this subregion. Although a description of the complexities of the many ancient land claims and seething rivalries in the former Yugoslavian countries and Albania is beyond the scope of this book, this section briefly summarizes them.

Fractures and Feuds: The Former Yugoslavia

Although this ancient land has a long history of settlement, the political entity called Yugoslavia has a relatively short history. With the disintegration of the Austrian–Hungarian Empire at the close of World War I, Yugoslavia was formed. Its early name was the Kingdom of Serbs, Croats, and Slovenes. This federation of loose kingdoms included Croatia–Slovenia, Bosnia and Herzegovina, Dalmatia, and Serbia and Montenegro. Even then, the groups didn't much like each other.

A harsh Nazi occupation characterized World War II, and from the ashes of the war emerged the ironfisted Marshal Tito, who brought a rigid unity to Yugoslavia. By elimi-nating all opposition, he imposed a communist system that the Soviet Union was never able to dominate. (Yugoslavia was not considered a Soviet satellite.) Although Tito's brand of communism offered more personal freedoms than the Soviet variety, Yugoslavia's economy still lagged behind its western counterparts. Communism also offered the illusion of unity; just below the surface, however, ethnic rivalries continued to seethe.

After Marshall Tito's death in 1980, a rotating form of government attempted to quell the hatred that had long existed between Yugoslavia's many peoples. It was not to be: By the early 1990s, autonomy movements were in full swing. Violence flared as Slovenia, Croatia, Bosnia–Herzegovina, and Macedonia all sought autonomy. Serbia and Montenegro possessed the bulk of the former Yugoslavian army—and used it against their former countrymen.

As the various factions struggled to scribe new borders on the map, the sides engaged in "ethnic cleansing" to solidify their claims and to vent historic hatreds. People were pushed back and forth across shifting borders, and, worse, entire villages were slaughtered in the name of ethnic purity. A tenuous peace now exists, and United Nations troops keep watch until the next stone is thrown.

What remains of the former Yugoslavia is Serbia and Montenegro. Slovenia, Croatia, Bosnia–Herzegovina, and Macedonia all have spun off to form independent countries. Even with the shifting and ethnic cleansing, the region remains an extremely complex patchwork of ethnic pockets. Rather than go into great detail about each country, this section provides an overview of the former Yugoslavian countries and then notes a few particulars.

Much of the development of the area's isolated ethnic pockets is due to the configuration of the land. The entire region is dominated by the Dinaric Alps, whose rugged mountains and isolated valleys provide the framework for the development of individual cultures. Because the mountainous divides hindered cultural mixing, groups tended to develop separately. The result is today's multiple groups, such as Serb, Croat, Slovene, Montenegrin, Macedonian, and Bosnian.

The languages also are scattered. Although Serbo-Croatian, Macedonian, and Slovene are the most common, smatterings of Albanian, Hungarian, and Turkic are also spoken. Religious affiliation is no less complicated: Eastern Orthodox, Roman Catholicism, and Islam are the most common, and pockets of Protestantism also exist.

Here's a rundown of the new countries that formed from the dissolution of the former Yugoslavia (in this area, be aware that generalizing is fraught with inaccuracies!):

Terra-Trivia

In 1984, the world held its Winter Olympics in Sarajevo, Yugoslavia (now Bosnia–Herzogovina). The world watched while smiling faces and elaborate ceremonies proclaimed the oneness of the human spirit and the cooperation of all nations. Who was to know that within ten years Sarajevo would be in ashes, the victim of Serbian artillery and ethnic hatred?

Terra-Trivia

The predominant language throughout the former Yugoslavia is Serbo–Croatian. It has a number of ethnic peculiarities and has failed to be a unifying factor in the area. Although the spoken language is the same, the Croatians and Slovenes use the Latin or Arabic text, and Serbs, Montenegrins, and Macedonians use the Cyrillic alphabet. Spoken the same, written differently: Is there no common ground in the Balkans?

➤ **Bosnia–Herzogovina:** Its capital is Sarajevo (Mostar is the traditional Herzogovinian capital), and its population is a mixture of Muslim, Serb, and Croat groups.

➤ **Croatia:** This country has mostly ethnic Croats of the Roman Catholic faith. Its capital, Zagreb, has almost one million people.

➤ **Macedonia:** Alexander the Great's old stomping ground is now a mixture of Macedonians, Albanians, Turks, Serbs, and Gypsies. Its religions are Eastern Orthodox and Muslim, and its capital is Skopje.

➤ **Slovenia:** Slovenia has mostly Slovenes and some Croats, Serbs, and Muslims speaking Slovenian and Serbo–Croatian. The country is largely Roman Catholic, and its capital is Ljubljana.

➤ **Yugoslavia:** Composed of Serbia and Montenegro (but not officially recognized by all countries), almost the entire country speaks Serbo-Croatian. Its religious mix is mostly Eastern Orthodox and Muslim. Belgrade is its capital and the largest city in the entire area, with more than 1.1 million people.

Terra-Trivia
For centuries the beautiful Adriatic Sea and Dalmatian coast have beckoned vacationers. The Roman Emperor Diocletian retired here, and his huge retirement villa later encompassed a medieval town and the modern city of Split, Croatia. Perched on the edge of the sea and ringed with ancient walls, Dubrovnik (also in Croatia) has long been a traveler's dream. Unfortunately, it was severely damaged during recent ethnic struggles.

Albania: Chinese Ties and Isolationism

Now you can leave the tortured states of the former Yugoslavia to move on to the last Eastern European country described in this chapter—tortured Albania. Although it missed much of the ethnic infighting that characterized its neighbor to the north, its people have weathered their own isolated nightmare. Although Albania gained independence after the Turks dominated the area until the twentieth century, that era has not been particularly kind to the country.

Albania's government followed a strict form of communism patterned after the Maoist Chinese model. The country then lapsed into a long period of isolationism that resulted in economic ruin. Albania, now the poorest country in Europe, suffers from political turmoil and a population explosion. Both Albanian and Greek are widely spoken, and the capital is Tirana. The country was largely Muslim, but state-mandated atheism prevailed for decades, so it's hard to gauge who's what anymore.

Plains, Mountains, and Rivers

Physical features of Eastern Europe.

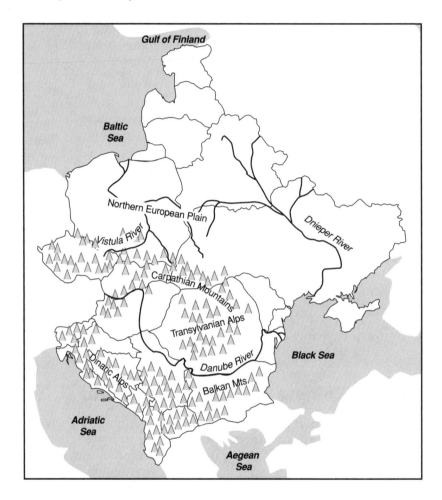

In addition to describing the people and culture of Eastern Europe, this chapter has taken you on a tour of the region's countries and noted some of its many problems. To conclude this overview, this section briefly describes the physical aspects that have helped to make this region unique, as shown on the preceding map.

The Landforms

To the north and east of Eastern Europe, the countries are dominated by large expanses of plains. Although this flatness has historically allowed armies to march across this portion of the region with depressing ease and frequency, the plains provide some blessing. With no intervening mountain range, the warmer air from the Atlantic Ocean and Baltic Sea help to moderate the climate. Although these plains areas are cold, they're warmer than might be expected for their latitude. The flat and fertile plains also support the agriculturally dependent populations.

Moving south and west through Eastern Europe, the plains give way to mountains. The Carpathian Mountains and Transylvanian Alps are in the center of the region, the Balkan Mountains are in the southeast, and the long chain of the Dinaric Alps are in the southwest. These mountains and their many valleys and basins gave rise to the diverse cultures that mark the region.

The Great Waters

Eastern Europe is sandwiched between three large bodies of water: To the north, it's bounded by the Baltic Sea; to the southeast, the Black Sea; and to the southwest, the Adriatic Sea. All except six of the region's countries share some coastline with one of these three seas. (Belarus, the Czech Republic, Slovakia, Hungary, Moldova, and Macedonia are all landlocked.)

The region is also laced by several important rivers. The northern plains are drained by Poland's Vistula River, which empties into the Baltic Sea. In the east, waters from the vast plains of Belarus and Ukraine drain south through the Dnieper River to the Black Sea. The region's most important river, the Danube, winds its way through most of the countries in the south to exit in the Black Sea.

The Least You Need to Know

➤ This region's 18 countries include former Soviet republics, former satellites, and new offshoots from the former Yugoslavia.

➤ The Balkan Peninsula dominates Eastern Europe's southern flank.

➤ The Danube River runs through much of the region.

➤ Mountains help to create ethnic divisions in the south.

➤ Slavic languages are the most common in the region.

➤ Although the region's people are highly literate, incomes typically lag behind those elsewhere in Europe.

The New Russia

This region is called the new Russia for good reason: The country is new! Okay, it's not exactly new; with the crumbling of the Soviet Union in 1991, however, Russia as a separate country reemerged. Perhaps more striking than the political distinctions is that almost every aspect of the society has undergone dramatic change in just the past ten years. The personal freedoms are new, a market economy is new, nuclear disarmament is new, the rebirth of religion is new—and fast-food restaurants in Moscow are new.

This chapter looks at Russia's origins, the rise of the Soviet state, the emergence of a new Russia, and its prospects for the future. Even without its former republics, Russia is a huge country. In any place as large, you would expect variety, and Russia has it. Varying landscapes, varying resources, and varying culture—each of these subjects is explored as this chapter describes the world's largest country.

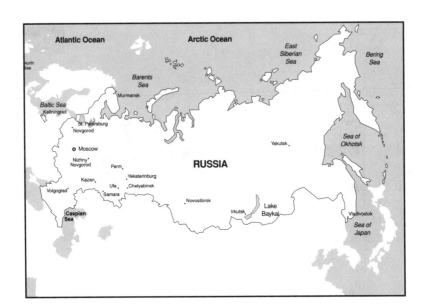

Tsars, Communists, and Free Markets

GeoJargon

Steppes are Russia's semiarid plains, where short grasses are the predominant vegetation.

The first Russian dynasty was started by Vikings who navigated Russia's rivers inland from the Baltic Sea to Novgorod—the oldest city in Russia, even though its name means "new town." By the eleventh century, eastern Russia was under the domination of Kiev, the most powerful city in the region at the time and now in Ukraine. Kiev's hold was not to last, however, as the Mongols swept across the *steppes*, or plains, to destroy the city. The Mongols were not expelled until the late fifteenth century.

The Tsarist Centuries

In the sixteenth century, Ivan the Terrible became the first tsar (the term used for a Russian ruler until 1917). Another noted tsar, Peter the Great, founded Saint Petersburg and began the process of westernization in a Russia locked in medieval ways. Later in the eighteenth century, the tsarina Catherine the Great maintained expansion of the empire and continued Peter's modernizations.

In the early nineteenth century, Napoleon led his armies onto the steppes in a foolhardy effort to subdue Russia. He was horribly defeated, as much by the Russian winter as by the Russian army (a history lesson Adolf Hitler failed to heed in the twentieth century). The last tsar was Nicholas II, who reigned over an increasingly corrupt regime until the Bolshevik Revolution of 1917. He and his family were subsequently shot by the Bolsheviks in 1918. The Communists, led by Vladimir Lenin and Leon Trotsky, persevered in a

144

violent civil war and by 1920 had complete control over Russia. The Union of Soviet Socialist Republics, or U.S.S.R., was born in 1922.

Union of Soviet Socialist Republics: The Age of Communism

After Lenin's death, a political power struggle began between Leon Trotsky and Joseph Stalin. Emerging victorious, Stalin eliminated opposition and expanded his power in a vicious manner. Through purges, starvation, collectivization, and work camps, Stalin was responsible for an estimated 20 million deaths.

As the Germans began their World War II expansionist policies in Europe, Stalin jumped on the bandwagon and signed a nonaggression pact with the Nazis in 1939. Under this measure of insurance, his armies invaded Poland and Finland and absorbed territories to the west. In 1941, Hitler double-crossed Stalin by invading the Soviet Union.

Although the Nazis made rapid initial advances across Russia's western plains, like Napoleon, they were slowed by Soviet resistance and the Russian winter. The Soviets made their stand in what became known as their "seven heroic cities." In 1942 and 1943, the tide of the "great patriotic war" turned against the Nazis in Stalingrad, later renamed Volgograd. Stalin pushed the Red Army west with a vengeance, grabbing territory, establishing satellites (politically subordinate countries), and exacting retribution along the way.

After World War II, more Soviet aggression toward the West prompted the Allies (Western forces) to respond by establishing the North Atlantic Treaty Organization (NATO). When Stalin attempted to isolate the Allies in west Berlin, they responded with the Berlin airlift, a massive airlift of food and supplies delivered from Western Europe and America to west Berlin. Satellite countries were cowed, boundary lines were redrawn, and the Iron Curtain was in place. The Cold War was in full bloom.

The Cold War was marked by several notable developments over its icy history. The Soviet Union embarked

Terra-Trivia

The Nazis exacted a horrible toll on the Soviet people. Some estimates place the Soviet battlefield deaths at more than 13.6 million. Civilian deaths, including in labor camps and concentration camps, totaled another 7.7 million. With more than 21.3 million military and civilian deaths, almost every family in the Soviet Union lost a loved one, and the survivors will never forget the "great patriotic war."

GeoJargon

NATO was established in 1949 and its headquarters located in Brussels, Belgium. The treaty's principal tenet is that an attack on any one member would be regarded as an attack on all. The Soviet Union responded with the formation of the Warsaw Pact, which included the former U.S.S.R. and its satellite countries. Russia is a signer of the NATO Partnership for Peace agreement but not a full member of NATO.

on a countrywide modernization program and stressed industrialization and collectivization through its planned economy. The modernization program was paralleled by an economically devastating arms race with the West (specifically, the United States). Both sides stockpiled nuclear weapons, and the entire world figuratively held its breath during the Cuban missile crisis of 1962. In 1979, the U.S.S.R. invaded Afghanistan, and U.S.–Soviet relations hit another low.

By 1985, change was in the air. Mikhail Gorbachev, the Soviet president, began plans for liberalizing personal freedoms and initiating elements of an open-market economy in his rigid country. The 1986 Chernobyl nuclear power-plant incident shook the world, and the catastrophic release of radiation had devastating effects on the region. Finally, after more than 70 years of harsh Communist rule, the U.S.S.R. was dissolved. In 1991, Russia joined with ten other autonomous and semiautonomous Soviet republics to form the new Commonwealth of Independent States.

The Loss of the Former Republics

Just how geographically different is Russia from the Soviet Union of yesterday? This section helps make a comparison between the old and the new. The Union of Soviet Socialist Republics consisted of 14 republics and one Soviet Federated Socialist Republic, Russia (as shown on the following map):

➤ Armenia	➤ Kazakstan	➤ Russia
➤ Azerbaijan	➤ Kyrgyzstan	➤ Tajikistan
➤ Belarus	➤ Latvia	➤ Turkmenistan
➤ Estonia	➤ Lithuania	➤ Ukraine
➤ Georgia	➤ Moldova	➤ Uzbekistan

Former Soviet republics.

Nowadays, this list reads like a United Nations roll because all the republics have formed their own countries (although Georgia has signed a cooperation treaty with Russia). Before the breakup, the U.S.S.R. had a land area of 8,649,489 square miles—the largest country in the world. Russia has a land area of 6,592,800 square miles—still the largest country, with more than 11 percent of the earth's total land surface. Nonetheless, Russia lost more than 2 million square miles, or almost a quarter of its former land area.

The losses in terms of population were even more dramatic. The population of the former Soviet Union was about 289,000,000. Russia's population is now 148,000,000, or almost a 50 percent drop. It's interesting that Russia's own population has declined in recent years. With a birthrate around 10 per 1,000 and a death rate around 25 per 1,000, the country is one of the few places in the world witnessing a negative population growth. Previously, ethnic Russians in the former U.S.S.R. represented only about 52 percent of the population. With the loss of its republics, the percentage of ethnic Russians has now soared to 82 percent of the remaining population of Russia.

Although Russia lost all its western republics, it still spans 11 time zones. The reason is that Russia retained a small, fragmented area, the Kaliningrad Oblast, located in the former area of East Prussia. This little parcel on the Baltic Sea, between Poland and Lithuania, extends Russia's 10 time zones by one to maintain the original 11 spanned by the U.S.S.R.

The result of the disintegration of the U.S.S.R. is that Russia remains the largest country in the world by land area. In terms of population, the U.S.S.R. was the world's third-largest country (behind China and India). Russia ranks sixth in the world in population (behind China, India, the United States, Indonesia, and Brazil).

Terra-Trivia
Running across the entire breadth of Russia is the famous Trans–Siberian Railway, or Baykal–Amur Mainline (BAM). The rail line connects Saint Petersburg in Russia's northwest area with Vladivostok in the far southeast on the Pacific Ocean. When you sign on for the journey, you spend more than a week on the line and cover more than 6,000 miles of track.

Terra-Trivia
Until 1867, the area that is now Alaska was Russian territory. For just $7.2 million, the United States, under the direction of Secretary of State William Seward, purchased the land in what became known as "Seward's Folly." (The gold and oil alone make the Alaska purchase a steal.) Had Russia retained Alaska, the country would have spanned 14 time zones.

An Unforgiving Climate

All it takes is a quick look at a globe to see just how far north Russia lies: Almost its entire mass is located above 50 degrees north latitude. Russia's latitude is roughly equivalent to Canada's. Saint Petersburg is located at about 60 degrees north latitude, placing it less than 7 degrees south of the Arctic Circle and north of Alaska's capital, Juneau.

Although Russia's latitude parallels Canada's, Russia is colder. Canada's northern area contains the huge Hudson Bay and consists of thousands of islands that spread north into the Arctic Ocean. Russia is located in the extreme north of the world's largest continental landmass, Eurasia, with no large, interspersed body of water to moderate the harsh temperatures.

The winter in northern Siberia consists of frigid temperatures and months in which the sun never rises above the horizon. High pressure is indicative of cold, dry air. The world's highest sea-level air pressure was recorded at Agata, Russia, in central Siberia.

With no major body of water to moderate temperatures, Siberia is also subject to tremendous ranges in annual temperature—the normal annual temperature covers a 100-degree range. In parts of central Siberia, the normal January temperature is less than –50 degrees F. The same area averages 50 degrees F in July.

Because the temperatures are so low, the air can hold little moisture. The result is precipitation totals of less than five inches between November 1 and April 30. Although the summer months get slightly more precipitation, as do the western parts of Russia, the bulk of Russia is quite dry. As though cold and dry weren't bad enough, much of Russia's northern lands are covered with permafrost.

Permafrost, or permanently frozen subsoil, poses all sorts of problems to permanent structures and transportation lines. If it remains frozen, it provides a hard surface. Beware, however, if it thaws. Road-building equipment, houses, and chunks of road surface have simply disappeared into the quicksand-like ooze of thawed permafrost. Rail lines laid across permafrost and warmed from the friction of train traffic or solar radiation look like they have been strung across a giant washboard.

So there you have it, cold, dry, permafrost—what more could a farmer dream of? Russia is not, of course, agricultural nirvana. Feeding its people has always been a problem. Even the U.S.S.R., with the more fertile lands of Belarus, Ukraine, and Moldova, had trouble producing enough food. The scenario in Russia is worse now that these lands are independent and no longer part of the country.

By Land or by Sea

As you might expect in a country as large as Russia, the physical realm, as shown on the following map, includes just about everything (except anything having to do with the tropics). Vast mountains, vast plains, vast forests—Russia is sort of a vast-fest. It also has some of the world's largest lakes and rivers. Russians are a little like Texans: They think that anything having to do with their homeland is the world's largest. Unfortunately, these braggarts are often right.

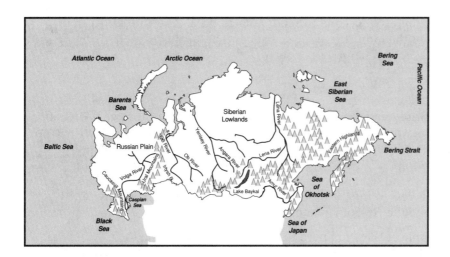

Physical features of Russia.

The Land—and Lots of It!

In the West, Russia's European territory is dominated by the Russian Plain, an extension of the Northern European Plain. The northern extent of the Russian Plain is primarily woodland, and the southern reaches represent Russia's primary agricultural lands. To the south, the plains terminate between the Black and Caspian seas in the Caucasus Mountains. The eastern edge of the plain (and the end of continental Europe) is marked by the long north–south chain of the Ural Mountains. The Ural chain is also extremely mineral rich and is a primary source of the fuels and minerals that supply Russia's industry.

Beyond the Urals lies the often frozen expanse of Siberia. The first thousand miles or so consists of the Siberian Lowlands. They're dominated by needleleaf evergreens and tens of thousands of square miles of swamps. East of the lowlands, the land begins to rise. First comes the Central Siberian Plateau, and, on Russia's Pacific coast, the mountainous Eastern Highlands. Much of the plateau and highlands are covered with needleleaf deciduous trees called *taiga*.

As inhospitable as this land may seem, it contains great wealth. Beneath the frozen surface lie tremendous energy reserves. Russia has more than 16 percent of the world's known coal reserves (second to the United States, with 23 percent) and more than 14 percent of

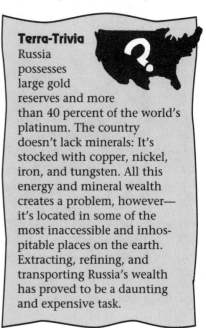

Terra-Trivia
Russia possesses large gold reserves and more than 40 percent of the world's platinum. The country doesn't lack minerals: It's stocked with copper, nickel, iron, and tungsten. All this energy and mineral wealth creates a problem, however—it's located in some of the most inaccessible and inhospitable places on the earth. Extracting, refining, and transporting Russia's wealth has proved to be a daunting and expensive task.

the world's petroleum reserves (second to Saudi Arabia). It's also a world leader in uranium reserves. Perhaps the most staggering statistic is that Russia has more than one-third the world's total known natural-gas reserves.

Frozen Waters

You can take a clockwise tour around Russia's coast to get a feel for the large bodies of water that bound it. Starting in the north are the Barents Sea and the frigid Arctic Ocean. To the northeast, Russia is separated from the United States (Alaska) by the Bering Strait. In the east are the Bering Sea, the Sea of Okhotsk, the Sea of Japan, and the Pacific Ocean. To the south is mostly land but also a coastline on the Caspian Sea and the Black Sea. In the west, Russia has access to the Baltic Sea via the fragmented bit of territory called Kaliningrad Oblast and the Gulf of Finland.

Russia has only a couple of lakes you need to remember. It borders on the Caspian Sea, which is the world's largest lake in terms of surface area. (Because the Caspian is landlocked, it's considered a lake, even though it's called a sea and contains saltwater.) Lake Baykal, known for its cold, crystal-clear water, is the earth's sixth-largest natural freshwater lake (in surface area). Lake Baykal is by far the earth's deepest lake, with a maximum depth of 5,316 feet (more than a mile deep!).

Russia also has its share of large, important rivers. Four of the world's ten largest rivers, in fact, are in Russia. First, however, consider Russia's most important river. The Volga River and its tributaries stretch through the heart of the populous Russian Plain before finally draining into the Caspian Sea. As Russia's most important commercial waterway, the Volga flows through its history, literature, and cultural heartland.

Geographically Speaking

If you have to remember just one Russian river, make it the Volga. As Russia's 5th-longest river (and the 16th longest on the earth), four other Russian rivers are longer (from west to east): the Ob'–Irtysh, 5th longest in the world, 3,362 miles; the Yenisey–Angara, 6th longest, 3,100 miles; the Lena, 9th longest, 2,734 miles; and the Amur–Shilka, 8th longest, 2,744 miles. Unlike the Volga, which flows from north to south, the first three of these rivers flow from south to north and drain into the Arctic Ocean. The last, the Amur, flows from west to east and forms part of the Russian–Chinese border before emptying into the Sea of Okhotsk and the Pacific Ocean.

The Search for an Ice-free Port

Russia has the world's second largest navy and one of the world's longest coastlines—so what's the problem? The problem is a port. Despite a huge ocean coastline, most of it is on the Arctic Ocean and lies frozen under ice much of the year. On the Pacific Ocean is the port of Vladivostok, but it's subject to freezing in the cold winter months. Although Russia has Baltic Sea ports and the Baltic Sea provides access to the Atlantic Ocean, the sea is also subject to freezing. The Black Sea has ports, but they present long journeys through NATO-controlled waters before ever reaching the Atlantic or Indian oceans.

Russia does have a port, however, that provides ice-free ocean access year-round—in a most unlikely spot. Far to the north, well above the Arctic Circle and abutting Finland and northernmost Norway, is Russia's Kola Peninsula. On this peninsula is the famous port of Murmansk. Despite its location in the far north, it's kept ice-free by the last gasp of the Gulf Stream. The same current that originates in the Gulf of Mexico and warms the British Isles and much of Europe keeps the ice from Russia's lone year-round ocean port.

Russia's Cities: The Big Two

Though Russia has 11 cities each with a population of more than one million, you have to remember only two of them: the capital, Moscow, and the former capital, Saint Petersburg. These two cities are Russia's largest and, more importantly, are the centers of Russian politics, history, and culture. This section briefly describes each one, all shown on the map of Russia at the beginning of this chapter.

Moscow is one of the world's largest cities. With 8.8 million inhabitants, it is by far Russia's largest city. In the sixteenth century, Ivan the Terrible established Moscow as the center of his newly formed empire. Although Peter the Great moved the capital from Moscow to Saint Petersburg in the early eighteenth century, the Bolsheviks moved it back to Moscow in 1918. In the heart of Moscow is the traditional seat of government, the red-walled buildings of the Kremlin. Adjacent to the Kremlin is the famous Red Square, with the spectacularly ornate Saint Basil's Cathedral.

Saint Petersburg is Russia's second city. Built by Tsar Peter the Great, it was constructed on marshes along the coast of the Gulf of Finland. The city's architectural beauty, in addition to the canals that were cut to drain the swamps, earned Saint Petersburg the title "Venice of the North." Because Peter imported architects from the West to design his prize city, Saint Petersburg has a distinctly western European flair. The city served as Russia's capital for a couple hundred years before its name was changed to Petrograd, and then Leningrad, and back to Saint Petersburg. The city now has almost 5 million people.

> ## ! Geographically Speaking
>
> In addition to Moscow and Saint Petersburg, Russia has nine other cities with a population of more than one million (in order from largest to smallest population): Novosibirsk, Samara, Chelyabinsk, Yekaterinburg, Nizhny Novgorod, Kazan, Perm, Ufa, and Volgograd (formerly Stalingrad).
>
> Four of these cities (Samara, Nizhny Novgorod, Kazan, and Volgograd) are located on the Russian Plain west of the Ural Mountains. Four (Chelyabinsk, Yekaterinburg, Perm, and Ufa) are located in or around the mineral-rich Ural chain. Only one of the cities, Novosibirsk, is located in southern Siberia far west of the Urals. The city locations alone can tell you much about Russia's population distribution.

People of the Steppes

When you talk about the people of Russia, think west and think south. The overwhelming mass of the Russian population lives west of the Ural Mountains. On this great Russian Plain is Russia's agricultural heartland and most of its people. The population in this area is most concentrated in the southeast and gradually dissipates as you move to the north. East of the Urals, the huge expanse of Siberia is largely uninhabited. The hardy souls who do subsist in this frozen realm live primarily along its southern reaches.

Although the Russian population is more than 80 percent ethnic Russian, the remaining population is composed of a variety of other ethnic groups. Ukrainians, Mongolians, Caucasians, Finno–Ugrians, Tartars, and other Turkic peoples are scattered across Russia's tremendous landmass. Although Russian is the official and almost universal language of the land, many people also speak in their native language.

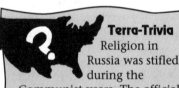

Terra-Trivia

Religion in Russia was stifled during the Communist years. The official religion of the state was atheism, and many Russians still claim no religious affiliation. Before the breakup of the U.S.S.R., a large concentration of Muslims lived in the southern republics. Most of them departed, however, as the republics gained autonomy.

Russia of the 1990s has seen a tremendous resurgence of religious interest. Most of the growth has occurred in the ancient faith of its forefathers, the Russian Orthodox church. Russian Orthodoxy on the steppes has a long history that dates back more than 1,000 years. The onion-domed churches that the Soviets converted to museums are now churches again, and the medieval icons and sacred rituals have new importance in people's lives.

Russia's people are highly literate, and although the country historically was known as an agrarian peasant society, only one of every 15 Russians now works directly in agriculture. Although Russia is a mineral-rich, industrialized country, its economy is suffering through difficult times. The transition to a market economy after almost 80 years of a planned economy is proving to be difficult.

Food shortages that have long been a problem in Russia continue, particularly in outlying areas. Other essentials are in short supply, or their prices have soared, especially medicine and medical supplies. Inflation, unemployment, and a declining growth rate all have been problems for Russia. The basic infrastructure has even degraded from a lack of funds. Foreign investment has not been as plentiful as was hoped, in part because of bureaucratic red tape and rampant organized crime.

Russia in the Twenty-First Century

In the process of industrializing Russia, the country's Communist leaders wrought environmental havoc across the land. Lakes of spilled oil spread across the fragile tundra, cedar forests fall as the last range of the huge Siberian tiger evaporates, the air is choked by coal-burning factories with scrubberless stacks, and the waters are fouled with radioactive and chemical spills. Much of industrialized Russia is an environmental nightmare that plagues its wildlife and has an untold impact on the health of its people.

Before Russia can hope to improve its lot, it must address the land that it has so long offended. The problem is cash: Cleanup costs are astronomical, and Russia is broke. The income Russia generates is based largely on the additional exploitation of its resources, which degrades the environment even more. Russia must reverse its long downward spiral. Its leaders are turning outside the country for technology and business expertise. One such potential partner is Russia's eastern neighbor, Japan. It has economic power, business savvy, and the technical resources Russia lacks, and Russia has the raw materials that Japan lacks. It's almost a perfect match, but not quite. At the end of World War II, Stalin grabbed Japan's Kuril Islands (despite signing an nonaggression pact with Japan). Japan wants these islands back, and Russia is reluctant to part with them. This political stumbling block must be resolved before the two nations can forge a happy marriage.

Terra-Trivia
In the Surgut area of east–central Siberia, massive oil and gas fields have been extensively developed since the 1970s. The harsh environment has wreaked havoc on pipelines that are leaking with depressing regularity. Oil pools the size of lakes have decimated bird species living in the tundra environment. Flared gas (a by-product of the oil drilling) contributes to the atmosphere's greenhouse gases.

Russia is no longer the superpower it was only a decade ago. Its military has been greatly weakened, and defense rubles have all but dried up. Russia's future must lie with its economy and the spirit of its people. Its economy, based on outmoded factories, a crumbling infrastructure, and machinery from the Cold War era, has a long way to go. The average Russian has tolerated a difficult climate, inept and corrupt leadership, bureaucracy, endless lines, and shortages. Still, the people's resolute spirit lives on and will continue to do so as the twenty-first century hopefully improves their lot.

The Least You Need to Know

➤ Russia's past rulers were called tsars.

➤ Joseph Stalin was the dominant figure during the formative years of the U.S.S.R.

➤ The U.S.S.R.'s 15 former republics have all spun off and become separate countries (including Russia).

➤ Russia is the world's largest country in area and sixth largest in population; Moscow and Saint Petersburg are its largest and most important cities.

➤ High latitudes mean that the climate is extremely cold in Russia.

➤ Most of the region's people speak Russian and are Russian Orthodox.

Chapter 13

Japan: The Pacific Dragon

In This Chapter

➤ Tracing the sword of the samurai

➤ Touring Japan's islands, mountains, and plains

➤ One people, 126 million individuals

➤ Sizing up the economic giant

➤ Knowing what lies ahead for the Land of the Rising Sun

Often called Nippon, Japan is more correctly called Nihon Koku, or Land of the Rising Sun. It's an appropriate name because Japan is one of the first major countries to see the sun rise each day. By the time the sun finally rises in New York, it has set in Tokyo. Just as Japan leads the world into each new day, it has also been a world economic leader. (Sorry—that may have been the most tortured segue in this book.)

Japan is an island country astride the Pacific Ocean and the eastern coast of Asia. Geographically, the country is well positioned to access the markets both of Eastern Asia and across the sea in the Americas. Although Japan is the world's 8th-largest country in population with 126 million people, its 146,000 square miles make it only the 59th largest in area. The result is that Japan's population density (people per square mile) is among the highest on the earth.

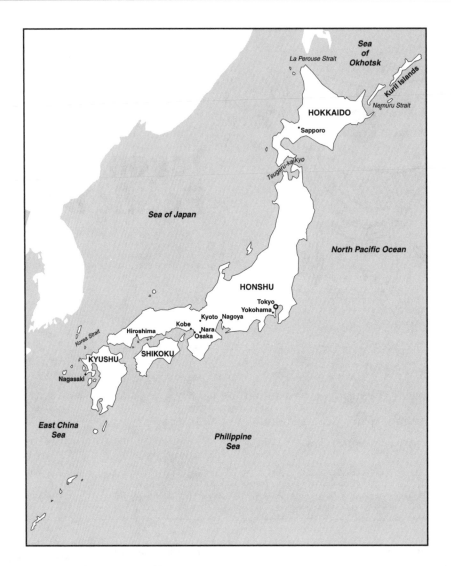

Song of the Samurai

Japan's earliest inhabitants were the Ainu, an obscure tribe that probably populated all the islands by the first millennia B.C. As Asians began to migrate into Japan, the Ainu were pushed farther and farther north, where a few now remain. Legend has it that, around 660 B.C., the migrants united under the first emperor of Japan, Jimmu Tenno.

The Chinese began to arrive on the Japanese islands about A.D. 400. They brought culture, writing, Buddhism, and, later, Confucianism. The Japanese readily absorbed all these elements and first demonstrated a desire to learn from others. In the mid-ninth century, as the emperors grew more reclusive, the Fujiwara family came into prominence and remained in power for three centuries.

In this era of local warlords, each one had his protecting band of professional soldiers, or samurai. The various samurai groups fought bitterly to achieve greater power for their warlord. Finally, in A.D. 1192, the Minamotos defeated the Tairas, and Yoritoma became the first shogun (the military commander-in-chief under the emperor) of Japan. The shogun ruled over the daimyo (200 to 300 feudal lords, each with his own samurai and fiefdom). The system, which lasted from the twelfth to the ninteenth centuries, seemed designed to defeat itself and to promote constant strife. One warlord versus another, the daimyo versus the shogun—the possibilities for intrigue were endless.

Geographically Speaking

From A.D. 794 to A.D. 1868, the splendid city of Kyoto was the capital of Japan and the inland center of this isolated country. Kyoto also was the country's cultural center, where theater, music, and the arts all flourished. In nearby Nara, a huge Buddhist shrine encloses one of the world's largest bronze statues of the Buddha.

The city of Kyoto boasts more than 2,000 shrines, temples, and magnificent miniature gardens. It's also the home of the former Imperial Palace and the Nijo Castle. Only small cottage industries are encouraged because the past is deliberately being preserved as a revered springboard to the future. Because of Kyoto's lack of war industy, the United States spared it from bombing in World War II.

In the thirteenth century, Kublai Khan's Mongols tried to invade twice, but were repulsed each time. In the mid-sixteenth century, Europeans began to arrive. First, Portuguese sailors arrived, looking for trade. Then came Saint Francis Xavier, the Jesuit missionary, to convert the country's people to Christianity. Despite disapproval and persecution, some 300,000 did convert before the shogun put a stop to the conversions and all things foreign in the fourteenth century.

This isolation lasted until the early eighteenth century, when the then shogun repealed the ban. In 1867, the last shogun resigned, and in 1868 the Emperor Mutsuhito reestablished himself as supreme ruler. He took the name Meiji (meaning "enlightened government"), moved the capital to Edo and renamed it Tokyo (meaning "eastern capital"), and by 1871 abolished all fiefs. The Meiji Restoration had begun, and Japan had entered the modern era.

Emulating Colonialism

For Japan, the modern era truly began with the Meiji Restoration, in 1868. The planners of that time took stock of what they had and what they needed in order to become a modern country, capable of a world leadership.

157

Japanese leaders looked to the West, where they saw that type of capability. They looked first to Britain, whose industrial revolution in the preceding century had helped the country advance from a primarily agrarian island country to become the world's leading manufacturer, exporter, and trade center. The Japanese sought technology, economic and educational reform, and advice in city layout, plant location, and communication and transportation. The West served as the model by which Japan's leaders built their modern nation.

Unfortunately, Japan lacked coal and iron and other raw materials necessary for product manufacture and export, and it also needed money for investment in plants and machinery. It realized that it could earn some money from the export of its present products (textiles, porcelain, wood, and small metal products) and could raise taxes on its own farmers. But how would Japan get coal, iron, and other raw materials? Although it could purchase the materials, that would take more money. Then again, it could use the old tried-and-true European colonial system: They could colonize and exploit.

Forging the Japanese Empire

First, Japan seized the Ryukyu Islands. China objected, but so what? Why the Ryukyus? Their people were ethnic Japanese and spoke Japanese and produced excess food. In addition, the islands, which offered homeland protection (a buffer zone to protect the main Japanese islands), were an easy place to begin building an empire. They were also a strategic jumping-off place for Japan's next move. With its new navy and new army, Japan took Taiwan from China, getting coal, petroleum, natural gas, and a market for its finished goods.

Japan defeated the Russians, seized Korea (getting iron and more coal and another market), and then grabbed Manchuria (more iron and coal and yet another market). The empire then expanded to several Pacific archipelagos by conquest or forced mandate (homeland protection). In 1937, the Japanese invaded China and within a year controlled almost all of eastern China—a huge market.

To fully establish itself as a superpower, the empire had to rid its sphere of the British and the United States. As the Western powers became ever more entangled in the World War II European and African conflict with the Axis powers, a window of opportunity opened for Japan. (See Chapter 24 for more information about Japan's role in World War II.) The Japanese apparently thought that they would have it made if they could destroy the American fleet with a surprise attack on Pearl Harbor; seize the Philippines, Malay, Singapore, and the East Indies; and then take a few more Pacific islands for protection of their homeland.

The Japanese empire almost pulled it off. Japan didn't destroy the American fleet at Pearl Harbor (because it wasn't all there). The Imperial navy lost two critical sea battles at Coral and Midway, and Japanese leaders completely underestimated the will of the United States and its ability to fight and win on two fronts. After undergoing months of homeland bombings and eventually two atomic bombs on Hiroshima and Nagasaki, Japan—morally and physically devastated—surrendered in 1945.

Land of the Rising Sun

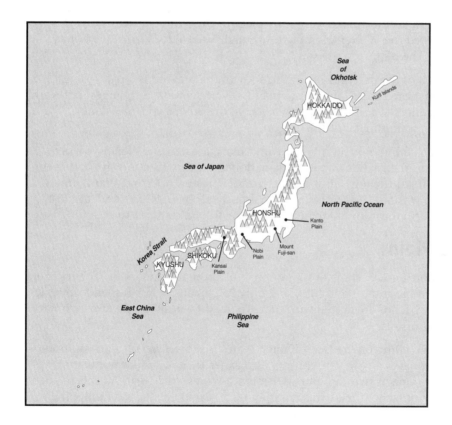

Physical features of Japan.

Millions of years ago, tectonic action caused the Pacific plate to smash into the Asian plate, and great mountains were formed on the edge of the Asian continental shelf. A deep gorge formed where the Pacific plate and Philippine plate folded under the Asian plate. Today, the mountains are Japan, and the gorge is the Japanese Deep (28,000 feet deep, off Japan's southeast coast). Japan's most beautiful mountain, Mount Fuji-san, is also its highest mountain (12,389 feet). This snow-capped volcanic cone was last active in 1707.

The Waters and the Islands

As an island country, Japan is bounded by seas and an ocean. To the north is the Sea of Okhotsk; to the west is the Sea of Japan and the East China Sea. To Japan's south and east is the endless expanse of the Pacific Ocean. Finding Japan on a map is easier after you throw in some land references: The Japanese islands are southeast of Russia and due east of the Korean Peninsula.

Japan is an archipelago, which means that it's made up of many islands. Although the country has about 1,000 islands, you have to keep track of only the "big four," which loosely form the shape of a huge dragon. To the north is the dragon's head, the island of Hokkaido. Its northern latitude makes it a chilly place in the winter, as witnessed in 1972, when this island hosted the Winter Olympics. Although Hokkaido is large, it's not as heavily populated as the islands to the south.

The next island south of the dragon's head (the body of the dragon) is Japan's largest island, Honshu. Considered to be Japan's mainland, it's home to Japan's largest cities and the bulk of its population and industry. Farther south, the other two large islands, Shikoku and Kyushu, form the tail of the dragon.

The Kuril Islands, which link northern Japan with Russia, are an issue of great concern. Although Japan claims the islands as part of its national territory (they always had been), Russia says "nyet"—it got them in a last-minute grab at the end of World War II and it occupies them. Even without the Kurils, Japan is almost 1,200 miles long and, at most, only 200 miles wide. No location in Japan is more than 100 miles from the sea.

Mountains and Plains

Japan has a major north–south spine of mountains called the Japanese Alps. The Japanese islands are 80 percent mountains and hills, and only 25 percent of Japan's land slopes less than 15 percent. The islands are dotted with more than 200 volcanoes, 60 of which have been active in recent history.

Japan is on the Pacific's Ring of Fire (see Chapter 2) and is subject to volcanic eruptions—and at least 15 earthquakes a year. The country also experiences a similar number of typhoons and at least one or two tsunamis (seismic sea waves) each year. All this happens right on top of the spot where the earth's Eurasian plate, Pacific plate, and Philippine plate all collide! (Plates are discussed in Chapter 2.)

Geologically, Japan is one of the world's most dangerous places to live. Tokyo sits largely on fill (land created artificially by filling in parts of Tokyo Bay) that could liquefy in the event of a major quake. Tokyo has narrow streets dotted with many skyscrapers built before stringent regulations were in force, and it's part of the world's largest population concentration. If the twenty-first century is born with a "big one," it could create a global catastrophe.

Japan's largest island, Honshu, has three major plains:

➤ **Kanto Plain:** Located in the east central part of the island, the largest and most important plain has much of Japan's agriculture, industry, and population. Its southern edge is dominated by the world's largest urban conurbation of Tokyo, Kawasaki, and Yokohama.

➤ **Nobi Plain:** To the southwest, it's centered around the city of Nagoya.

➤ **Kansai Plain:** Even farther southwest, it's home to Kyoto, Kobe, and Osaka.

The plains and their cities all are connected by superfast trains known as "bullet trains." Japan has as many small valleys as mountains and hills, almost all with short, fast-flowing, steep streams and rivers (only two rivers in Japan are more than 200 miles long).

Lakes also are minute and few; Lake Biwa near Kyoto is the largest. Japan's most dominant body of water is its Inland Sea (240 miles long and from 8 to 40 miles wide), connecting Honshu on the north with Kyushu and Shikoku on the south. The sea is known for its islets and beauty and is valued for its fishing.

Terra-Trivia

In 1995 a quake of magnitude 7.2 on the Richter scale struck near Kyoto, Osaka, and Kobe. The results were catastrophic: 5,100 people were killed, and another 27,000 were injured. Buildings couldn't withstand the seismic forces and crumbled. Emergency rescue procedures broke down. Temporary housing was nonexistent in many places. Economic damage from the quake was placed at around $100 billion!

Terra-Trivia

For centuries, the storms of the Pacific and the Philippine Sea have been battering Japan's southeast coast and Kyushu's west coast. The result is a heavily carved and eroded smattering of islands, inlets, bays, and peninsulas—an indented coast giving Japan more than 15,500 miles of coastline.

Japan's Temperate Climate

Latitudinally, Japan is the same as the east coast of the United States from mid-Maine to north Florida. The two areas have much in common. Because they're both temperate, they generally have similar temperatures and rainfall patterns, usually about 40 to 50 inches. Unlike the United States, though, Japan has two annual rainy seasons and gets an occasional coastal pocket of precipitation of as much as 100 inches.

Both the United States and Japan have eastern warming currents (the Gulf Stream and the Japanese Current, respectively). Both countries have prevailing cold winter winds blowing from the northwest, from Canada and Siberia, respectively. Both get heavy winter snows in their northern areas as these winter winds pick up moisture across water expanses (the Great Lakes in the United States and the Sea of Okhotsk and Sea of Japan). Both are in belts of tropical storm paths, from hurricanes and typhoons, that prevail in late summer and early autumn. Other similarities are discussed later in this chapter.

The People of Japan

Perhaps nowhere on earth do the values and traditions of the past thrive so strongly alongside present ideas as they do in Japan. The country is extremely cohesive: It has one race (Mongoloid, similar to the Chinese and Korean people but smaller in stature), one language (Japanese), one culture, and one highly developed—and highly technically oriented—system of education that produces virtually 100 percent literacy.

Until 1947, the state religion was Shinto, in which the emperor is God. Now all religions are permitted. The Shinto religion, indigenous to the Japanese islands, involves the worship of the divine forces of nature. Ancestor worship also factors strongly into Shinto practice.

Although most Japanese are still Shinto, almost 35 percent are Buddhist. Many adhere to both religions because in practice the religions largely overlap. The dominant form of Buddhism in Japan, as in most of the Buddhist world, is Mahayana. Unique to Japan is a particular Buddhist discipline called Zen Buddhism. Beautiful and serene rock gardens are in Zen temples across Japan.

Terra-Trivia

Mutsuhito (of Meiji Restoration fame) was only 14 when he became emperor in 1866. His grandson, Hirohito, became regent at 20 and emperor at 25. He reigned as God and then as man until his death in 1989. Legend has it that Hirohito's son, Akihito, who became emperor at 56, is the 125th direct descendent of Jimmu, Japan's first emperor (660 B.C.).

U.S. occupation after World War II brought westernization. Core values, however, are the base on which change is founded. Although rural life—now modernized in agricultural techniques and equipment—has changed little, 77 percent of Japan's people live in cities. In the cities, more change is evident: For example, smaller families living alone (rather than with parents), condos, Western dress (but still kimonos and obis for special events), fast food (but still rice and raw fish), Western movies and music, more leisure time, and skiing and baseball.

The Japanese had a brief baby boom after World War II but brought the birthrate down from 34 per thousand to 11 per thousand by the end of the 1980s. With a drastic decrease in the death rate (life expectancy is now up to 79 years), the natural increase rate is only 0.3 percent, one of the lowest in the world.

Stretching about 350 miles across the three major plains of the island of Honshu is the Tokaido megalopolis, a long population belt with more than 50 million people. It includes three major conurbations: the Tokyo–Yokohama–Kawasaki sprawl of 30 million people (the world's largest); the Nagoya area, with 5 million people; and the Osaka–Kobe–Kyoto complex, with another 15 million. The parallel to the east coast of the United States (the Boston-to-Washington megalopolis of similar latitude, length, and population) is hard to miss.

An Economic Giant

With a devastated manufacturing segment after World War II, but with massive financial aid from the United States and a willingness to learn and plan ahead, Japan began its

economic miracle all over again. Again, it looked to the West, this time to industrial buildups that had enabled Germany and Italy to prepare for World War II, including the cohesion of government, business, and labor that had made this miracle possible. Japan, Inc., was just being born.

The goal was the best-quality product made right the first time. The Japanese aimed to gain a world market share in their export trade. Price based on cost became obsolete—only long-term profits mattered. Earnings were invested in research, and expansion was financed through low-interest, government-sponsored bank loans (made possible by high personal savings due to a deliberate scarcity of consumer goods). High-cost equity financing requiring dividend payments was avoided. The Japanese also began studying the work of Deming (an American quality statistician) and Drucker (an American economist).

Terra-Trivia

Japan is the second-largest consumer of electricity on the planet. With only 2 percent of the world's population, it consumes more than 7.5 percent of the world's total electricity. The number-one consumer of electricity is the United States, which, with 4.5 percent of the world's population, consumes more than 25 percent of the world's generated electricity.

It took Japan until the mid-1950s to get back to pre-war production levels. By the 1960s, its economy had skyrocketed. From 1965 until 1980, economic growth averaged almost 10 percent a year. During the next decade, growth "slowed" to 7 percent a year but Japan became number one or number two in the world in shipbuilding and in producing steel, automobiles, trucks and buses, watches, VCRs, TVs, cameras, microwave ovens, refrigerators, computers, and copying machines.

In 1991, Japan's economy began to falter, and recession hit the economy in many countries to which Japan exported its goods. Competition from those same countries resulted in improved quality and equally priced goods. The country's Asian neighbors began to produce lower-priced goods. Japan felt increased pressure to open its own markets to lessen its imbalance of trade with other nations.

All these situations began to take a toll on Japan's economy. Thousands of small-company bankruptcies combined with political, stock, and money scandals. It also experienced its worst stock-market crash, which helped cause the country's deepest and longest post-war recession and its highest unemployment.

A Look at Japan's Future

Japan has begun to come back, however. As it comes out of its recession (with two reasonably good years in a row), it must again reevaluate its current strengths and weaknesses. The country is still one of the world's three greatest economic powers. Its populace is one of the best-educated and most cohesive in the world. The country has a proven record of success. It is wealthy and can invest in anything it wants or buy anything it needs.

Japan is almost out of usable land, however, for farming, industry, and people. It can supply only 71 percent of its food needs. Its standard of living, both economically and socially, is changing: It's more difficult and less popular to cling to values of the past. Many of Japan's competitive advantages are now lost. The countries of the world will no longer let Japan sell in their markets if its own aren't more open.

If Japan's keen analysis of the world scene is as astute now as it has been in the past, its prospects are rosy. If the Japanese people's fantastic work ethic can still apply in this ever increasing materialistic and hedonistic society, its prospects are rosy. In all likelihood, Japan will find its way toward continuing its goal of world economic preeminence. At least it will remain as one of the leaders.

The Least You Need to Know

➤ Japan is one of the world's most densely populated countries.

➤ The Meiji Restoration of 1868 initiated planning and reform that started Japan's thrust into the modern economic era.

➤ Ownership of the Kuril Islands is still a sticky subject between Japan and Russia.

➤ Japan's militaristic approach to world supremacy ended with its defeat in World War II.

➤ The aftermath of World War II brought democratic reforms and capital infusion and began Japan's global economic leadership.

➤ In less than half a century, Japan has attained economic leadership, but now competition threatens its future.

Chapter 14

Australia and New Zealand: Looking Down Under

In This Chapter

➤ Looking back at kiwi history

➤ Inspecting the land of volcanoes

➤ Getting to know the kiwis

➤ Living in the city: A tour of New Zealand's big cities

➤ Digging into Australia's sordid beginnings: Outback parole

➤ Trekking into the outback

➤ Introducing Australia's inhabitants

➤ Organizing the cities, states, and territories

Australia and New Zealand, which are neighbors despite being separated by 1,300 miles of Tasman Sea, form a region in the far South Pacific. Both countries are firmly based in English tradition, language, and rule of law and government, despite their 12,000-mile distance from Europe. Young relative to many other nations, literate and prosperous, and developed in an underdeveloped part of the world, both New Zealand and Australia have many similarities.

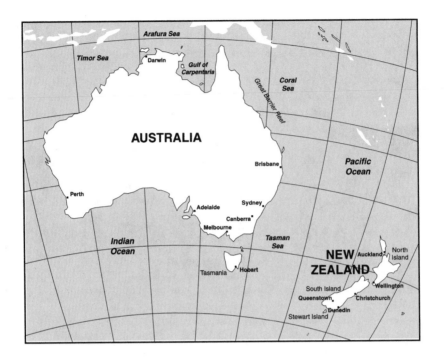

Don't be deceived by their whimsical outlook, odd-looking animals, informality, and macho cowboys and surfers. The people of this area are serious, hard-working, independent souls not to be pushed around. As in other nations in the South Pacific, the "no nuke" policy is real: Neither nuclear-powered nor nuclear-armed vessels are allowed ashore—period.

Despite a number of islands, only two countries are in this region. Although the countries have many commonalities, this chapter describes them separately, for clarity's sake. Although Australia is by far the larger of the two in both area and population, let's start with the little guy, New Zealand.

Land of the Kiwi: New Zealand

If New Zealand were placed in the Northern Hemisphere, it would be at approximately the same latitude as the stretch from Los Angeles, California, to Seattle, Washington. The country is farther south than Australia and cooler—closer to the South Pole. New Zealand's Stewart Island is, in fact, just 20 miles off the coast of South Island and hosts lots of penguins (to say nothing of the world's best oysters.)

First Peoples: The Maoris

Before the fourteenth century A.D., Maoris paddled their dugouts filled with seeds, children, plants, dogs, and rats to Aoearoa, "land of the long white cloud." The Maoris saw the beautiful islands of New Zealand, where misty mountains rise from the seas. The moa, a ten-foot flightless bird, supported them until it was hunted to extinction. (Smaller but similar, the kiwi, New Zealand's mascot, enjoys protected status.)

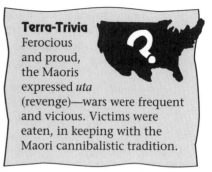

Terra-Trivia
Ferocious and proud, the Maoris expressed *uta* (revenge)—wars were frequent and vicious. Victims were eaten, in keeping with the Maori cannibalistic tradition.

The Flood of Europeans

The Dutch explorer Able Tasman found an agricultural people, albeit unfriendly, in 1642 in the place he named New Zealand. Facing an unfriendly welcoming party in the Maoris, he sailed away, leaving New Zealand unpenetrated. For 127 years it remained that way, until the Englishman James Cook revisited and explored the islands. His *Endeavor* crew included Sir Joseph Banks, who recorded detailed facts about the flora and fauna and noted the "...almost certainty of being eaten as you come ashore!"

Braving the Maoris, settlers flocked to New Zealand for its whales, flax, and timber. Guns and rum flourished in the colorful frontier towns of North Island, where cemeteries speak of the New England whalers who finished their lives far from their homes in Plymouth or New Bedford. Curiously, some are recorded as dying of smoke from sulfur burned in the ship holds to control rats and the plague.

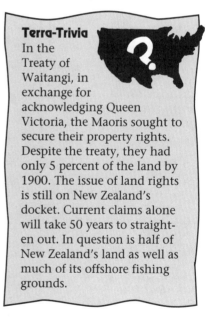

Terra-Trivia
In the Treaty of Waitangi, in exchange for acknowledging Queen Victoria, the Maoris sought to secure their property rights. Despite the treaty, they had only 5 percent of the land by 1900. The issue of land rights is still on New Zealand's docket. Current claims alone will take 50 years to straighten out. In question is half of New Zealand's land as well as much of its offshore fishing grounds.

By the nineteenth century, European woes and over-crowding sent thousands of English people to New Zealand seeking a better life. In an effort to create cooperation with the British, 400 Maori chiefs signed the Treaty of Waitangi in 1840 but didn't understand the subtleties of the British concept of land ownership. Although Edward Wakefield's New Zealand Company was started to organize the new country and the treaty mandated respect for Maori property rights, settlers had plans of their own.

Self-government began in the 1850s, when the population of New Zealand had reached 50,000. Six provinces formed, with a governor to plan for peaceful European–Maori development of the country. Alas, a decade of warring over land finally ended with the *pakeha* (white man) firmly in control. A gold rush in the 1860s flooded South Island and was followed by a jade rush.

The advent of refrigeration in the 1880s enhanced the sheep–mutton business. Trade prospered, and welfare programs and universal education followed as attempts to integrate the Maoris continued. By 1907, New Zealand was granted independence from Britain and became a member of the Commonwealth of Nations.

Although the intervening years were fairly quiet in this distant corner of the world, 1985 brought an important milestone in the Treaty of Rarotonga. This document established the South Pacific nuclear-free zone: No nuclear-powered ships from the United States or anywhere else are allowed to dock on participants' shores.

Trembling Land

As shown on the preceding map, New Zealand is more than 1,000 miles long and about 1,200 miles southeast of Australia. The country is sandwiched between the Tasman Sea on the west and the South Pacific on the east. It's a mountainous nation, in two major parts: North Island and South Island. (A third island, Stewart, located south of South Island, is significantly smaller than the two main islands.)

Physical features of New Zealand.

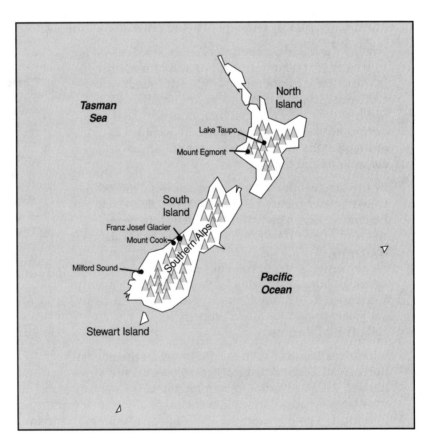

Although dwarfed by Australia in size, New Zealand is still slightly larger than the United Kingdom, the European source of the bulk of New Zealand's population. Separating the two islands is Cook Strait, which thins to 16 miles at its narrowest point. North Island is slightly smaller than South Island but is more geologically active.

Unsteady Ground on North Island

Volcanically speaking, North Island is active indeed. Three major volcanoes are in the island's center, and one of them, Ngauruhoe, erupted in 1995. All three volcanoes are part of Tongariro National Park. To the west in Egmont National Park, extinct Mount Taranaki (Egmont) rises more than 8,000 feet above the rich surrounding plains and vineyards.

With geysers shooting 100 feet high and fumaroles (a surface hole from which volcanic gases gush out) spewing steam and gas, geothermal plants produce 10 percent of New Zealand's power. Another major source of power is hydropower from South Island's rushing streams, sent across Cook Strait by undersea cable.

In the heart of North Island is Lake Taupo, the island's largest lake, teeming with fish. To the far north is a long peninsular proruption (long extension of land) called the Northland. This largely low-lying land is famous for its 90-mile beach north of Auckland. The Northland also has the remnant groves of the enormous kauri tree. Most of these ancient forests were turned into fishing fleets by whalers and were the early source of the huge Maori war canoes.

The Alpine South

South Island, measuring 58,000 square miles, is larger and more varied than North Island and is dominated by the Southern Alps (Fjordland). In this chain is the country's highest mountain, Mount Cook, or Aorangi, at 12,349 feet. Mount Cook is central to the massive chain that follows the west coast of South Island and provides alpine scenery and the 16 miles of uninterrupted ski runs famous the world over.

The Southern Alps are cloaked in glaciers that shrink and expand according to the snowmelt. The Franz Josef Glacier is unique in that it ends in a rain forest. In the southwest portion of the island are Milford Sound's Mitre Peak and matching fjord, in which 6,000-foot mountain faces rise from the water, high valleys display thundering falls (Sutherland Falls is fourth highest in the world at more than 1,900 feet high), and seals play ashore. The mountainous terrain has served as an excellent source of hydropower that is transported by underground cable to electrify North Island.

GeoJargon
Mount Cook is called Aorangi by the Maoris. The name means "cloud piercer."

South Island's west coast was the base for the kiwi gold rush, begun when a Maori found dust in a dog's fur while saving him from drowning. More than half a million ounces of gold were extracted before 1860, when the greenstone, or jade, trade took over.

The Kiwis

North Island has two-thirds of New Zealand's 3.5 million people. Most of New Zealand's indigenous Maoris also live in the north. North Island also boasts the country's largest city (Auckland), its capital (Wellington), three-quarters of its industry, most of its dairy, and more than half of its huge sheep population.

Terra-Trivia

New Zealand has more than 55 million sheep, which amounts to more than 15 times its number of people. It also leads the world in wool exports. This small country has almost 5 percent of the world's sheep and accounts for nearly 40 percent of the world's total wool exports.

Because of the mountainous nature of New Zealand's interior, the majority of its population lives in coastal cities and towns (which is where the majority of rail lines and roads are also found). As with the language (English is the most widely spoken), the population is 90 percent European-based, with a sizable Maori minority.

Protestant religions prevail in New Zealand, although a sizable minority of people (15 percent) are Catholic. The Catholic contingent is accounted for by Irish immigrants who arrived in the 1800s and brought their religion, as did many new refugees from World War II. Also in New Zealand are a few hardy Mormons (Church of Latter–day Saints) who live a long way from Utah.

The Maori

The indigenous Maori people who make up 12 percent of New Zealand's population are unrelated to Australia's indigenous people.

Maoris are a Polynesian group rather than Australia's indigenous Austroloid group. With the arrival of the *pakeha* (white man), the 250,000 Maoris shrank to 42,000 because of disease and more efficient methods of killing. Maoris have had full rights since 1865 and have fairly successfully integrated into white society. Well-known and talented Maoris include the novelist Keri Hulme and the opera star Dame Kiri Te Kanawa.

The 1990s have brought efforts to rectify past mistakes and restore the homeland to native peoples by enacting the true intent of the Waitangi Treaty. Although the treaty was supposed to protect the Maoris' interests, it failed in its application. Efforts are being made to preserve indigenous heritage in school so that all kiwi children will appreciate their bicultural roots.

Livelihoods and Pastimes

New Zealand is highly agricultural. The fertile Canterbury Plains on the eastern side of South Island are the country's breadbasket. It has enough good land to grow 90 percent of the country's grains. Hops, apples, and raspberries thrive. When you talk about New Zealand fruit, the indigenous kiwi fruit must be mentioned, of course, because it's an export that's popular around the world.

The spectacular natural beauty of the islands makes backpacking a favorite pastime. Queenstown, on South Island, is the trekking center of the island, where hikers stock up, load their swags (backpacks), rest, and scramble for transport to the mountains beyond. In addition, the lawn bowling tournaments and rose gardens of Queenstown are equal to those in Christchurch and entertain the less energetic.

Terra-Trivia
The kiwi reforestation project has introduced new species, such as the California pine, Japanese cedar, and European larch, to replace the ancient kauri and mingle with the native rimu, totara, and beech. Twenty-five percent of New Zealand is now forested to hold the soil in place and to provide for pulp and timber for export.

New Zealand's Cities

New Zealand is highly urbanized: 85 percent of its people live in the cities. This section gives you a rundown of New Zealand's three most important cities.

Auckland

By far New Zealand's largest city, Auckland has more than 900,000 people. The population includes the largest Polynesian population in the world—one in every seven of Auckland's residents. This bustling North Island city of "California-style commercialism" is a major port, built on an isthmus that separates the Pacific Ocean and the Tasman Sea.

Wellington

Wellington is New Zealand's capital, seat of government, and second-largest city. Located on the southwestern tip of North Island, it straddles a fault, much like San Francisco. Wellington enjoys hilly terrain and gusty winds as well as the ever-present threat of an earthquake. The city was named after Napoleon's British Waterloo nemesis, the Duke of Wellington.

Christchurch

South Island's "English city" (New Zealand's Garden City) is built around the Gothic cathedral in the central square. Built by Anglicans, Christchurch houses several fine universities and, from its port in Lyttleton, exports Canterbury Plain's wheat, meat, and wool in addition to products from its other industries.

The Land Down Under: Australia

Unlike the Maoris, Australian Aborigines walked the land bridge from Asia in the north with their dingoes (dogs) about 38,000 years ago. Representative of Stone Age man, Aborigines lived and hunted in isolation, using stone (not metal) tools. More than 300,000 Aborigines in 500 tribes and speaking 300 languages inhabited Australia when Captain Cook arrived in the 18th century. As English settlers pushed the Aborigines farther into the interior, starvation resulted in bloody raids over food.

Unlike New Zealand with its selected Anglican settlers, Australia's addition to the British Empire began with the First Fleet. This flotilla consisted of 11 ships and 1,487 people, of whom 736 were convicts. After landing in Botany Bay, they moved a week later to Sydney and "the finest harbor in the world." They claimed all of Australia and Tasmania as British territory while the Aborigines looked on but did not protest.

Most of the new inhabitants were male, giving some credence to Australia's being a "man's land." Although many of the new arrivals were convicts, the crimes were invariably minor: an 11-year-old boy transported for seven years for stealing ten yards of ribbon, a West Indian for taking 12 cucumber plants, and a 9-year-old chimney sweep for petty theft—not exactly hardened criminals.

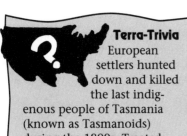

Terra-Trivia
European settlers hunted down and killed the last indigenous people of Tasmania (known as Tasmanoids) during the 1800s. Treated with contempt by whites, the Tasmanoids died from disease, starvation, and outright butchery. Six thousand in 1804 dwindled to 300 by 1830 and to none in 1876, when the last Tasmanoid died and an entire group of humans was then erased.

Tasmania (formerly Van Diemen's Land) also became a penal colony centered in Hobart. A total of 161,000 convicts settled down to farm and trade in Australia. The introduction of a hardy Spanish breed of sheep with superior wool had a huge and positive impact, unlike other species that were introduced, such as the Australian rabbit and the pestiferous (irritating!) New Zealand possum. Farmers pushed into the interior across Australia's eastern mountains to "squat" on land for eventual ownership.

The gold rush and the settlers who provided supplies to the miners increased Australia's sparse population. Friction with Chinese settlers in the gold fields led to a restrictive immigration policy and "white Australia." Following the two world wars, however, a more multicultural society developed that included Vietnamese, Cambodians, and Lebanese. Australia has become a melting pot: Half of all Australians are foreign or have a foreign-born parent.

The Big Dry: Australia's Land

Although Australia is a huge place (the world's sixth-largest country), it's the world's smallest continent. Roughly the size of the United States minus Hawaii and Alaska, Australia is the land of superlatives. It's the lowest and the flattest continent. Except for Antarctica, Australia is also the driest continent and has the sparsest population. Covering nearly three million square miles, it's surrounded by one sea after another (Coral, Tasman, Timor, and Arafura) and the Pacific and Indian oceans, as shown on the following map.

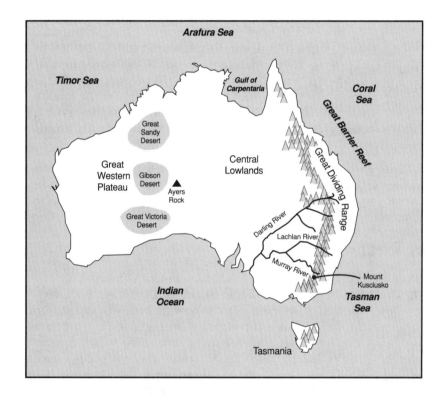

Physical features of Australia.

Mountains and Outback

The Great Western Plateau, Central Lowlands, and Eastern Highlands make up the main physical regions of Australia. All except one-third of Australia is part of the Western Plateau, which is characterized by no trees or rivers and lots and lots of sand. Sand ridges that reach 60 feet high, sand hills, sand dunes, and sand plains make up the Great Sandy Desert, the Gibson Desert, and the Great Victoria Desert, which cover the bulk of central and western Australia. Also in central Australia is the famous Uluru (Ayers Rock).

Terra-Trivia
Ayers Rock, the largest monolith in the world, is 1,143 feet high. The huge, red, sandstone outcrop adds interest to "gibber" plains (huge, flat stretches covered with pebbles). This unmistakable landmark in the outback has been a sacred place to the Aborigines for countless centuries.

The Central Lowlands extend from the Gulf of Carpentaria (the big notch in the north) through the pastureland of the Great Artisan Basin. The area is laced with intermittent streams that sometimes swell with the help of runoff from mountains to the east. Much of this water is too salty for crops and is used only to irrigate grazing land. In the southeast is the best farmland, thanks to the Murray, Darling, and Lachlan rivers; sediment in this area has also yielded opals.

The Eastern Highlands, or Great Dividing Range, separate the dry interior from the fertile coast. The area is tropically sticky and hot in the north and humid and subtropical in the south. Steep mountains (such as 7,310-foot Mount Kusciusko, Australia's highest) rise in the southeast and are home to a great deal of hydroelectric activity. The mountains in the south feed the long, gentle rivers that water the area and make possible the "fertile crescent" filled with apple and pear orchards, farms, and sheep stations.

Tasmania, Australia's island to the south, lies 150 miles across the Bass Strait and marks the mountainous southern extension of the Eastern Highlands. Dairy, wool, and oil deposits and hydroelectric programs are big there. The few remaining wombats and the Tasmanian wolf are remarkable examples of wildlife in this wet and isolated region.

The Great Barrier Reef

The Coral Sea has one of the world's most splendid natural wonders: Off Australia's northeast coast lies the Great Barrier Reef. Six hundred islands and atolls run for 1,250 miles, 10 to 100 miles from shore in shallow water to form the base of this vast reef. The size of England and Scotland combined, this immense living mass of coral reef, the earth's largest, covers 80,000 square miles. It's the home of 400 types of coral and 1,500 varieties of fish. Three million terns and boobies cluck and squawk above, and green sea turtles, sharks, and giant clams live below.

Terra-Trivia
Bommie (from the Aboriginal word *Bomboara,* meaning "submerged coral head") coral, which results from primitive polyps that have secreted a skeleton, can be wispy or hard. Oil drills, strip mines, phosphate guano deposits, polluting runoff, and silt from farming all help to ruin these fragile living reefs.

The Great Barrier Reef Marine Park was established in 1981 in an attempt to save the marine life from the depredations of (turtle soup) canning factories, oil drilling, bombing practice by the Australian navy, and reckless tourists. If environmental care is taken, tourists can still be fascinated and enchanted by exploring Wistari Reef or watching mantas leap from the Coral Sea. Beware, however, of the great white sharks.

Australia's Climate, Flora, and Fauna

Up is down, south is cold, and north is hot. The top third of Australia is tropical, and the seasons are opposite of the Northern Hemisphere. Santa sweats in his beard and red suit, and revelers eat plum pudding with brandy sauce on a hot December day, with "beer and barby" (barbecue) on the beach the next day. Tasmania and the south coast of the mainland are marine. Otherwise, Australia's climate is fairly mild—perfect for tennis and surfing all year. Most of this vast country is arid or at least semiarid.

Here's a word about the amazing plant life of the land down under. An incredible variety of native plants grows only here—honey flower, spear lily (12 feet high with red blossoms), 600 kinds of orchids, aromatic eucalyptus trees, more than 500 types of gum trees, and acacia, the sweet-smelling plant known as the "wattle tree." (Settlers used acacia saplings for wattling, or interlacing, the frame for mud and plaster walls.)

I can't describe Australia without mentioning the "joey" (kangaroo) because marsupials make up half the continent's native mammals. Marsupial females carry their young in a pouch until the babies are fully developed. Many types of kangaroos hop with surprising speed across the open plains. Koala bears cling to eucalyptus trees, their favorite food. An egg-laying mammal, the duck-billed platypus, is the oddest of all the animals. (New Zealand can claim only the bat and eel as native critters.)

People of the Southern Continent

People are extremely urbanized in Australia: 85 of every 100 Australians live in the cities. The population distribution is extremely peripheral; this coastal settlement also occurred in New Zealand, but the mountainous interior caused it there. In Australia, the deserts (not the mountains) limit interior settlement.

Australia's population is almost completely literate and English-speaking. (Later in this chapter, you read about some peculiarities of Australian English.) About one-quarter of the population is Anglican, one-quarter is Roman Catholic, and another quarter is of mixed Christian faiths. Aussies have a high standard of living, with a gross domestic product per capita of more than $20,000.

The Aborigines

After the arrival of the Europeans, Aboriginal life was wretched. Aborigines eventually worked at menial jobs for food. State policy, assuming the Aboriginal race to be doomed and dying out, wrested 100,000 children from their parents between 1910 and 1970. The light-skinned children were placed with white families and the black-skinned children put into orphanages.

Protective programs and extra social benefits are attempting to assist the Aboriginal people and assimilate them into the general structure of Australian society. Citizenship is recognized, segregated schools are closed, welfare benefits have increased, and most

Aborigines live and work as grazers (sheep farmers) near sheep or cattle stations in the willie woolies (outback, or back country) or in the mines. Many Aborigines still cling to tribal ways, and "going walkabout" (trekking into the outback) is not unheard of.

Geographically Speaking

The Aborigines were nomadic hunters—*woomeras* (spear throwers) rather than the Polynesian farmers of New Zealand. They used stone tools, and their rock paintings illustrate how they tracked the animals that helped them survive in a harsh land. Aboriginal trackers are now used to locate people who are "bushed," or lost in the bush.

Spiritual life evolved along with practical survival techniques. Aboriginal beliefs were illustrated in caves, and their music told tales of the coming of fire to earth or the sun's rising. The distinctive digeridoo, a drone pipe five feet long and hollow, is still used to create this other-worldly music of the outback. Tapping boomerangs set the scene for special dancing celebrations called *corroborees*.

Terra-Trivia

Both Australia and New Zealand make territorial claims to portions of Antarctica. The claims are pie-shaped wedges that radiate from the South Pole and are due south of their respective countries. Australia's claims amount to almost 2.4 million square miles, and New Zealand claims more than 160,000 square miles of Antarctic territory.

Speaking Like a Native

Unlike the English, Australians have few regional differences in speech, though they use words differently from other English speakers. They put the billie on for tea and might talk about a drongo (the Aboriginal word for "fool"), an ocker (a loud, rude person), a silvertail (a member of high society), or tall poppies (successful people). A pom is an Englishman (taken from convict uniforms labeled P.O.H.M., or *prisoners of Her Majesty*). Indicative of their brash friendliness, colorful and fun-loving Australians call immigrants "new chums."

The Australian States, Territories, and Cities

Australia is politically divided into eight states and territories. To the west is—you guessed it—Western Australia, in north central Australia is Northern Territory, and south of that is South Australia. In the northeast is Queensland, and south of that, in the southeast, is New South Wales. To the far south is Victoria. Off the southern coast is Tasmania, and the small governmental area carved out of New South Wales is called the Australian Capital Territory.

Because Australia is a country of city dwellers, its cities are important. This section describes the largest and most important.

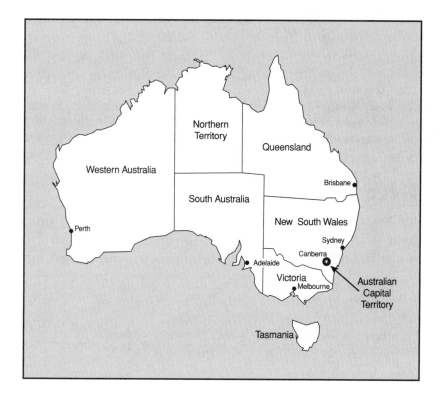

Australian states, territories, and cities.

Canberra

Australians couldn't decide on whether to put their capital in Melbourne or Sydney, so they built it between the two cities. Canberra comes from the Aboriginal word meaning "meeting place." This major city is the only one not located on the coast. Full of administrative buildings, Canberra is home to government, science, academia, the Australian War Memorial, and the Australian American Memorial commemorating America's help defending Australia in World War II.

Sydney

Australia's largest city has 3.7 million people and, according to Captain James Cook, "the finest harbor in the world." Sydney is nestled beside the mountains and the sea. In a city sprawling over rocky terrain, Sydney's residents live in houses clinging precariously over the harbor. The Sydney Opera House echoes the harbor's many sails and will provide a dramatic background for the Summer Olympics in the year 2000.

Melbourne

Australia's second-largest city, with more than 3.2 million residents, Melbourne boasts a new Cultural Center to compete with Sydney's Opera House. Melbourne is located at Australia's southern tip across the Bass Strait from Tasmania.

Perth

Perth (1.2 million residents) is the capital of Western Australia and was described by a 19th century writer as a resort "of rogues and drunkards." Although it's remote from the rest of the country, it's a thriving metropolis, a port city, and a gateway to the Indian Ocean. Perth gained global exposure during recent America's Cup yachting races held off its shores.

The Least You Need to Know

➤ Both Australia and New Zealand have indigenous populations, the Aborigines and Maoris, respectively.

➤ Modern societies of both countries are based on English heritage: language, law, and church.

➤ The countries in this region are isolated by distance from the rest of the world but are not isolationist.

➤ The population is urbanized and on the edges of both countries, thanks to Australia's red desert center and New Zealand's mountainous middle.

➤ Australia has extremely varied flora and fauna that, because of their remote location, are found nowhere else in the world.

Chapter 15

Pacific Rim Economic Giants

In This Chapter

➤ Tying in to China's colonial past

➤ Taking a look at the countries

➤ Journeying across urban landscapes

➤ Exploring the cities, from Seoul to Taipei

➤ Meeting the people of the Pacific Rim

➤ Understanding the dangers of the future

As the chapter title indicates, this region is an economic one, not a typically geographic one. The countries in this region, however, have certain geographic similarities: The countries and territories making up the region are all in Asia; they're all on the western "rim" of the Pacific Ocean or its adjacent waters; and they're all islands, peninsulas, or both. These giants include South Korea, Taiwan, Hong Kong, and Singapore. Although the list could include Japan, because of its size and world importance and for other reasons, this book treats it as a separate region.

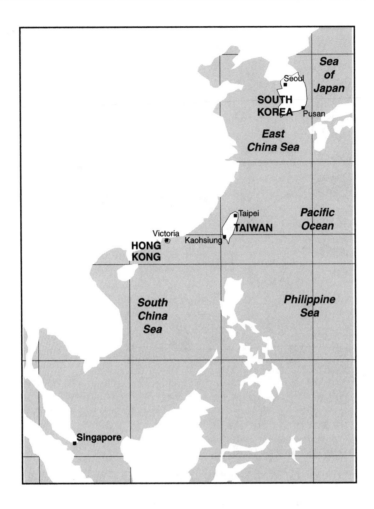

As you can see on the preceding map, South Korea occupies the southern half of the Korean peninsula, just about 100 miles west of Japan. South Korea is twice as far from China as it is from Japan, and it's separated by North Korea on the northern half of their peninsula. Taiwan (formerly Formosa) is off the southeast coast of China. Hong Kong occupies the tip of a peninsula and several islands in southeastern China. At the southern tip of the Malay Peninsula, 1,500 miles southwest of China, is the island state of Singapore.

All the Pacific Rim countries have high levels of economic development, and all are highly industrialized and urban. As city states, both Hong Kong and Singapore are virtually 100 percent urban. South Korea and Taiwan are about three-fourths urban. Although they all export a large amount of goods to China, the vast majority of their exports cross the Pacific to America and Europe.

The export totals of each country vary, of course. Each country annually exports about $100 to $125 billion worth of goods. As a bloc, the region represents almost 40 percent of all the United States' imports, more than all the NATO countries combined, and also more than Mexico and Canada combined. As a bloc, it's the world's third largest exporter, behind only the United States and Germany but ahead of Japan, which is fourth.

Because the Pacific Rim region is not well endowed with natural resources, it must import its raw materials, perform the necessary manufacturing and assembly operations, and then export its finished products. In many cases, the region acts as a trade agent and merely ships other countries' goods, having some of the best, well-located harbors in Asia. The economies all are built on an ample supply of highly skilled labor working for very low wages.

Chinese Ties, Colonial Past

The Pacific Rim doesn't form a region in the typical sense: Geography is secondary to history and economics in this region. Each country or territory is developed independently of one another, though Chinese cultural influences and colonial domination permeate the region.

South Korea

South Korea has experienced a seesaw history. Until 108 B.C., it was known as (old) Chosun. Then China took it over—and then Mongolia. Then, South Korea became independent again as (new) Chosun. In 1910, Japan took it over as a colony and mistreated it miserably until the end of World War II. After the war, the peninsula was split at the 38th parallel: The north went to the Soviet Union, and the south to the United States. (If you were paying attention in Chapter 4, you might remember that *parallels* are lines of latitude that run east–west around the globe.)

The northern half became the Democratic People's Republic (usually called North Korea), and the southern half became the Republic of Korea (usually called South Korea). In 1950, North Korea invaded South Korea and attempted a military unification, which precipitated the Korean War. In 1953, a truce was signed and a no-man's land was set up just north of the 38th parallel, replacing the border between the two states.

Terra-Trivia
Although the Korean peninsula is an extension of the Chinese mainland, the Korean language is more closely related to Japanese.

Since the war, South Korea has had an unstable political life and continued tension with North Korea. Supplied with tremendous U.S. and Japanese aid through the years, South Korea has built the necessary infrastructure and manufacturing facilities to spawn one of the most remarkable economic miracles of modern times.

Taiwan

Taiwan's history is recorded back to A.D. 603, when the Chinese took over. Then Taiwan was conquered by the Japanese, then the Portuguese, then the Spanish, and then the Dutch. At the end of the 17th century, China regained control for the next 200 years, until 100 years ago, when Taiwan was conquered by Japan. At the end of World War II, Taiwan reverted to China, which still claims it as a province. (Got that?)

Just as South Korea owes its modern existence to World War II, Taiwan owes its existence to the Communist Revolution in China. In 1949, Nationalists under Chiang Kai-shek who were fleeing the Communists under Mao Zedong took refuge on the island of Formosa and renamed it Taiwan. The United States' naval forces prevented an invasion by China and have offered protection ever since. In addition, military and economic aid and this infusion of capital sparked yet another miracle of economic growth.

Terra-Trivia

With China's emergence on the world scene, friends have been few and far between for Taiwan. (A great deal of animosity exists between China and Taiwan.) When China entered the United Nations, Taiwan was expelled. China pressures its trading partners to sever links with Taiwan. And, because China has a potential market of more than one billion people, it usually gets its way. The island of Taiwan has become a metaphor for its position among the world's nations.

Hong Kong

Before the British took over, Hong Kong was a small fishing community and a haven for opium smugglers and pirates. At the end of the first Opium War in 1842, Hong Kong was ceded to the British in perpetuity. Both the Chinese and the British soon began to realize the importance of this trading port as the only good harbor between Shanghai and Indochina. In 1898, the additional area (adjacent to Hong Kong) of the New Territories was ceded for 99 years.

In the 1940s, when the Communist leader Mao Zedong was conquering all of China, he easily could have seized Hong Kong but chose not to because it was his gateway to the world for the goods his country desperately needed for economic survival. In 1984, the British agreed to include all of Hong Kong in the mid-1997 giveback, with a guarantee from China that it would preserve its people's civil liberties.

Hong Kong's entrepreneurs and bankers had become extremely wealthy and were investing all over the world, especially in China. By the mid-1990s, they accounted for 75 percent of all foreign investment in China. The former British territory recently became a special administrative region (SAR) of China, to operate under a "one country, two systems" policy.

Singapore

Little is known about the early history of Singapore. The first known information about it was that it was (as it is now) a trading center. In the fifteenth century, it was claimed by the Malacca Sultanate. In 1819, present-day Singapore was founded by the British colonial administrator, Sir Thomas Raffles, and five years later deeded by the sultan to the British.

Because of Singapore's strategic location at the mouth of the narrow passage connecting the Indian Ocean to the South China Sea and because it became a "free port," it soon became an important trading mecca. During World War II, Singapore was occupied by the Japanese, who treated Singapore citizens brutally. In 1945, the British liberated Singapore and the next year made it a crown colony.

In 1959, Singapore became a self-governing state. Independence didn't last long, however, because in 1963, only four years later, Singapore joined with Malaya, Sabah (Northern Borneo), and Sarawak to form a new country, Malaysia. Again, this situation didn't last long because only two years later, in 1965, Singapore separated from Malaysia and became an independent country and a member of the United Nations. Senior Minister Lee Kuan Yew, who was prime minister for 31 years, is a dominant political figure in this tightly controlled state.

> **Terra-Trivia**
>
> In May 1994, a United States teenager was caned in Singapore for vandalizing a car. Corporal and other harsh punishments are the norm in Singapore, a country with an extremely low crime rate. The caning raised a debate in the United States about whether corporal punishment is inhumane or might help to reduce notoriously high crime rates in the States.

The Big Players in the Game

From South Korea to Singapore, the Pacific Rim region spans about 2,900 miles (the approximate width of the contiguous United States). The region's four small countries and territories lie scattered in this vast expanse, each with its own politics, economics, and character.

South Korea

South Korea is blessed with the prime agricultural land on the Korean peninsula (which became more apparent with the catastrophic food shortages North Korea recently experienced). Despite being blessed with the better agricultural lands, South Korea was not blessed with the bulk of the peninsula's raw materials—North Korea got most of those. It's unfortunate that complementary North and South Korea can't find common ground.

South Korea has long had the reputation as a producer of semiconductors. Although it does make television sets and other electronic products, it has recently increased its heavy industry to produce cement, pig iron, crude steel, and passenger cars (Hyundai). South Korea is also one of the world's largest shipbuilders.

Taiwan

In 1988, when Lee Teng-hui won the national presidential election, he became the first Taiwanese (rather than Nationalist) to hold the office and the first person in history to be elected head of state in any Chinese country. With the help of American money and political stability, the country's gradual economic miracle fed on itself. The accent in Taiwan went from farming to manufacturing.

Geographically Speaking

Taipei's National Palace Museum is considered to be one of the four greatest museums in the world, ranking up there with the Louvre in Paris, the British Museum in London, and the Metropolitan Museum of Art in New York. The National Palace Museum houses more than 700,000 pieces of art, some dating back 5,000 years, collected by Chinese emperors and stored in Beijing's Forbidden City.

This collection represents only about one-fifth of the total that had been crated up and moved all over China and finally to Taiwan (under Chiang's orders) to keep it away from first the Japanese and then the Communists. Because only 15,000 objects can be displayed at a time, the priceless bronzes, porcelains, jades, and paintings are changed every three months. If you visited the museum quarterly, it would take you almost 12 years to see the entire collection.

Although Taiwan does produce steel, ships, paper, and cement, its emphasis recently has been on textiles, clothing, and high-tech items, such as personal computers, telecommunications equipment, radios, TVs, VCRs, and calculators. Most of these goods go to Western markets, but more than $6 billion per year goes through Hong Kong to China. Although China is more than 260 times the size of Taiwan, with more than 50 times Taiwan's population, China must take second place as a trading nation.

Hong Kong

Hong Kong is not really a country, even though, for the sake of convenience, this book refers to it that way. Until 1997, Hong Kong was officially identified as a Chinese territory under British administration. Now it's just plain part of China. Nonetheless, this book refers to it as a separate entity based on its unique history and economic status.

Some people couldn't care less about Hong Kong as a trading giant or its political aspirations. What really matters is that it's the tourist capital of the East. As the world's mecca for shopping, Hong Kong's free port status invites wares from all over the world to be sold at possibly the world's lowest prices.

Singapore

Space, or a lack of it, is a key to understanding Singapore. Much of its food and most of its water must be imported because *desalinization* (the conversion of saltwater to freshwater) is a decade away. Most housing is high-rise, and Singapore doesn't have enough of it, especially for middle- and low-income families. Because of Singapore's location, excellent harbor, and a lack of space, being an *entrepôt* (a port that ships goods from one country to another) is an ideal source of wealth.

Crude oil from Southeast Asia is imported, refined, and then transshipped to other Asian countries. Malaysian products (rice, spices, and timber) are also exported through Singapore, as are Malaysian and Indonesian rubber. The list of these types of products, goods, and raw materials being transshipped for lesser-developed neighbors with poorer port facilities is endless. Singapore also imports products from developed countries, however, for transshipment to Asian countries (machinery and autos, for example).

Local manufacturing tends to be limited because of a lack of space. Although some shipbuilding, oil refining, and food processing takes place, most products

> **GeoJargon**
> An *entrepôt* is a port that specializes in the *transshipment* of goods from one country (not its own) to another country (or other countries). Hong Kong plays this role, and Singapore does it in spades.

made in Singapore don't require massive manufacturing facilities, including computer disks and other electronic items, chemicals, pharmaceuticals, plastics, clothing, and rubber products. Most of these items are exported to Japan, the Americas, and Europe. Singapore has also become a financial center and has a prominent stock market.

Singapore is a clean, modern city with no slums and an almost zero percent crime rate. Attributable partly to the moral values of the people, much of this situation also has to do with the dominant role the government plays under the direction of Prime Minister Lee. Singapore has countless laws governing behavior, almost religiously enforced with heavy fines, caning, or imprisonment.

In Singapore, you cannot chew gum (outlawed because youths attached it to subway doors to make them continually reopen), pick flowers in public parks, eat cashew nuts, or eat anything on the subway. You must flush public toilets, and men cannot have long hair. Singapore's strict rules pertain to political opposition and almost everything else.

Urban Landscapes

The countries and territories of the Pacific Rim all are located on islands or peninsulas. Although this chapter discusses some of South Korea's and Taiwan's physical geography, it is limited by the urban nature of Hong Kong and Singapore (they have little vegetation or wildlife). This section looks at the natural side of these economic giants.

GeoJargon

A *typhoon* is the Pacific Ocean equivalent of an Atlantic Ocean hurricane. This massive storm system swirls counterclockwise, with winds more than 74 m.p.h and sometimes even as high as 180 to 200 m.p.h. It forms in the summer over the ocean (or sea) and brings to any landmass unfortunate enough to be in its way very high tides and rains along with devastating winds.

South Korea

In area, South Korea is by far the largest country in this region—it's almost three times the size of the next-largest country, Taiwan. Even so, it's only about the size of Indiana. Here's one difference between the two, though: Indiana has a population of less than 6 million people, and South Korea has more than 45 million people. South Korea is a rugged, mountainous country, especially on its eastern side. As you might guess, because of the mountains, the population tends to be distributed primarily on the western side of the peninsula.

South Korea has cold winters and warm-to-hot summers; the waters surrounding the peninsula tend to moderate South Korea's climate. The country gets most of its rain on its southern coast, where agricultural land is most plentiful. Because of the temperate climate, more than one crop per year can be grown, and rice cultivation is of primary importance. South Korea is also vulnerable to an occasional typhoon (hurricane-like) deluge.

Taiwan

The island of Taiwan, formerly called Formosa, is only a little more than 100 miles off the southeast coast of China. Located on the edge of the Philippine tectonic plate and in this seismically precarious position, it's extremely vulnerable to earthquakes. As though earthquakes weren't enough, Taiwan also gets its share of tropical typhoons.

Although Taiwan measures only 13,900 square miles, it has almost 22 million people. Hilly and mountainous in the east, it has a long coastal plain in the west. The Tropic of Cancer crosses the island, which means that it's at the edge of the tropics. The result is minimal seasonal temperature variation—Taiwan has hot summers and warm winters. Rainfall is heavy throughout the year.

Because of the country's warm temperatures, multiple cropping is practiced on the western half of the island. The fields are intensively farmed, largely in small subsistence plots. The dominant crop in Taiwan, as in South Korea, is rice. Tropical rain forests are on the eastern, mountainous side of the island.

Hong Kong

On approximately the same latitude as southern Taiwan, on China's southeastern coast, is Hong Kong. The former British territory is made up of the New Territories and Kowloon on the tip of a peninsula and several islands, chiefly the island of Hong Kong, across one of Asia's busiest and best harbors.

Although Hong Kong is in the tropics, the southwestern monsoon makes the weather subtropical. This monsoon is a warm, moist equatorial wind that gives Hong Kong a rainy season in early summer. The temperature range is 59–82 degrees F. The area is naturally hilly and rugged, which affords from Victoria Peak a view that's unbelievably spectacular; whether it's best during the day or night is a toss-up, however.

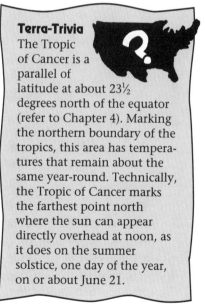

Terra-Trivia
The total area of Hong Kong is now about 415 square miles. I say "now" because it continually dredges up new land by adding fill to shallow bays to the point that it becomes dry land. (The Dutch apparently have not cornered the market on creating land from water.)

Singapore

Singapore is a city–state, a republic, and an island off the southern tip of the Malay Peninsula. Although Singapore is one of the largest ports in the world, it's the smallest of the Pacific Rim states in both size (only about 250 square miles) and population (3 million). If Singapore were in Europe, it would vie for microstate status, as described in Chapter 10.

Unlike the other countries in the Pacific Rim, Singapore is not mountainous. Despite years of dredging, in fact, it's still somewhat swampy. Because Singapore is only one degree north of the equator, it lies in the heart of the tropics; because it's almost on the equator, it's tropical, wet (averaging 95 inches of precipitation annually), hot (81–86 degrees F), and humid. If you don't like it hot and sticky, don't go to Singapore!

Terra-Trivia
The Tropic of Cancer is a parallel of latitude at about $23\frac{1}{2}$ degrees north of the equator (refer to Chapter 4). Marking the northern boundary of the tropics, this area has temperatures that remain about the same year-round. Technically, the Tropic of Cancer marks the farthest point north where the sun can appear directly overhead at noon, as it does on the summer solstice, one day of the year, on or about June 21.

The Cities: From Aberdeen to Victoria

Major cities of the Pacific Rim.

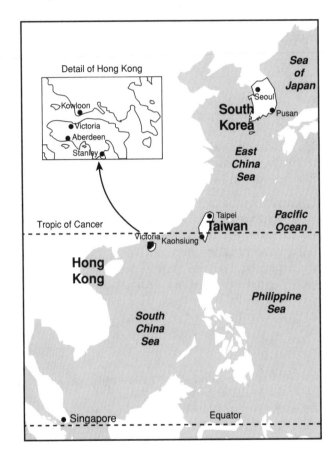

South Korea's soul is Seoul! Between one-fourth and one-third of the entire population of the country lives in or around this capital city. The next-largest and next-most-important city is Pusan, South Korea's largest port city and also a major manufacturing center.

In Taiwan, the country's capital, Taipei, is also its preeminent city. Six million people live in the Taipei metropolitan area, on the northern part of the island. The second-largest city is the southwestern port of Kaohsiung.

The chief population centers in Hong Kong have their own names. Victoria, the capital, is on the north coast of Hong Kong island. Kowloon is across the harbor from Victoria on the mainland. Hong Kong reveals its British history with its other English-sounding centers, such as Aberdeen and Stanley, which are both on the south end of the island of Hong Kong.

Singapore's primary city is an easy one to keep straight: It's Singapore.

People of the Pacific Rim

South Korea's people are 99.9 percent Korean, and all speak Korean. More than half are atheists, one-quarter are Buddhists, and about one-fifth are Christians. Although the economy used to be agrarian, now the dominant sector is manufacturing, which employs more than one-third of all workers. More than three-fourths of South Korea's people are urbanites.

In Taiwan, most of the people live in the north and the west. Highly urban, the bulk of the population is Taiwanese (84 percent), and almost all the rest are Nationalists (or their descendants) who fled China. Mandarin Chinese is the official language. A mixture of Confucianism, Taoism, and Buddhism dominates the religious scene, with about 5 percent Christian.

The people of Hong Kong are 97 percent Chinese, and both English and Cantonese Chinese are its official languages. The prevalent religion is a blend of Buddhism, Confucianism, and Taoism. About 8 percent of Hong Kongians (Hong Kongers? Hong Konganese? Hong Kongites? Hong Konganavians?) are Christian.

Seventy-eight percent of Singapore's people are Chinese; 14 percent, Malay; and 7 percent, Indian. Interestingly, Singapore has four official languages: English (the language of administration), Mandarin Chinese, Malay, and Tamil. Its primary religions are Buddhism, Islam, Hinduism, Confucianism, Christianity, and mixtures of all five.

Terra-Trivia
If an armed conflict with North Korea were to break out, South Korea's capital and its huge population cluster would be extremely vulnerable. Seoul is only about 25 miles from the demilitarized zone (or DMZ), which marks the border between North Korea and South Korea. Although this zone is one of the most heavily defended lines on earth, geographic proximity is enough to make a resident of Seoul quite nervous.

Terra-Trivia
The smallest country in the region, Singapore is also the most ethnically diverse.

The Dangers of the Future

This chapter has examined each of the Pacific Rim states and given you a glimpse of their histories, locations, climates, and peoples. Although you have seen some of the keys to the great success stories of these states, this section gives you a look at some of their concerns for the future.

All these states have based their economies and their success on enterprising people willing to work for low wages. With success, however, comes higher standards of living and greater demands for the goodies of the world—and, therefore, higher wages. Then the competitive advantage is lost, right? Yes and no. The first part is true, and indeed wages have gone way up. Pacific Rim countries have not lost their advantage, however; they have *increased* it.

Although both the countries and territories in the Pacific Rim have had wage increases, so has the United States, and now the difference between the low and high ends of the scale is larger than ever.

South Korea

The big concern in South Korea, of course, is its neighbor to the north, Communist North Korea. Earlier, North Korea wanted unification and invaded South Korea to try to get it. Although that effort failed, the situation has been up and down since then, with unrest, suspicion, and a smoldering desire for unification.

South Korea has good reason to be concerned because North Korea possesses one of the world's largest standing armies and is a very poor nation. South Korea is a rich nation in comparison. Is the plum too tempting for North Korea not to try to pick it?

Taiwan

Taiwan's worry, of course, is a takeover by Communist China: The strait separating the two is narrow, and China openly considers the island one of its provinces. Taiwan is also losing many of its official friends. Most countries want so much to trade with China (and a share of the world's potentially largest market) that they are breaking official ties with Taiwan and opening them with China.

Considering this situation, is anyone really willing to take on China in its own backyard if it steps across the 110-mile strait? Will China take back the island and enhance its riches with Taiwanese trade?

Hong Kong

Will China successfully operate its promised "one country, two systems" policy and its other promise to maintain the personal freedoms of the Hong Kong people for at least 50 years? Or will it break these promises and divert Hong Kong's wealth and earnings to solve other needs elsewhere in its own vast country? China knows well the value of Hong Kong's port and the impressive results of its economic system.

China could have choked off Hong Kong at any time, yet it chose not to. Will China kill its new golden goose? Only if China doesn't fully understand or appreciate what makes free people in a free market society really work.

Singapore

You have read about how all the other Pacific Rim states have concerns involving a Communist-country neighbor—South Korea with North Korea, Taiwan with China, and Hong Kong with China. At one time, Singapore had concerns about another Communist country, first North Vietnam and later Vietnam. Because Singapore is so small and defenseless and yet so valuable a prize, it feared that it was at the mercy of Vietnam, which appeared to want to play a more dominant role in the South China Sea.

This fear doesn't seem relevant anymore (indeed, if it ever was). Vietnam is very poor and concerned with its own economy rather than with political domination. Singapore has other concerns, however. Senior Minister Lee (and now perhaps his son) have exercised such tight control and so strongly restrained all opposition that the question has been raised: What's next after the Lees? Will Singapore open itself up to becoming a more truly democratic state, or will it continue or enhance its anti-opposition policy?

Another Singapore concern is the development of its neighbors' economies. Both Malaysia and Indonesia are indeed moving forward; as they do, they will want to do more within their own countries and avoid having to ship goods through Singapore. Doing so will enable them to not only avoid costs but also capture profits now going to Singapore.

Will Singapore have to replace this loss with more domestic manufacturing or face a curtailment in economic growth? Singapore is so limited in space that it will not be able to fully replace its transshipment profits. Singapore may be facing, therefore, a reduced economic growth rate.

The Least You Need to Know

➤ Four small, very industrialized major trading states located on the western rim of the Pacific comprise the Pacific Rim: South Korea, Taiwan, Hong Kong, and Singapore.

➤ Each state exports about $100 to $125 billion annually, and, as a bloc, the region is the world's third-largest exporter.

➤ South Korea is the largest country in the region; its capital is Seoul.

➤ Taiwan owes its birth to the Communist Revolution in China, when the Nationalists fled to the island of Formosa.

➤ Hong Kong's future is in the balance as it rejoins its huge neighbor, China.

➤ Singapore is the smallest country in the region, a city–state whose economy is based on transshipping.

Part 3
A Regional Look: The Developing World

It's time to set sail for the unknown—and this time, I mean it! The places you visit in this part of the book are often a little off the beaten path. Although these developing countries of the world have more than half the world's population, news about them is often relegated to pages in the back of the newspaper. The areas discussed in Part 3 are not the world's economic powerhouses: Industry lags behind the industry in developed countries, and technology is often primitive. Theses regions don't always stand up and demand to be recognized—you have to search them out.

Because many of the places discussed in Part 3 are likely to be unfamiliar to you, you might want keep an atlas at hand. Although Bhutan, for example, may not occupy a place on the United Nations Security Council or be one of the world's largest steel producers or a leader in semiconductor technology, if you want to see an exotic place that's off the beaten path, Bhutan is for you!

As global communications and transportation shrink our world, culture becomes more and more homogeneous. You can go to essentially the same restaurant in Moscow or Los Angeles or listen to the same music in London or Tokyo. Because that sameness can start to wear on you, thank goodness that a few Bhutans still remain in the world.

Middle America: A Bridge in Time

Between the two great continents of North America and South America is a bridge. This bridge (which is really two bridges) is Middle America, which connects and separates these two giants. As you can see from the following map, the first bridge is a funnel-shaped mainland in the west, and the second is an arc of islands in the east, separated from the mainland by the Gulf of Mexico and the Caribbean Sea.

Middle America has been the stage for empires, bold conquests, and political upheavals. This chapter describes the geographic importance this region plays as a land bridge between continents and demonstrates that it's not just a bridge but also an important destination.

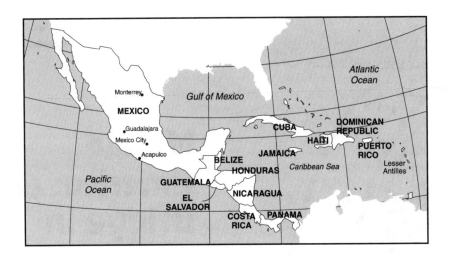

Empires of the Past

The map shows that the Middle America mainland region does indeed look like a bridge, or what physical geographers refer to as a *land bridge*. At one ancient time, another land bridge existed between Asia and Alaska. Both these bridges most likely played a vital role in enabling humans and animals to spread from Asia eastward into North America and then later southward into South America.

The Amazing Maya

The Mayan civilization developed in the lowlands that are now in Honduras, Guatemala, Belize, and the Yucatán Peninsula in southern Mexico. The Mayan population was at its peak somewhere between two to three million people. Although no one can be certain about when the Mayan civilization began, its formative period probably began as early as 1500 B.C. Its classical period was between A.D. 300 and 900.

The Mayan empire was a group of related tribes speaking variations of the Mayan language. Agriculture was the basis of the empire's economy. Although maize was the principal Mayan crop, they also grew beans, squash, manioc, and cotton. The Maya were also known for their fine cloth, pottery, jewelry, and, perhaps their most remarkable accomplishment, the Mayan calendar. It was the most accurate one used by humankind until the advent of the Gregorian calendar the world now uses.

Perhaps the Maya are best remembered for their architecture. Many of the ruins, some amazingly well preserved, are scattered all over their vast empire. These ruins originally were huge centers for religious ceremonies. The usual plan was a number of pyramidal mounds surmounted by temples and palaces and other buildings grouped around open plazas.

About A.D. 900, the many Mayan vast pyramidal centers of temples and palaces were, strangely, completely abandoned. To this day, geographers still don't know why, although speculation abounds. The remaining Maya emigrated north and formed small groups in what is today's Yucatán Peninsula. Many descendants, mixed and pure, now live in Yucatán, Guatemala, and Belize, and about 350,000 still speak Mayan.

> **Terra-Trivia**
> Maize was important to the Maya. It was their chief food crop, and geographers believe that the Maya thought that it was the raw material from which their creator made them.

The Mighty Aztecs

About the time that the Maya civilization had reached its peak, another civilization was beginning to form around present-day Mexico City. This Toltec nation spread out and ranged into the southern lowlands of the Yucatán Peninsula and conquered the deteriorated Maya. The Toltec civilization was relatively short-lived; by A.D. 1200, it was in decay.

Another Amerindian group, the Aztecs, arrived from the north and developed into a powerful force that had no difficulty in conquering the weakened Toltecs. The Aztecs founded their settlement on an island in one of the many lakes that lay in central Mexico. Their city, Tenochtitlán, was to become the greatest city in the early Americas and the capital of the mighty Aztec empire.

The Aztec religion called for human sacrifice and even cannibalism. As the Aztec empire expanded, the Aztecs acquired slaves to work their fields and become fodder for the seemingly insatiable needs of their gods for human sacrifice. Over the 200 years of Aztec domination, more than 100,000 slaves climbed the steps of Aztec pyramids to their slaughter. Those conquered by the Aztecs naturally feared and hated this group.

Hernán Cortés capitalized on this hatred when he led the Spanish into the area in A.D. 1519. The defeated neighboring Indian tribes became his army when he set out to conquer the Aztecs and their king, Montezuma II (Cortés' horses, cannon, and armor didn't hurt, either, of course). Cortés eventually conquered the Aztecs, and the long period of the Spanish and Catholic Church domination of mainland Middle America had begun.

Middle America's Mainland

The mainland bridge starts in the south with the narrow Panama *isthmus* (a narrow strip of land) jutting out from South America—in this case, only 40 miles wide at its narrowest point. This isthmus gradually widens as it winds northwesterly toward North America. It

GeoJargon

An *isthmus* (try to say it three times fast) is a narrow strip of land with water on both sides that connects two larger land bodies—which pretty much sums up Panama.

has two major bulges on the east: the Yucatán Peninsula and the eastern Nicaraguan and Honduran hump. The Baja California peninsula on the west juts south for more than 750 miles from the point where the Middle America bridge (now about 1,400 miles wide) joins with North America.

The mainland funnel has as many contrasts as it does similarities. The massive South American Andes Mountains continue through Central America and Mexico under many local names (Sierra Madre is the most famous) until they reach the great Rocky Mountain chain in North America. Most mountains in the Andes are volcanic in origin, and many are still active.

Earthquakes are frequent in this region, and, combined with volcanic eruptions, make it one of the more dangerous areas in the world. The local mountain chains almost always run through each country with highland plateaus in the middle and lowland coastal forested belts or plains to the west and east. The climate is affected by the altitude: The lowlands and coastal belts are tropically hot and extremely rainy (as much as 150 inches a year), and the highlands are temperate, much cooler, and less rainy.

Terra-Trivia

In 1993, Mexico took a giant step toward becoming a truly developed nation. Along with Canada and the United States, it signed the North Atlantic Free Trade Agreement (NAFTA), which removed tariff barriers between the trading partners. In addition to bolstering business for *maquiladoras* (assembly plants of U.S. companies along the border that often have a reputation for labor violations and environmental problems), NAFTA was expected to stimulate foreign investment even more and create thousands of Mexican jobs.

Spanish is the official language (except in Belize, where it's English). The countries' populations are generally of mixed race, either mestizo (white and Amerindian) or mulatto (white and black African). Guatemala is the exception, where Mayans make up 55 percent of the country's population. In almost all the countries, the people are Roman Catholic. (Belize is again the exception, where almost half are Protestant). The populace is typically about half rural and half urban.

The mainland funnel of land, more than 3,800 miles long, has the giant Mexico in the north. At 762,000 square miles, it is by far the largest country in Middle America—almost twice the size all the others combined—and the 14th largest in the world. Because of its size and world importance, Mexico is considered in this book as one of four subregions in Middle America, along with Central America, the Greater Antilles, and the Lesser Antilles.

Mexico: The Wakening Giant

Mexico, officially The United Mexican States, is composed of 31 states and a federal district. The country has about 95 million people. In addition to the huge capital of Mexico City, Guadalajara and Monterrey are its largest cities. For

300 years, during the period of Spanish colonization, the Spanish-Amerind culture that now prevails was slowly but unalterably developed. In 1821, Mexico gained its independence from Spain; since then, the country has undergone the difficult process of becoming a developed nation.

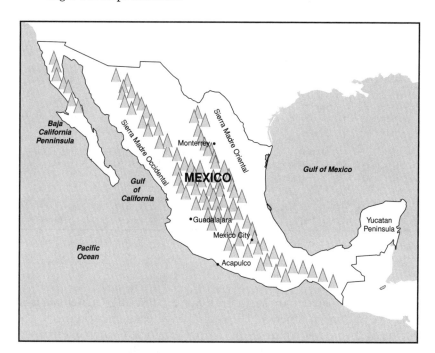

Mexico.

The Turbulent Central American Countries

South of Mexico on the funnel is the subregion of Central America, made up of seven countries, as shown on the nearby map: Guatemala, Belize, Honduras, El Salvador, Nicaragua, Costa Rica, and Panama.

This chapter has quickly looked at much in these seven countries: their topographies, climate, peoples, language, and religion.

This section outlines significant characteristics of these seven countries, starting from the north and moving south. (Only cities with a population of at least one million people are mentioned.)

Central America.

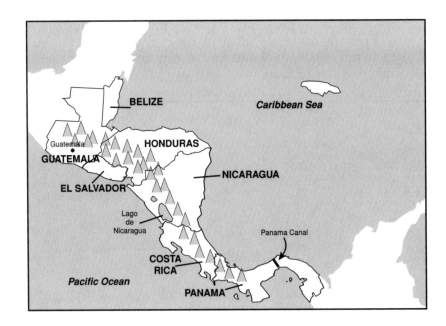

Guatemala

With about 10 million people (55 percent are Maya or of Mayan descent), Guatemala is the most populated country in Central America. The capital, Guatemala, is also the country's largest city. One of the poorer countries in Central America, Guatemala has had a constant series of revolutions, coups, wars, and military dictators. Guatemala is now arguing with Belize about their common border.

Belize

Belize is the only Central American country with a British background (it was called British Honduras until 1981) and the only one with no Pacific coastline. The second-smallest country in the subregion, it's also one of the wealthiest and least populated, with only 210,000 people.

Honduras

Honduras, the second-largest country in Central America, is also near the bottom in terms of personal income. Since its independence from Spain in 1821, it also has had a history of dictators, revolutions, and coups. It had a war with El Salvador in 1969 and served as a base for Nicaraguan rebels in that country's civil war.

El Salvador

El Salvador, the smallest country in Central America, is the only one with no Caribbean coastline. It became independent in 1841 and had several wars with its neighboring countries during the rest of the century. The first few decades of this century saw some stability and growth in the economy, only to be followed by a series of strong but oppressive dictatorships. El Salvador is now only tenuously democratic.

Nicaragua

With only $340 gross domestic product per capita (the value of domestic goods and services divided by the number of residents), Nicaragua is not only the largest country in Central America but also the poorest—one of the poorest in the world, in fact. Much of its poor economy results from extreme political unrest, almost civil war, that has prevailed for the past three decades. Before the recent unrest, 150 years of post-Spanish colonization were filled with dictators and coups and three different U.S. interventions.

Costa Rica

Costa Rica is one of the wealthiest countries in Central America. The indigenous natives resisted colonization, and, with no obvious wealth, Spanish subjugation came late and then only indirectly through Nicaragua. Independence came in 1821, and democracy settled in with few hitches. With no army and a thriving middle class, the entire country enjoys a standard of living that's quite high by Middle American standards.

Panama

Panama has a more typical Central American story: dictators, coups, and revolutions. The only real difference from the rest of Central America is that it didn't become independent until 1903, and that was from Colombia, in South America. Panama's most recent turmoil was the late 1989 U.S. invasion and subsequent arrest of its dictator, General Manuel Noreiga, who was tried and convicted on drug charges and money-laundering.

Of economic and historic interest to Panama, the entire region—and, for that matter, the entire world—is the Panama Canal. One of the greatest engineering feats of all time, it was built in 1905–1914 by the U.S. Army Corps of Engineers. The canal is 40 miles long and has lock chambers that lift and lower ships 85 feet during the passage between oceans. In one recent year, more than 12,500 commercial ships passed through the canal. Although the original U.S. lease

Terra-Trivia
Many citizens of Central America and some geographic historians don't consider Panama part of Central America because Panama was, for most of history, part of South America's Columbia. Also, Panama runs east–west, whereas the rest of the funnel runs northwest–southeast. This book takes the more common view of including it in Central America.

was "in perpetuity," in 1977 the United States agreed to give control of the Canal Zone back to Panama in 1979 and the canal itself in 2000.

Sorting Through the Enchanted Islands of the Caribbean

Many people are confused by the various names of island chains in or around the Caribbean. To help clarify, use these groupings while you look at the map of the Caribbean:

➤ **West Indies:** Includes the Bahamas, the Turks and Calcos Islands, Greater Antilles, Cayman Islands, Lesser Antilles, and Netherland Antilles

➤ **Greater Antilles:** The big islands of Cuba, Hispaniola (Haiti and the Dominican Republic), Jamaica, and Puerto Rico

➤ **Lesser Antilles:** The Leeward Islands, Windward Islands, and Netherland Antilles

➤ **Leeward Islands:** The northern half of the island arc (protected from the trade winds): the Virgin Islands, Saint Eustatius, Montserrat, Saba, Antigua–Barbuda, Saint Kitts–Nevis, and Guadelupe

➤ **Windward Islands:** The southern half of the island arc: Dominica, Martinique, Saint Lucia, Barbados, Saint Vincent and the Grenadines, Grenada, and Trinidad and Tobago.

➤ **Netherland Antilles:** Just off the northern coast of South America, the ABCs of Aruba, Bonaire, and Curaçao.

Islands of the Caribbean.

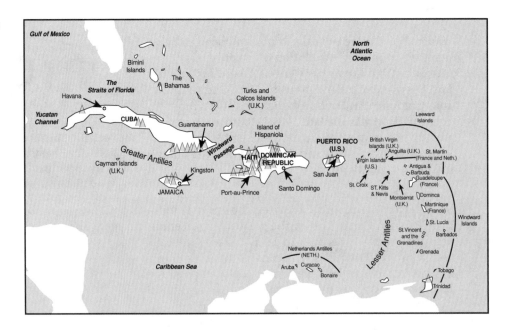

Geographically Speaking

A common problem the islands of the Caribbean share is their annual exposure to nature's dreaded hurricanes. These storms, spawned in the tropical waters of the southern Atlantic, the Caribbean Sea, and the Gulf of Mexico, are rated from 1 to 5: 1 means minimum winds of 74 m.p.h., and 5 means winds exceeding 155 m.p.h.

From a center eye of about 15 miles in diameter, hurricane winds circle out to about 150 miles in diameter. Gale winds (32 m.p.h. and higher) at their perimeter enlarge the circle to 300 miles in diameter. These storms, combined with the high seas they create, can cause severe loss of life and untold damage. In 1988, Hurricane Gilbert devastated the Gulf region and claimed more than 260 lives. For these islands of little wealth, hurricanes can devastate economies for years.

The Greater Antilles: The Big Islands

Another subregion of Middle America is the Greater Antilles, which includes Cuba, Hispaniola (Haiti and the Dominican Republic), Puerto Rico, and Jamaica. Cuba is by far the largest island in the Greater Antilles group. On the east, Cuba is separated from Hispaniola by the Windward Passage. Hispaniola, the second-largest island, is the home of two countries: Haiti on the west and the Dominican Republic on the east. South of Cuba is the third-largest island, Jamaica. To the east of Hispaniola is the smallest island, Puerto Rico. To the east, Puerto Rico is separated from the Lesser Antilles by the Virgin Passage.

These four islands have in common their large size (compared to the Lesser Antilles) and their relative adjacency (they're all close together in the north Caribbean). They also share an important historical fact: All were discovered by Christopher Columbus and therefore started their New World life under the Spanish flag and the Roman Catholic cross.

Peaceful Caribs in the islands were conquered by warlike Arawaks, who in turn almost became extinct in the early days of Spanish rule because they were subjected to cruel enslavement and exposed to diseases to which they had no immunity. Varadera, Cuba, has classic Maya sculptures, indicating that these venturous canoesmen at least visited (if not settled) this land.

The islands also have many contrasts. Cuba is only one-quarter mountainous, although the others are mostly mountainous. Haiti, Puerto Rico, and Jamaica have tropical climates, complete with rain forests; Cuba and the Dominican Republic are semitropical. In Jamaica's rain forest, the amount of rainfall can exceed 200 inches per year. Parts of Haiti, however, get only 20 inches of rain annually.

This section looks at each country separately but not in great detail; many books have been written about their turbulent, rebellious, and even violent political histories, involving mostly intervention, occupation, or war with the United States. Only Jamaica came somewhat peacefully into the present age.

Cuba

Cuba has 11.5 million people, all Spanish-speaking and most with a mixed Spanish, black African, or Amerindian heritage. Its capital and largest city is Havana. Until the current Communist regime, Cubans were almost all Roman Catholic, but now only one-third are, and more than half declare themselves nonreligious. The cultivation and production of sugar is king, and Cuban tobacco (in cigars) is the world's favorite.

The real story is Fidel Castro and his Communist state, one of the last on the planet. Before the downfall of the Soviet Union, the U.S.S.R was giving Cuba aid worth $5 billion a year. This amount is down to nothing—a true disaster for Cuba and Castro, and undoubtedly the end of Communism there—the only question is when.

Haiti

Terra-Trivia

Voodoo is a religion (practiced primarily in Haiti) that combines elements of Roman Catholicism with tribal religions of western Africa. Voodoo cults worship the high god Bon Dieu, ancestors, the dead, twins, and spirits called Ioa (frequently identified with Roman Catholic saints). African elements include dancing, drumming, and the worship of ancestors and twins. The priest is called a *houngan;* the priestess, a *mambo.*

Poverty is the problem in Haiti, with one of the world's lowest gross domestic products per capita, only $370. Haiti has 7.3 million people, 95 percent black who speak French or Creole. Although most Haitians are Roman Catholic, many practice voodoo, which on Haiti is a blend of Catholicism and African religions.

Haiti has few cash-producing exports, and coffee leads the trickle. History's bad guys were the Duvaliers: "Papa Doc" and his son, "Baby Doc." In 1994, the political situation got so bad that the United States sent in troops to restore order.

Perhaps the worst problem facing Haiti is its ecological destruction. For centuries, plantation owners and Haitians have stripped the land of its trees for agriculture and firewood. Trees, however, bind the soil; with 99 percent of the country now deforested, tropical rains have washed away much of the soil. The result is a land that's denuded—in many areas, down to bedrock level. What little soil that remains is overworked and is used for growing export crops rather than life-sustenance crops.

Dominican Republic

The Dominican Republic shares the island of Hispaniola with Haiti. The border between the countries is one of the few national borders that can be seen from the air. The

Dominican Republic's forests stop, and the denuded slopes of Haiti start at a fence line marking the border.

The Dominican Republic has 8.3 million people who speak Spanish, have mixed blood, and are Roman Catholic. Its capital and largest city is Santo Domingo. Although it's one of the world's poorer countries, it's more than twice as rich as Haiti. Sugar is its main source of income.

Jamaica

With more than 300 years of English rule following more than 100 years of Spanish rule, Jamaica didn't finally attain independence until 1962.

The island has 2.7 million people, mostly black with mixed blood, in addition to its established English, Asian, and Chinese minorities.

Most Jamaicans speak the official language, English, or a local English dialect that incorporates African, Spanish, and French elements. The people are Christians of many sects, with Protestants most prevalent. The popularity of Rastafarianism, a religious sect whose members believe in a prophetic return of blacks to Africa, is increasing. Bauxite and alumina are keys to the economy, and tourism is close behind.

Terra-Trivia
As the homeland of Bob Marley, Jamaica is perhaps best known for its favorite export, reggae music.

Puerto Rico

Puerto Rico is by far the wealthiest nation in the group. Almost all its four million people are Roman Catholic, speak Spanish, and have some small amount of pure Spanish blood (mostly mixed blood). Sugar is no longer king of the economy: Now it's chemicals and tourism.

Puerto Rico was under the control of Spain until the Spanish–American War in 1898, when it was ceded to the United States, under whose control it has remained. Although Puerto Rico is now a commonwealth (a political unit but not an independent nation), the ongoing question for decades has been its preferred form of rule. In 1993, the popular vote was 48 percent to remain a commonwealth, 46 percent to become the 51st state of the United States, and 4 percent for independence. Now the question of political status is in the air again.

The Lesser Antilles: The Small Island Arc

The fourth subregion of Middle America is an *archipelago* (a large group of islands) that extends in a grand arc all the way down to South America. Composed of the Lesser Antilles and the Netherland Antilles, it has nine independent countries, many possessions and territories, and almost countless islands, islets, and cays (tiny islands of rock, sand, or coral).

The Lesser Antilles, all small in both population and size, share a Carib–Arawak pre-European history. Most are lush, some have rain forests, and all have golden white (or in some cases, black) sand beaches and coral reefs. Some have still or recently active volcanoes, and all are the tops of underwater mountains.

All the islands of the Lesser Antilles have peoples of mixed blood, usually dominated by that of the black Africans brought over as slaves to replace the decimated Amerindians. The islands are a tourist paradise for Americans, Europeans, and almost every other population that craves the islands' warmth and beauty.

> ## Geographically Speaking
>
> The Lesser Antilles are the fabled pirate islands. *Pirates* (such as Edward Teach, the notorious Blackbeard) plundered land and ships for personal gain. They had no letters of marque (legal authorization) from any country and attacked whomever and wherever they chose. Although *privateers* always had marque with some country, they generally used their own vessels to do their warring and plundering.
>
> In the 17th century, pirates called *buccaneers* almost always worked without marque but often "donated" a part of their booty to a favorite crown and usually attacked only the enemies of that crown (the Englishman Sir Henry Morgan, "the Scourge of the Spanish," was such a man). The headquarters of the buccaneers was the little island of Tortuga, off the north coast of Haiti.

You can group the islands of the Lesser Antilles by their European ancestry:

➤ **British:** The British Virgins, Antigua–Barbuda, Saint Kitts–Nevis, Montserrat, Barbados, Saint Lucia, Saint Vincent and the Grenadines, Grenada, and Trinidad and Tobago. Although the Virgins and Montserrat are still dependencies, the others are independent countries who still look to Queen Elizabeth as their sovereign.

➤ **Danish:** The U.S. Virgin Islands, until 1917.

➤ **Dutch:** St. Maarten (half an island), Saba, and Saint Eustatius, all still under Dutch rule.

➤ **French:** St. Martin (the other half of the island shared with the Dutch), Guadeloupe, and Martinique. All are parts of France, not colonies or dependencies. Dominica has been a true political football: It started out Spanish, then became French, then English, and finally independent. It also was one of the Leeward Islands but then switched to the Windward Island group.

> **!** **Geographically Speaking**
>
> Tourists visit the Caribbean because of the region's warm climate and natural beauty. Tourism is sometimes embraced by its local people because of their financial need. Sugar is no longer a big moneymaker on the small islands, farming barely feeds its people, and industry is essentially nonexistent.
>
> Tourism provides employment (primarily service jobs) or markets to get that little bit of extra cash needed to improve the standard of living. Tourism can also be demeaning for local residents and a constant reminder of a lavish lifestyle they can never attain.

Middle America's Urban Nightmare: Mexico City

This chapter has not discussed cities in Middle America in much detail because the region doesn't feature many cities of international stature—except one.

Mexico City has about 24.5 million people and is growing by more than a half-million people every year. The Tokyo conurbation (a continuous network of urban areas) is possibly larger; by the year 2000, however, it will be a poor second. Considering its current growth rate, by 2010 Mexico City will have between 45 and 50 million people. Many other smaller cities are facing the same problems on a reduced scale.

Mexico City's concerns are legion. Water must be imported from hundreds of miles away. Sewage and other wastes must be exported. Food must be imported, stored, and distributed. Housing must be provided, jobs created, and transportation networks installed.

Mexico City's severe air pollution is exacerbated by thin air at its 7,350-foot elevation, with a constant flow of noxious fumes from millions of vehicles and the daily spewing of chemical pollutants by tens of thousands of area factories. On any given day, Mexico City's air pollution can be 100 times worse than levels deemed safe to breathe.

To make matters worse, much of Mexico City has been built on a filled lake bed. Its location on the edge of the Cocos tectonic plate adjacent to the Pacific Ring of Fire means that the city is subject to violent earthquakes. Hit with sufficient seismic force, the land created by filling the lake can potentially liquefy (called *liquefaction*) and cause devastating structural damage to Mexico City's buildings.

The Least You Need to Know

➤ The Middle American region has four primary subregions: Mexico, Central America, the Greater Antilles, and the Lesser Antilles.

➤ Middle America was home to the great empires of the Maya and the Aztecs.

➤ Mexico is the largest and most politically and economically important country in the region.

➤ The Panama Canal connects the world's two largest oceans via the thin Isthmus of Panama in Central America.

➤ The Greater Antilles are composed of the four largest Caribbean islands: Cuba, Hispaniola, Puerto Rico, and Jamaica.

➤ The Lesser Antilles consist of many small islands, where tourism and pirate lore are big news.

South America: Colonial Legacy

South America is the fourth-largest continent, with 12 percent of the earth's landmass. It's south of North America and much farther east than most people realize. Longitudinally, virtually the entire continent lies east of Florida. Its eastern bulge is less than 1,700 miles from the western bulge of Africa, whereas the Atlantic Ocean measures more than 3,500 miles from New York to Lisbon.

This more easterly location of South America extends the Pacific Ocean. The distance from western South America to Australia is almost twice as much as from California to Japan. The equator passes through the northern part of South America. From east coast to west coast, the continent measures 3,000 miles; from north to south, it's 4,600 miles. South America is 900 miles south of Florida but only 600 miles north of Antarctica.

The Human Succession

How and when did humans get to South America? The most commonly held theory is that they crossed the Bering land bridge (when it connected Asia with Alaska) and then fanned out east and south. Perhaps 200 centuries ago, humans passed down through the Middle American land funnel and reached South America. This belief is only a theory, of course—plenty of scientists have their own ideas about South America's human origins.

Before the Europeans "discovered" the Americas, hundreds of native tribes lived, hunted, and herded in South America. The most famous, the Andes peoples, lived all the way up South America's west coast. In the fertile valleys, protected on both sides by high mountains, the Andes Indians developed as farmers, perhaps starting as far back as 500 B.C. They grew potatoes, maize, squash, peppers, some fruits, and even cotton.

These people eventually produced crop surpluses, which allowed for the development of permanent settlements and cities in which an advanced culture began to develop. Stone buildings, roads, altars, and even gold all began to appear. After many centuries of gradual development, something rather momentous began to happen.

The Incan Empire

It began with a small, warlike band of Quechua-speaking peoples living in the central Andean highlands of Peru. In about A.D. 1100, they moved south into the valley of Cuzco, conquered the peoples there, and remained more or less peaceful for the next 300 years. In the next century, they formed the vast Incan empire (2,500 miles long and 500 miles wide with almost 20 million people) despite extremely difficult terrain—coastal desert, mountain rain forests, and high mountain plateaus.

The Incan empire was a centrally controlled monarchy. Politically, it was arranged like a pyramid, with a godlike Inca (king) on top; a court and advisors, priests, and administrators were next in the hierarchy; and farmers, artisans, and other producers were on the bottom. The empire had fortresses, palaces, temples, aqueducts, canals with balsa boats, thousands of miles of paved roads for people and llamas to tread (they had no carts), amazing suspension bridges (more than 300 feet long), and grain storage towers (in case of famine).

The Inca built dry stone walls (which still stand) to terrace the hillsides. The terraces were called *andenes* by the Spanish (perhaps the origin of the word *Andes*?). Archaeologists still look with amazement at Incan masonry: Despite the use of only massive stone blocks and no mortar, the joints are so fine that an index card cannot fit between the stones.

Although the Inca were sun worshippers, they did believe in an unknown god who was the creator of all things. Although some question remains, most historians believe that the Inca didn't practice human sacrifice. They had little personal freedom; almost everything was state-controlled. Citizens were required to marry, work was obligatory, and the common people were told where to live and what occupation to pursue.

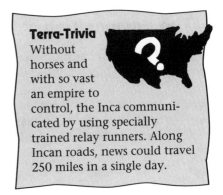

Terra-Trivia
Without horses and with so vast an empire to control, the Inca communicated by using specially trained relay runners. Along Incan roads, news could travel 250 miles in a single day.

One-third of all goods produced went to the king, one-third went to administrators, and the other third was left for family support and sustenance. The Inca were great weavers and made exquisite pottery. They mined gold but only for decorating temples and the use of the king. This massive empire is the one a Spanish adventurer named Francisco Pizarro found when he landed on the coast of Peru in 1531.

The Coming of the Conquistadors

After establishing a base on the coast, Pizarro sent emissaries to ask for a meeting with Atahualpa, the Incan god–king. Assuming that the fair-skinned Spaniards were returning Incan gods, Atahualpa agreed to the meeting with hordes of unarmed Incas in attendance. Pizarro asked Atahualpa to become a Christian and swear allegiance to the Spanish king. Atahualpa refused.

Pizarro arrested Atahualpa, massacred 2,000 Incas, marched into Cuzco, and imprisoned the king in his own palace. To ransom his freedom and throne, Atahualpa offered a 17-by-22-foot room filled with gold as high as a man could reach (and then the same room filled twice again with silver). Although Pizarro accepted the ransom, a few weeks later he had Atahualpa strangled to death.

After a few years, another king, the last male in the line, was beheaded by the Spaniards. With his death, the Incan Empire was history, although evidence of its culture still remains—along with two to three million Incan descendants living in the Andean plateaus in much the same way as they did 500 years ago.

Geographically Speaking

At age 50, Pizarro was a minor Spanish official living in Panama. He had little money and owned only a small parcel of land. After going to Spain to get the king's permission to conquer the Incan empire, he traveled to South America with 180 men and 40 horses to do just that. At age 57, Pizarro conquered the mightiest and wealthiest nation South America ever produced.

The Famous Line

Portugal and Spain didn't seek to colonize for the enhancement of the peoples they found as much as they did for the wealth they could extract. In 1493, Pope Alexander VI wisely realized that Spain and Portugal would be competing for the wealth of the new lands just "discovered" by Columbus, so he drew a north–south line to divide the continent. West of the line was Spanish territory; to the east, Portuguese. The next year, the Treaty of Tordesillas modified the line to about 51 degrees west, as shown on the following map.

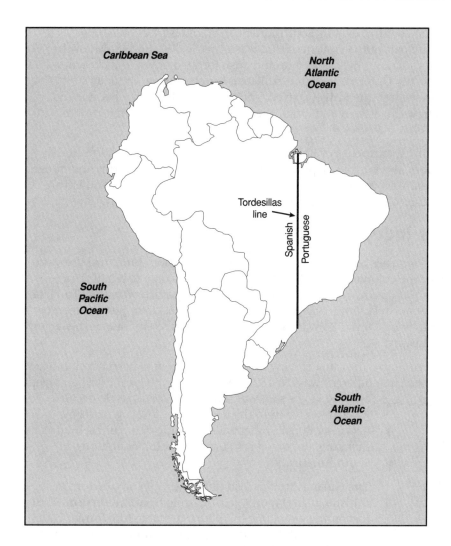

Tordesillas line.

On his third voyage, in 1498, Columbus sailed along the coast of South America far enough to be convinced that it was a continent, but he didn't land or make any territorial claims. In 1499–1500, Amerigo Vespucci also sailed by and also made no claims. In 1500, the Spanish under Vicente Yáñez Pinzón (the former captain of Columbus' ship *Nina*) discovered the mouth of the Amazon and landed at Recife. Because of the Tordesillas line, however, the Spanish didn't push his claim to Brazil. In 1520, Ferdinand Magellan (an ex-Portuguese who was by that time sailing for Spain) stopped in for repairs before he set out to circumnavigate the world. On his way, he discovered the strait that now bears his name.

Spain began to exercise control over its South American colonies by establishing viceroyalties. The only part of the continent that wasn't settled by the Spanish or the Portuguese were the Guianas. This hot and humid area, forested down to its beaches, had little to loot. In 1621, the Dutch came and established Dutch Guiana, or today's Suriname. Shortly thereafter, the British arrived, and their British Guiana became today's Guyana. The French finally arrived and established French Guiana, which is now a department of France.

In the Guianas and the vast reaches of Brazil to the south, the Europeans didn't find any Incan empire to loot or any massive population to conscript for "their" farms and mines, so they turned to what was working well up in the Caribbean. The Spanish and Portuguese imported millions of black Africans and Asians to labor on their plantations.

A Tidal Wave for Independence

It took almost three centuries of mineral and economic exploitation, human abuse, and political injustice for the colonial pots of Spain and Portugal to come to a boil. They finally did in the early 1800s. In 1810, the locals deposed the Spanish viceroy of La Plata, and by 1825 all the Spanish colonies had declared independence. Great heroes emerged, including the most famous, Simón Bolívar. It took only 15 years to end one of the largest colonial systems in history.

In Portugal-dominated Brazil, the same craving for independence had been developing, but it all came to a head in a different way. In 1822, Dom Pedro, the local Brazilian regent and the king's son, was recalled to Portugal. He decided instead to give in to the demands of local agitators by declaring Brazil's independence and setting himself up as its first king. (So much for Portugal's colonial holdings in South America.)

Terra-Trivia

In the seventeenth century, the Dutch and British both occupied what is now Suriname. In 1667, the British ceded their half as a trade for New Amsterdam, which is now New York City—nice trade!

The Guianas story is also different. Guyana (the former British Guiana) didn't gain self-rule from Britain until 1961 (becoming a republic in 1970). Suriname (the former Dutch Guiana) didn't become independent from the Netherlands until 1975, and French Guiana never did get its independence from France. Who would have guessed that France, which owned the smallest area of South America, would own the last?

The Countries and Their Big Cities

Brazil is the world's fifth-largest country, with more than 160 million people and half of South America's total population. The country has ten cities with a population of more than 1 million. The two largest are São Paulo (South America's largest) with more than 16 million people, and Rio de Janeiro with more than 10 million. Brazil's capital, Brasília, with a population of 2.5 million people, is hardly worth mentioning. Brazil, the world's largest producer of coffee, abuts every country in South America except Ecuador and Chile.

With 38 million inhabitants, Colombia is South America's second most-populated country. Uniquely situated, the country has a coastline on both the Pacific Ocean and the Caribbean Sea. Its capital, Bogotá, has more than 5 million people. Cali, Medellín, and Barranquilla each has a population of more than 1 million people. Although Columbia is big in the world of emeralds and tin, it's best-known for refining much of the world's cocaine.

Terra-Trivia

Machu Picchu is an abandoned ancient Incan city on a high ridge in the Andes about 50 miles north of Cuzco. It wasn't discovered until 1911. Despite its unknown history, the speculation is that it may have been the last refuge of Incas fleeing from the Spanish.

Argentina has 35 million citizens and is South America's second-largest country in land area. Its capital, Buenos Aires, has about 3 million people and almost one-third of the country's people living in the area around it. Cordoba is another significant city, with more than 1 million people. Although Argentina still smarts from its military loss of the Falkland Islands (Malvinas) to Britain in 1982, it takes pride in the renewed attention given to Evita Perón, the strong wife of its former dictator.

Peru is home to about 24 million people, about half of whom are Incas and other Amerinds living in mountain plateaus. The other half of the population are mostly mestizos living on the coastal plain. Peru's capital is Lima, also its largest city, with more than 6.5 million residents. Peru's President Fujimori gained initial renown for his attacks on the Shining Path, a Marxist rebel group, and again for the successful conclusion of the Tupac Amaru hostage affair in the Japanese embassy. Peru is the world's largest grower of coca, from which cocaine is refined.

Venezuela is a country of 22 million people, with 3.5 million living in the modern capital of Caracas and more than a million living in Maracaibo. The least important area to the gold-seeking Spanish, Venezuela is now South America's wealthiest country.

Chile is 2,650 miles long from north to south, but averages only 110 miles wide from east to west. It has 15 million people, almost all mestizo and some Native Americans. Chile's capital, Santiago, has almost 5 million people and one-third of Chile's total population. The country has long been big in fishing, nitrates, copper, and, more recently, manufacturing.

Ecuador (with a *c*) straddles the equator (with a *q*). The country has 11 million people and two cities each with a population of more than one million people apiece: Quito, its capital, and Guayaquil. The famous Galapagos Islands, where the naturalist Charles Darwin first began to formulate his theories on natural selection and evolution, are a part of Ecuador. Tourism and bananas are big business in Ecuador.

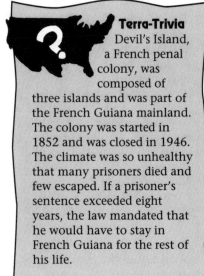

Landlocked Bolivia has eight million people and two capitals: one legal (Sucre) and the other administrative (La Paz). Each city has a population of less than one million. Bolivia, with mostly mestizo and Native American inhabitants, is South America's poorest and historically most unstable country. It lost its only corridor to the sea to Chile in one of its many wars.

Paraguay, also landlocked, has a population of five million with more than one million living in the capital of Asunción. Almost all of Paraguay's people are mestizo.

Uruguay has a population of just more than three million, half of whom live in or around the capital of Montevideo. The population is largely white, and raising stock is important.

The Guianas (ABC—Asian, black, and Creole) were all plantation colonies. Characterwise, the Guianas are more Caribbean than South American. Guyana, the largest, has Dutch and English roots, has about 800,000 people, and its capital is Georgetown. Suriname is next-largest with Dutch and English roots and about 450,000 people; its capital is Paramaribo. French Guiana (still part of France), with only about 120,000 people, has its capital in Cayenne and is South America's smallest state. Of the three, Suriname has some bauxite; the others have few mineral resources.

Physical Facets

A glimpse at the following map of South America's physical features shows that the continent is dominated by two geographic wonders: the mighty Amazon River and the Andes Mountains. With its main trunk just south of the equator flowing from west to east, the Amazon dominates the northeastern portion of the continent. Along the entire western edge of South America lies the spectacular Andes Mountain chain, which acts as a huge wall that divides the low interior areas from the vast Pacific Ocean.

The World's Largest River

Although the Amazon is probably second to the Nile as the world's longest river (the Amazon is about 4,100 miles to the Nile's 4,145), it's by far the world's largest river. The drainage basin of the Amazon is more than 2.7 million square miles, 40 percent of all South America. Its discharge into the Atlantic Ocean measures more than 50 billion gallons per second. In one day, that's enough to meet the needs of all U.S. households for five months.

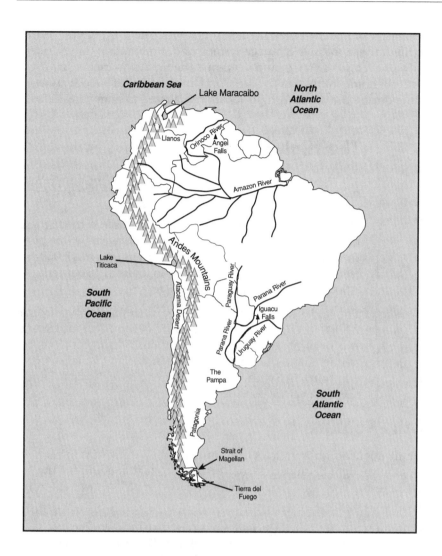

Physical features of South America.

The Amazon was discovered in the mid-sixteenth century, when Francisco de Orellana sailed down the Napa River into what is now the Amazon and sailed it from west to east into the Atlantic Ocean. The source wasn't discovered, however, until the late twentieth century, when Loren McIntyre, an American explorer, author, and photographer, traced the main trunk to the Ucayali River, and then to the Tambo River, and then to the Ene River, and finally to the Apurímac River. There, the huge Amazon begins as a small trickle on the northeast slope of the 18,363 Andean Mount Nevado Mismi. This source in southern Peru (about 65 miles south of Cuzco) establishes the Amazon's total length as about 4,100 miles.

The Amazon's exact length can't be measured because it changes course every year. With its more than 2,000 tributaries, a massive drainage system, and its abundant rainy season, the river rises annually an average of 65 feet and overflows the jungle on each side as much as 30 miles. This phenomenon makes the Amazon almost 60 miles wide at spots. When the river subsides during the dry season, its location is altered in many areas along its 4,100-mile course.

GeoJargon

The *snow line* is the point on the landscape above which snow remains year-round. The warmer the average temperature at a mountain's base, the higher the snow line's elevation. The *tree line* is the point on the landscape above which no trees will grow. It becomes too cold, and the growing season is too short to sustain tree life above this line.

The High Andes

The Andes chain is one of the world's greatest mountain systems. It's 4,500 miles long and ranges from 50 to more than 400 miles wide. It reaches heights of as much as 22,831 feet (Mount Aconcagua, in Argentina, is the highest mountain in the Western Hemisphere). The snow line of the Andes ranges from 4,000 feet in the south to 17,000 feet at the equator. The tree line ranges from a southerly 3,000 feet to an equatorial 12,000 feet.

The Andes have many lofty peaks, including 50 higher than 20,000 feet. Located at the juncture of the Nazca and South American plates, the Andes Mountains are subject to frequent, and often violent, earthquakes and volcanic eruptions. In South America, only Paraguay, Uruguay, and the Guianas are considered non-Andean.

Environments of the Continent

When most people think about South America's environment, they think of tropical rain forests. It's not an illogical thought because the tropical rain forests that dominate the Amazon basin are the largest in the world. South America has other well-known environments, a few of which are discussed in this section.

In northern South America is an area called the *llanos* of Colombia and Venezuela; a similar environment exists throughout much of southern Brazil. These areas of grassland and open dry forests support much of South America's cattle industry. Another famous cattle area farther south is the *pampa*. This landscape of tall grass is located primarily in Argentina around Buenos Aires and throughout Bolivia. In addition to being prime cattle country, the pampa is also a major grain-producing region.

Southern Argentina has another livestock area called Patagonia, whose higher elevations, cooler temperatures, and drier climate lend themselves well to sheep ranching. The final area of note is along central South America's west coast. In this narrow strip west of the Andes is the Atacama Desert, one of the driest places on earth. Because the desert is on a 2,000-foot plateau and is favored by cool winds blowing off the Humboldt Current, it averages only 65 degrees F year-round.

The Waters

Venezuela has, in addition to Lake Titicaca, Lake Maracaibo. At 5,000 square miles, this huge body of water is larger than Lake Titicaca, but it's really not a lake! Lake Maracaibo is really a bay because it's connected to the Gulf of Venezuela by a narrow channel. Other large lakes dot the continent, many as a result of human-made river dammings.

In addition to the Amazon, South America has several long and locally important rivers. The Orinoco River in the north, more than 1,700 miles long, ends in a vast delta on the Atlantic Ocean. In the south, the 2,450-mile Parana River flows from south–central Brazil. It has a confluence with the Paraguay River and then unites with the Uruguay River to form the Río de la Plata estuary on the Atlantic Ocean.

A few more physical features are important. In one of Venezuela's remotest spots, almost impossible to get to by land, are the Angel Falls. At 3,212 feet, they're the world's highest falls (compare them with Africa's Victoria Falls at 400 feet and Niagara Falls on the United States–Canada border, at 167 feet). On the Brazil–Argentina border are the Iguaçu Falls, one of South America's great natural wonders. In the dry season, the river drops in two 200-foot-high crescent-shaped falls. In the wet season, however, it becomes one gigantic fall that's two miles wide (compare this with Victoria's one-mile width and Niagara's combined width of ⅔ mile).

To South America's extreme south lies the Strait of Magellan, "discovered" by Ferdinand Magellan in 1520. The strait connects the Atlantic and Pacific oceans by winding through South America just north of the southerly archipelago, Tierra del Fuego. It measures 331 miles long and only 2 to 15 miles wide. Although the strait is difficult to navigate, sailing it avoids the hazardous winds, rains, currents, and waves off Cape Horn.

Terra-Trivia

The South American continent's largest lake is the famous Lake Titicaca, on the Peru–Bolivia border. What makes the lake famous is its elevation. Titicaca is the world's highest navigable lake: It's about 3,900 square miles in area and almost 13,000 feet above sea level.

GeoJargon

An *estuary,* as used in this chapter, is an inland arm of the sea that meets the mouth of a river. It also refers to the area where the river's current meets the sea's tides.

> ## ! Geographically Speaking
>
> Chile owns the southernmost island in the Tierra del Fuego archipelago, Cape Horn (making it the southernmost country in South America). The name Cape Horn is a misnomer because a cape is usually defined as a strip of land jutting out into a large body of water—hardly a name for an island.
>
> The largest island in the archipelago is the Isla Grande de Tierra del Fuego. The passage between the large island and the mainland forms much of the Strait of Magellan. The western half of the large island is owned by Chile, and the eastern half is owned by Argentina. On the Argentine half is the city of Ushuaia, the world's southernmost city.

A Look at South America's Weather

South America's northern area is all tropical; southern Brazil, half of Paraguay, almost all of Argentina, much of Chile, and all of Uruguay lie south of the Tropic of Capricorn (about 23½ degrees south latitude) and are temperate. For most of South America's tropical zone, the yearly temperature is 75 to 85 degrees F; it's cooler in the Brazilian Highlands, however, and even cooler in the Andes. In the temperate zone, winters range from the 60s to the 30s as you head south. Again, it's always cooler in the Andes. Down near Cape Horn, it's less than 32 degrees F year-round.

GeoJargon
The *Tropic of Capricorn* is a latitude parallel 23 degrees 27 minutes south of the equator, marking the southern boundary of the tropics, where the temperature remains about the same year-round. Technically, it's the farthest south where the sun appears directly overhead at noon on or about December 22.

South America's People

As a continent, South America has 12 percent of the earth's landmass but only 6 percent of its people. The large majority of these people live within 500 miles from a coast, particularly the east, central east, and northwest coasts. The vast interior is sparsely inhabited. Today's ethnic breakdown is largely a result of South America's history.

Pure Native American peoples live primarily in former Incan lands. Mestizos (people of Native American and European blood) mainly surround Incan land and live in Chile, Paraguay, Bolivia, Peru, Ecuador, Colombia, and Venezuela. Asians, who were brought there as slaves, now live in Guyana and Suriname.

African blacks, who also came as slaves, live primarily in the Guianas and Brazil, with smaller populations in Peru, Ecuador, and Colombia. Mulatto peoples (those of black African and European stock) inhabit primarily Brazil and also live in Colombia and the Guianas. Whites dominate only in Argentina and Uruguay, where they represent more than 85 percent of the total population.

Language also parallels the region's colonial history. Portuguese is spoken in Brazil, English in Guyana, Dutch in Suriname, French in French Guiana, and Spanish everywhere else. Not surprisingly, many other indigenous languages are also spoken. In Paraguay, 90 percent speak Guarani. In Bolivia, Quechua and Aymara are also official languages, and 40 percent of the people don't speak Spanish. In Peru, although Quechua is also official, 70 percent speak Spanish. Quechua is also common in Ecuador. In Guyana, you hear Hindi and Urdu, and in Suriname, taki-taki.

Roman Catholicism is dominant everywhere in South America, except in Guyana, where it's third to Hindu and Anglicanism. Ethnicity and language are blends of European, African, Asian, and Amerind heritages. Because of this diverse mix, Roman Catholicism forms the most universal unifying cultural force throughout South America.

Post-Independence Political Instability

Unfortunately, all the countries of South America have something else in common: They haven't been able to sustain long-term democratic governments. Military strongmen, coups, boundary squabbles, wars, congressional takeovers—you name it, they've had it. The only difference from country to country is the degree of the unrest, from sporadic to constant. Bolivia has counted 180 coups and wars—an average of more than one a year since gaining independence.

Although all the states are now officially democracies, no president in this region has any true assurance against a coup. In 1996, all 13 South American countries suffered from at least three of these problems: government scandal, insurgent rebels, drug traffic, extreme poverty, attempted coups, police brutality, soaring taxes, government cutbacks, recession, high unemployment, major or general strikes, monetary fund problems, or hyperinflation.

A great disparity exists between the rich (few) and the poor (many). South America faces the usual big-city problems you have read about the developing world, as in Mexico City. The countries of the region are all poor—Bolivia has a gross domestic product per capita of only $2,400. Most hover between $3,000 and $5,000.

Terra-Trivia
Brazil is one of the most ethnically mixed and balanced countries in the world: mulatto, 22 percent; mestizo, 12 percent; Italian, 11 percent; black, 11 percent; and German, Japanese, Native American, and others, 19 percent. Pure Portuguese comprise only 15 percent, and Spanish (surprisingly) only 10 percent.

The Least You Need to Know

➤ South America has the world's largest river, the Amazon.

➤ The continent is home to one of the world's greatest mountain chains, the Andes.

➤ The region boasts the world's highest falls—Angel—and the widest— Iguaçú.

➤ South America featured the great civilization known as the Incan Empire.

➤ Brazil has Portuguese, black African, and Native American roots and represents half of South America's size and population. The rest of the region has primarily Spanish and Native American roots.

➤ All of today's South American countries have had unstable political histories and aren't yet out of trouble.

Saharan Africa: The Sands of Time

The region of Saharan Africa is vast; in fact, it encompasses the whole of northern Africa. It's bounded on the west by the Atlantic Ocean, on the north by the Mediterranean Sea, and on the east by the Suez Canal, the Red Sea, and the Indian Ocean. To the south is the valley of the Niger River and the semiarid and slightly wooded plains and forests that are typical of sub-Saharan Africa, as described in Chapter 19.

Although the region is made up of 14 countries, it is named for and dominated by the gigantic Sahara Desert, by far the world's largest. Its other major geographic feature is the lifeline of Africa, the Nile River. The region is united by the Islamic religion—in particular, the Sunni Muslim sect. Islam is not, of course, limited to Saharan Africa. With few

GeoJargon

An *oasis* is a fertile patch of land within a desert. It's often lush, with ponds, grass, and trees, usually date palms. The most common cause of the fertility is the sudden surfacing of an underground river flowing down from the mountains, often hundreds of miles away.

exceptions, it pervades all the region's countries; it is practiced by 95 to 99 percent of all the people regardless of whether they're Arab or Berber or from other African tribes.

The Sahara is even larger than this region. The desert extends beyond the Red Sea, across Arabia and the Persian Gulf, through Iran, and into Iraq. It measures more than 3.5 million square miles and includes about 80,000 square miles of oases, or fertile patches. Its daytime temperature in the shade can reach 130 degrees F, with nights falling into the 30s. Average yearly rainfall is less than five inches, and some areas get no rain for several years. (It's not a place where you would want to take a walk in the park.)

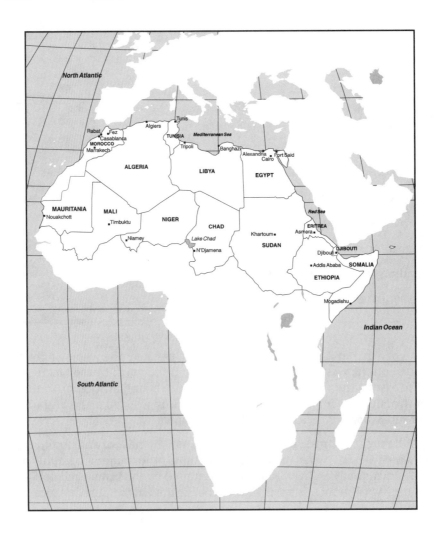

The Lands of Pharaohs and Berbers

The Nile River and its delta, rising and falling yearly for centuries, has nourished and sustained people along its banks since the beginning of recorded time. Great civilizations and kingdoms thrived there: Egypt was the land of the pharaohs. To the west lie thousands of miles of shifting sand, on which ancient caravans and desert people, the Berbers, traveled.

The Ancient Pharaohs

The Egyptian kings were the master architect–builders of the Pyramids and the Sphinx, the conquerors of the then "known" world, and the supreme rulers of highly sophisticated societies. Early Egyptians not only developed advanced agriculture but also domesticated horses, pigs, and sheep. They were proficient in calendrics (the study and development of calendars), astronomy, government, engineering, and metallurgy. The Egyptians also left the world enough ruins of temples and tombs to keep archaeologists busy for umpteen centuries.

The secrets of ancient Egypt, the tales of their wars, and the succession of their pharaohs, their gods, and even their everyday lives were locked up in the indecipherable hieroglyphics carved on their monuments and written on papyri. In 1799, Napoleon's forces found the Rosetta stone, a black basalt slab with inscriptions in hieroglyphic, demotic (a simplified form of Egyptian writing), and Greek. After about 15 years of international puzzle-solving, the code was finally broken by a British physicist, Thomas Young, and a French Egyptologist, Jean François Champollion. The silence was ended: Ancient Egypt could speak.

The Berbers: People of the Sand

While all the wonders of the Egyptian empire were going on around the lush and fertile irrigated valley of the Nile, nomadic tribes of Berbers began to settle the sandy wasteland to the west, particularly the Atlas Mountains and the valleys in the northwestern corner of Africa.

The nomadic Berbers lived in peace for centuries. Then came the invasions—first, the Phoenicians, and later

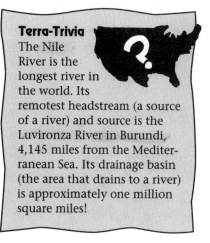

Terra-Trivia
The Nile River is the longest river in the world. Its remotest headstream (a source of a river) and source is the Luvironza River in Burundi, 4,145 miles from the Mediterranean Sea. Its drainage basin (the area that drains to a river) is approximately one million square miles!

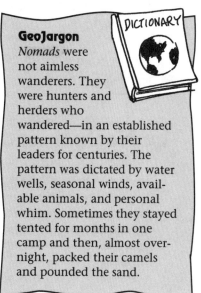

GeoJargon
Nomads were not aimless wanderers. They were hunters and herders who wandered—in an established pattern known by their leaders for centuries. The pattern was dictated by water wells, seasonal winds, available animals, and personal whim. Sometimes they stayed tented for months in one camp and then, almost overnight, packed their camels and pounded the sand.

225

the Carthaginians, the Romans, the Vandals, and the Byzantines. Later, the Arabs invaded, who became the Moors when they fraternized with the Berbers. As though these conquests weren't enough, the European nations finally got colonial and sought to form spheres of influence and eventually colonies in northern Africa. The English, French, Spanish, Italian—they all left their mark on the region.

Geographical Versus Political Divisions

A mountain range or a river makes a good border. Where one culture ends and another, different culture begins—that makes a good border. The line between two spoken languages, even if it's somewhat blurred because of bilingual people, makes a good border.

Lines drawn on a map by third-party countries segregating Arab from Arab, Muslim from Muslim, and one sand dune from another aren't likely to make good or lasting borders. This region is Saharan Africa, however. To make some sense in the study of these countries, this book groups them geographically rather than politically.

! Geographically Speaking

As the center of trade for nations and trading states, the Mediterranean Sea had always been of great strategic importance. It links the Atlantic to the Indian Ocean via the Suez Canal and the Red Sea. Because the southern shore of the Mediterranean is North Africa, whoever controlled this region was able to control the Mediterranean.

During World War II, the Axis and Allied powers duked it out for control of the region in the African Campaign. Although France and Britain maintained a presence in this region after the defeat of the Axis forces, nationalism was growing. By the mid-1950s and early 1960s, the colonies in this area were becoming independent states.

The Maghreb and Its Eastern Neighbor

The *Maghreb,* as shown on the following map, is an Arab word used to signify the Arab countries sharing the Atlas Mountains, their included valleys or coastal plains (all having abundant or ample water) to the north, and the Sahara Desert to the south. Moving south from the Mediterranean Sea (which all these Arab countries border), the amount of annual rainfall goes from 30 to 40 inches to less than 5 inches. Ninety-five percent of its people, of course, live "up north."

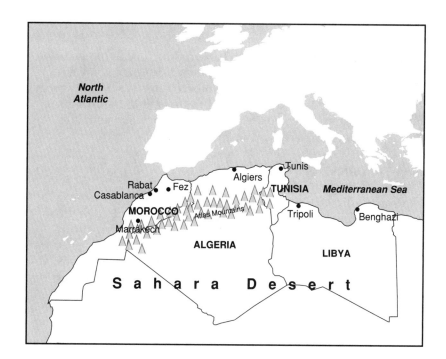

The Maghreb.

These people originally all were Berbers. Now they're mostly Moors (of mixed race) or Arabs. They all speak Arabic and are Sunni Muslims. Some European nationals, of course, still cling to their colonial roots here. The Maghreb is made up of four countries with a total population of 73 million.

Morocco

Morocco was formerly under the influence of Spain and particularly France. Recently, Morocco absorbed the Spanish Sahara to the south. Because the country borders the Atlantic and the Mediterranean, fishing is a big industry, as is mining the minerals of its mountain regions. Sheepherding is still important, as is growing cereals, vegetables, and fruits.

Morocco is the last of the African monarchies; by constitutional law, its monarch must be male. Tourism from Europe, the Americas, Japan, and various cruise ships is important and growing more important. Rabat (population 1.4 million) is its capital. Fez and Marrakech each have about 800,000 inhabitants; Casablanca (known for the famous movie) is its largest city, with a population of almost 3 million.

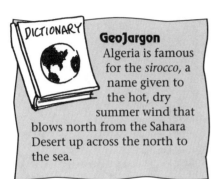

GeoJargon
Algeria is famous for the *sirocco*, a name given to the hot, dry summer wind that blows north from the Sahara Desert up across the north to the sea.

Terra-Trivia
The lands of the Berbers—primarily Tunisia but also Morocco and Algeria—were called the Barbary States and were infamous for their pirates who preyed on the lucrative Mediterranean Sea shipping trade. In 1805, 1806, and (finally) in 1815, the United States navy attacked the pirate bases, especially in Tunis, and ended this government-sponsored means of revenue.

Algeria

With primarily a French background, Algeria has had major difficulty in gaining independence. Tourism is a factor there, but less so than in Morocco. Algeria's climate, geography, and lifestyle are similar to those of Morocco's. In Saharan Africa, Algeria is second only to Sudan in area. Its capital, Algiers, has two million people.

Tunisia

In ancient times, before Tunisia was a French colony, Carthage was here and was the base for the Carthaginian empire. Its ruins now are the base for the ever-growing European tourism industry. In addition to the usual endeavors of farming and fishing, petroleum is a major industry—both on land and offshore. Tunisia's capital and largest city is Tunis, with a population of one million.

Libya

A former Italian colony, Libya is water-poor but extremely oil-rich. To solve the water problem, Libya started in the 1980s to spend a planned $25 billion of its oil revenue to bring water from the southern aquifers (subterranean water supplies) to the northern population. Some more of the oil wealth has been used by Colonel Mu'ammar Qaddafi, the head of state by coup, to sponsor international terrorism. A resultant international trade embargo seems to have lessened this activity. Libya's two major cities are its capital, Tripoli (population 600,000), and Benghazi (population 500,000).

The Sahel, or Transition Zone, Countries

As shown on the following map, south of the Sahara Desert is a transition zone where the alternately expanding and shrinking desert meets the *savanna* (grass plains) and finally the forests of tropical central Africa. Annual rainfall increases from less than 5 inches in the desert to more than 50 inches in the forest. This area is also a transition zone from Arab to black African and from Islam to Christianity and tribalistic religions. The countries in this zone, predominantly still desert, are called the Sahel; they cross all of Africa from west to east.

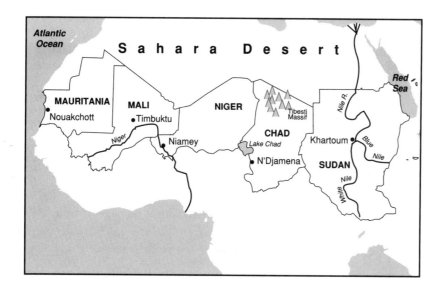

The Sahel.

Mauritania

Of Mauritania's two million inhabitants, 85 percent are Moorish Arabic-speaking nomads, and 15 percent are urban Muslims living in Nouakchott, the capital. Although only 17 percent of the people are literate, the percentage is high compared to Mauritania's easterly interior neighbors.

Mali

Almost all of Mali's ten million inhabitants are black Africans, although some are Berber nomads. Everyone is Muslim. Long ago, when caravans and camels were the only method of transportation and trade, the inland, oasis city of Timbuktu had great renown; now it's thought of as an outpost to nowhere. Irrigation from the Niger River is the lifeblood of its people; because the river is unpredictable, however, drought and famine are all too often their lot. The literacy rate is only 9 percent, and the annual PCI (per capita income) is just $200.

Niger

Of the ten million residents of the former French colony of Niger, most are black and 80 percent are Muslim. Niamey is its capital, with about 400,000 people. Uranium deposits were hoped to be an economic boon; as the nuclear star began to fade, however, foreign aid became the country's main support for subsistence. Drought and famine, a 10 percent literacy rate, and a $280 PCI characterize the country.

Chad

In this former French colony, half the six million people are Arab, and half are black. French and Arabic are both official languages. Although subsistence agriculture, goatherding, and some fishing in Lake Chad support life, drought and famine are prevalent. A 6 percent literacy rate, a life expectancy of only 39 years, and a PCI of $150 make it one of the poorest countries on earth.

! Geographically Speaking

Although the freshwater Lake Chad is fed by two rivers, its size is gradually decreasing because of evaporation and underground seepage. It now ranges from 10,000 square miles in the rainy season to only 4,000 square miles in the dry season. Lake Chad is also probably the world's shallowest lake, ranging from 3 to no more than 20 feet deep.

By way of contrast, in northern Chad is the Tibesti Massif, a mountain range in the middle of the desert with peaks rising to 11,200 feet. Clouds are occasionally trapped, and snow can be seen on the mountaintops.

Sudan

At 968,000 square miles, Sudan is Africa's largest country. It also has a sizable population of 29 million. Sudan's northern third is desert, its middle third is steppes (dry, level, and treeless plains) and low mountains, and its southern third, called the Sudd, is made up of vast swamps and rain forests. Up north, most people are Arabs and Muslim and speak Arabic. Down south, they're black, and they practice various traditional religions or Christianity and speak Arabic, English, or native tongues.

Though plagued with strife and civil war, Sudan is blessed with the Nile and its two major tributaries, the White Nile and the Blue Nile. These rivers provide fertile, cultivatable land that enables 70 percent of the country's people to farm on what is only about 5 percent of its total land area. In the south, where jungles prevail, all the animals associated with Africa (hippos, elephants, monkeys, leopards, lions, and even the dreaded tsetse flies) begin to make an appearance.

The population center is where the White Nile joins the Blue Nile to form the Nile itself. Almost two million people reside in and around the capital, Khartoum. Despite the desert and swamps and the nomads, coups, and civil wars, literacy is still 31 percent; life expectancy, 49 years; and the PCI, $870.

Egypt and Its Lifeline, the Nile River

Because Egypt is arguably the leading nation in the Arab world, this book puts it in a group by itself in Saharan Africa. Egypt is not a rich nation by the world's standards; because of its history and current position in world politics, however, it's a giant. This chapter has already taken a peek at Egypt's ancient greatness—now you can read a little about the Egypt of today.

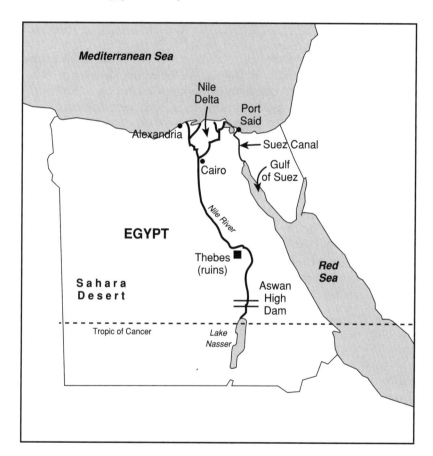

Egypt.

Ninety-nine percent of Egypt's population of about 65 million people live on 4 percent of the total land area—almost within eyesight of the Nile or its delta. Most of the people are descendants of pre-Arab Egyptians, and many are indeed Arabs. Others are of Berber descent, Nubian, and black African mixtures, and some are even descended from the British and French. Sunni Muslims account for 90 percent of all Egyptians; the rest are mostly Copt (a branch of the Christian Church centered in Egypt) or Copt Orthodox. Arabic is the national and official language. Most of the better educated also speak English or French as a second language.

> **! Geographically Speaking**
>
> The terms *upper* and *lower* invite confusion when they're applied to Egypt. Upper is north, and lower is south, right? Wrong. These terms refer not to direction but to elevation. Although upper Egypt is in the south, when *upper* refers to the Nile, it means upriver (toward the source). Lower Egypt is north because it means downriver (toward the delta).

Half the Egyptian population is engaged in agriculture and another 10 percent in fishing. With new discoveries of oil reserves, primarily in the Gulf of Suez area, petroleum and natural gas have become quite significant. Although manufacturing got off to a slow start because of Britain's trade preferences, it's now beginning to come of age. Tourism, always big business and getting bigger, has slowed its growth because of fundamentalist Muslim attacks. The Egyptian government is sparing no effort or expense in trying to stem this nationally self-defeating terrorism.

General Gamal Abdel Nasser (1954–1970) recently put an end to British control, nationalized just about everything (including the Suez Canal), opposed Israel in every way (including three wars), and brought Egypt into the sphere of the Soviet Union. Nasser was followed by Anwar Sadat (1970–1981), one of the truly great Arab leaders of modern times. Under Sadat, Egypt began by warring with Israel, but he soon realized that war was bringing financial and moral ruin to his people. He began to distance Egypt from the U.S.S.R. and drew it closer to the United States. He also began to reverse the trend of nationalization.

The world mourned when Sadat, while at a parade reviewing stand, was gunned down by Muslim extremists from his own army. Hosni Mubarak became the next president and started a major program for economic reform. He has maintained his country's peace with Israel and has also tried to mend fences with the other Arab states. Although Mubarak has followed "positive neutrality" with the world's leading powers, he has actively worked for Israel–Arab peace and even committed 38,500 troops to the anti-Iraq coalition in the Persian Gulf War.

Egypt is divided west from east by the Nile River. The Nile enters Egypt from Sudan in the south and flows north for about 960 miles to the Mediterranean Sea. For its entire length from Sudan to Cairo, the Nile valley is surrounded by cliffs on both sides. At the southern border with Sudan is Lake Nasser, formed by the Aswan High Dam. The lake is 300 miles long and 10 miles wide, and about two-thirds of it is in Egypt. The Nile Valley is almost 2 miles wide all the way north to Idfu, where it widens to about 14 miles. It generally remains at that width all the way to Cairo, with most of the arable land on the western side of the river. Just north of Cairo, the Nile fans out to form its delta, which is about 155 miles wide when it reaches the Mediterranean Sea.

Geographically Speaking

Historically, the Nile has flooded annually, bringing not only water but also silt from upstream. The flow was walled off into basins along the way; in these basins, silt was collected and crops were grown. Later, the walls or dams were broken, and the flow continued downstream to the delta. This one-crop-per-year system is called *basin irrigation*.

With the advent of the Aswan High Dam, the system has changed. Now the water flow is controlled: Much of the silt is trapped in Lake Nasser, and perennial irrigation is used in Egypt's farmlands. This process has resulted in a 50 percent increase in the Nile valley's arable land and has permitted two crops a year in most areas. On the down side, salinization and the disease schistosomiosis are on the rise in irrigated areas, and, because the silt isn't getting to the delta, a saltwater takeover threatens this precious fertile area.

Despite its heavy emphasis on agriculture, Egypt is also quite urban. Almost ten million people, about one-sixth of the entire country, reside in greater Cairo, its capital city. Cairo is a city of contrast: old versus new, poor versus rich, sand versus cement, and desert versus river, for example. This old city has *suqs* (bazaars), where much of the day-to-day trading goes on—to the utter fascination of tourists. In the City of the Dead, more than 250,000 squatters live in the tombs of the ancient cemeteries. Another city of illustrious past—Cleopatra, Caesar, Napoleon, and many more—is Alexandria (about three million). At the entrance of the Suez Canal is another important city, Port Said (population 450,000).

The Horn of Africa

The Sahara Desert finally ends its eastward sprawl at Sudan, where it meets the high tableland of the Ethiopian Plateau. It's also the beginning of the Horn of Africa, a huge eastern outcropping from the continent shaped much like the horn of a rhinoceros jutting up and out into the Indian Ocean. The Horn is made up of four countries with about 72 million people.

Terra-Trivia
The Suez Canal links the Mediterranean Sea to the Gulf of Suez, an arm of the Red Sea. Because the seas are at approximately the same elevation, the canal has no locks, the sealable portions of a canal that can be flooded or drained to raise or lower a ship in passage. It is 101 miles long and was built between 1859 and 1869 for $100 million. The Suez Canal reduces the crossing time from the Atlantic Ocean to the Indian Ocean by almost 50 percent.

The Horn of Africa.

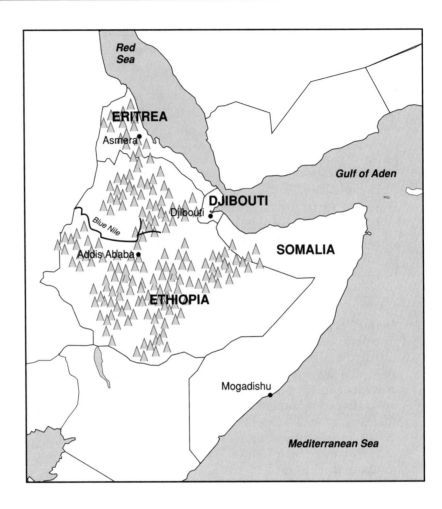

Eritrea

Eritrea is the "newest" country, having split away from Ethiopia in 1993 and thereby rendering it landlocked. Eritrea is bounded on the northwest and north by Sudan, on the east by the Red Sea, on the southeast by Djibouti, and on the south and southwest by Ethiopia. It has about 3.5 million inhabitants, and its capital is Asmera. The country has a provisional government but seems to be on its way to becoming a democratic republic. In all other characteristics, it's similar to Ethiopia.

Ethiopia

Ethiopia is a rugged country, with the high tableland Ethiopian Plateau separated from the coastal plain by 4,000-foot-high cliffs. The plateau is crossed with rivers and deep gorges or rifts. It's also mountainous, with peaks rising as high as 15,000 feet. Temperature and rainfall vary with altitude but hover between tropical and temperate. Animals and vegetation are profuse.

Terra-Trivia
Although people all over the world now drink coffee, it is believed that Ethiopians were the first to drink it.

Throughout most of the past 2,000 years, Ethiopia has had monarchs and emperors. It shunned all European overtures during colonization but did, with Portuguese help, defeat Arab invaders. The only exception to its self-rule was an Italian conquest and occupation during World War II. In 1974, a 13-year military coup ended the monarchy of Haile Selassie. Then Ethiopia flirted with a Marxist Communist regime, only to replace it recently with what seems be a democratic republic.

The Ethiopian population of about 57 million is extremely diverse: The Amhara and Galla tribes each make up a third, and the balance is composed of numerous other tribes and clans. The capital city of Addis Ababa has about 1.5 million people; most of the country is rural, with inhabitants engaged in herding or self-sustenance farming. Although more than 70 languages are spoken in Ethiopia, its official language is Amharic, spoken by some 60 percent. Half the people are Christian and belong to the Ethiopian Orthodox Church; the other half are mostly Muslim, with some Jews and African traditional faiths.

The population is extremely poor, with a PCI of only $120. The country has periodic droughts and famines, and civil wars still plague the land.

Djibouti

About half a million people live in the country of Djibouti, mostly in the capital city of the the same name. The people, who are mostly Muslim, speak Arabic or French or both. Because Djibouti is the terminus (the end point) of the Addis Ababa railroad, the economic situation isn't so bad: Its PCI is $1,070.

Somalia

East of Djibouti and Ethiopia and bounded on the north by the Gulf of Aden and on the east and south by the Indian Ocean is Somalia. This land is rugged and diverse in topography, climate, vegetation, and animal life. Most of its ten million inhabitants speak Somali or Arabic and are Muslim. The capital is Mogadishu, with 500,000 people. Civil war among the clans and warlords, combined with drought and famine, made the situation so bad that in 1992 the United Nations had to step in with humanitarian help. Although the money was desperately needed, getting it to the scattered, distrusting populace proved difficult.

What Might the Future Hold?

Racial and religious strife are deep-rooted in the Saharan African region and can easily flare up and get out of hand. When conflict happens, the struggles can be long and bloody, and Saharan Africa has several possibly in the making. The Sunni Islam-versus-Fundamentalism conflict is beginning to send off serious sparks throughout the Muslim countries, particularly Egypt, Sudan, and the Maghreb. Another concern is the unrest between the Christian Amhara and the Muslims in Ethiopia. Still another is the race and culture of the Arabs versus that of the African tribes throughout most of the Sahel. Although proven democratic traditions or strong leadership can control or even quiet these burning embers, neither of these qualities is Northern Africa's strong suit.

Desertification (the expansion of the desert) has many ecologists increasingly concerned, particularly in the Sahel of Saharan Africa. This process usually occurs as a result of two causes, both present at the same time: extra-dry desert conditions and overgrazing or overcultivating grasses or soils in the areas neighboring the desert. The results, of course, are drought and famine and loss of arable or life-supporting land for the future.

Desertification can be extensive. Between 1980 and 1984, the desert expanded 15 percent, or 500,000 square miles, which was devastating. Fortunately, the process isn't a one-way street: Because of an unusually moist summer in 1985, the desert "gave back" 300,000 square miles. This desert seesaw probably has been going on for centuries.

The Least You Need to Know

➤ The Sahara Desert, the largest in the world, is the dominant physical characteristic of northern Africa.

➤ The Muslim religion prevails throughout Saharan Africa and acts as a unifying cultural link throughout the area.

➤ The Nile River is the longest river on earth, and the Nile valley and delta enable life and agriculture in an arid land.

➤ The Pyramids, the Sphinx, the temples, and the tombs of the ancient pharaoh civilizations are all in this region.

➤ Despite the technology, wealth, and high standards of living that characterize much of the world, millions of people in North Africa are still scratching for mere existence and fighting the problems of drought and famine.

➤ At the edge of this region is the Suez Canal, one of the greatest manmade waterways in the world.

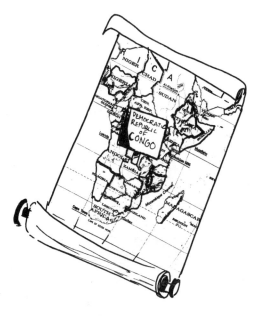

Sub-Saharan Africa: Where We All Began

Although Africa is the world's second-largest continent (Asia is the largest), it can truly be called the heart of the earth's landmass. If you look at a globe and think back about 100 million years, you can perhaps imagine all the continents fitting together as one massive landmass, called Pangaea (refer to Chapter 2). In the center of its southern portion, called Gondwanaland, is Africa.

Eurasia broke off from the north and northeast, leaving the Mediterranean and Red seas. India broke off from the east, leaving the Indian Ocean. Antarctica broke off from the southeast, leaving the Arctic Ocean. South America broke off from the west, just south of Africa's western bulge, and North America broke off from the western bulge, leaving the Atlantic Ocean. What remains is the core, or the heart: Africa. This book divides Africa into two realms: the area of the Sahara and the area north of the Sahara (Saharan Africa, covered in Chapter 18) and south of it (sub-Saharan Africa).

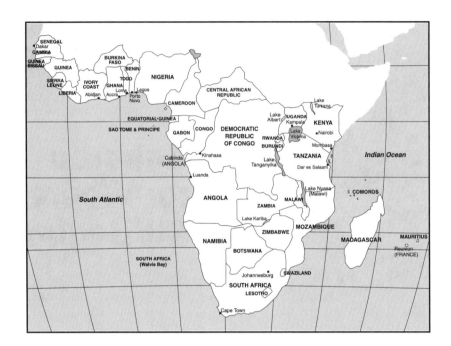

This chapter focuses on the region of sub-Saharan Africa. It has more than 40 countries whose combined population comprises more than 70 percent of Africa. Sub-Saharan Africa has one-tenth of the world's people, who speak one-third of the world's languages. That complexity stems from many millennia of human existence. In Africa, more than five million years ago, all humankind as it's now known is believed to have begun.

Ancient Roots

Scientists aren't sure whether collective human birth was by creation or evolution or some combination of both (or even by some other way not yet dreamed of or discovered). About 5 million years ago, geographers believe, some type of hominid inhabited southern and eastern Africa. They also believe that about 1.5 million years ago, this toolmaking hominid developed into the more advanced forms of Homo habilis and Homo erectus.

The earliest true human beings in Africa, Homo sapiens, apparently date from more than 200,000 years ago. As hunter–gatherers, they banded together with others to form no-madic groups. These nomadic peoples eventually spread throughout all of Africa.

Early humans left no written history, of course, for geographers to follow their adventures and development throughout the millennia that ensued. Not until the Egyptians began to record in hieroglyphs in about 5000 B.C. did geographers even get a chance to open the book on ancient civilization. Before then and outside of Egypt, they can only speculate from the few finds that have been unearthed.

Geographers do know that the peoples of the forests south of the great Sahara needed and produced goods quite different from those of the dry, distant north. Because the savanna peoples in between had the opportunity to solve these differences by establishing trade, caravan routes that ran east and west and north and south began to thrive, and trading "cities" were created.

> ## ! Geographically Speaking
>
> Pygmies now live in Africa, the Malay Peninsula, the Philippines, and New Guinea. They average about 60 inches in height and are either hunter–gatherers or farmers who usually adopt the language of their neighbors. Modern blood typing and other studies have found these various groupings of pygmies to be distinct from each other and to have had independent origins.
>
> The largest grouping, the African pygmy, is called the Bambuti. They're also the shortest, averaging only 51 inches tall. About 300,000 of them live in the Congo Valley, and it's now believed that they were the first to settle or herd this area.

Peoples of the region would have to band together for mutual help and protection. States also would begin to develop. Geographers are beginning to find traces of many of these states, including Ghana, Bantu, Mali, Songhai, Hausa, Fulani, Ibo, Kanem–Bornu, Benin, and Kango. All these states and others broke up into substates. Probably several hundred of these small-to-large groupings existed, along with roving warlike tribes, such as the Mossi, the Zulu, the tall Masai, and the short Pygmies.

Recent Stirrings

The subtle cultural fabric of Africa was both complex and remarkably intertwined. It's important to note that, as complicated as this developing culture, trade, and history were, it took place largely in interior Africa. In the fifteenth century A.D., all this changed.

European Involvement

With land routes to the riches of the Orient (Asia) becoming too dangerous and costly in the fifteenth century, the Portuguese were seeking a water route to the east by going around Africa. Each sailing took them farther south, and finally their gropings took them around the Cape of Good Hope, into the Indian Ocean, and on their way. Other

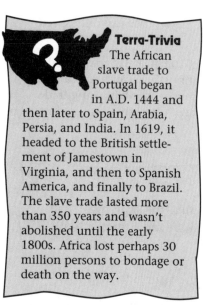

Terra-Trivia
The African slave trade to Portugal began in A.D. 1444 and then later to Spain, Arabia, Persia, and India. In 1619, it headed to the British settlement of Jamestown in Virginia, and then to Spanish America, and finally to Brazil. The slave trade lasted more than 350 years and wasn't abolished until the early 1800s. Africa lost perhaps 30 million persons to bondage or death on the way.

European countries began to ply African waters, and the need for coastal stations and forts began.

Although most of this western contact was in western Africa, it was also taking place in southeast Africa, in what is now Mozambique. The Arab states to the north were coming south down the east coast of Africa and seeking trade. Soon the way stations became more than just stopping points en route: They became centers for exporting Africa's own most valued product: human beings bound for slavery in the Americas, Arabia, Persia, and India.

Slavery not only ravaged the interior of Africa, from which the slaves were "mined"—destroying families, entire villages, and even cultures—it also caused untold misery to those affected and their descendants for centuries to follow. It also shifted prominent African trade routes from the interior to the coasts, where they would remain for the next few centuries.

European Lines on the Map

Not until the second half of the nineteenth century did the European countries begin to penetrate the interior of the great continent of Africa. It began with missionaries and explorers but soon moved on to serious attempts to establish "spheres of influence." To make the process more orderly, representatives from the European countries (not even one from native Africa) met in Berlin in 1884.

From the toeholds they already had on the coasts, the Europeans literally carved up all of Africa, drawing and redrawing boundary lines through known and unknown expanses. The final lines took little account of African peoples, settled areas, hostile tribes, local cultures, languages, or in many cases even natural geographic barriers, such as rivers, mountains, deserts, and lakes. This situation just begged for disaster—which for the past 100 years it has received.

Within 15 years of the Berlin conference, all of sub-Saharan Africa was under some sort of subjugation, except Liberia, where freed Afro-American slaves had established Africa's first independent state in 1847. The British generally followed their policy of indirect rule (where local rulers were made representatives of the crown) and almost succeeded in their Cape-to-Cairo quest, except for German East Africa, which interrupted the vast 4,900-mile stretch. They also picked up Nigeria and a few other smaller areas in West Africa.

The French followed their usual policy of creating an "overseas France," one of acculturation (cultural change resulting from the adoption of a culture from the the more dominant society) of Africans into the French way of life in all the rest of West Africa, or what is now the Central African Republic, Gabon, and Congo in equatorial Africa, and Madagascar off the east coast. The Belgians were paternalistic, attempting to convert the Africans to more "civilized" western ways. Belgian Congo, which became Zaire and is now the Democratic Republic of Congo, was their victim.

Portugal exploited its colonies of Angola and Mozambique, where the locals were forced to give almost their entire income to their colonial masters. Germany followed the British system, but not quite to such extremes, in Cameroon, Namibia, and German East Africa.

By 1950, 40 years later, little had changed. World War II had split up Germany's colonies—Cameroon went to France, German East Africa to Britain, and Namibia to South Africa, which had gained full independence in 1931. By 1970, another 20 years later, only Portuguese colonies and today's Zimbabwe weren't independent. By 1991, they all had their own flags.

> **Terra-Trivia**
>
> In the eighth century, Arab merchants established today's Zanzibar as their base to trade with the Zenj (Arabic for "black") states. Later, it became a sultanate, and in the 1860s the sultan founded Dar es Salaam (Arabic for "haven of peace") as his summer palace.

Independent Africa: Little Stability and Lots of Turmoil

Most of the sub-Saharan countries have had their independence for at least 25 years. Six years ago, only 5 of the 40 could be termed stable: Gambia, Senegal, Zimbabwe, Botswana, and Mauritius. Since then, 3 more have probably joined the stable list: Ghana, Gabon, and Côte d'Ivoire.

What about the other 32 countries, however? During 1996, 17 national elections were held—some blatantly unfair or militarily controlled and some mere affirmations of former coup leaders. The others may have been truly democratic. During their brief era of independence, almost none of these countries has escaped coups, military rule, governmental corruption, civil or neighbor wars, famine, widespread disease, tribal unrest, or riots—you name it, they all have experienced at least a few of these situations.

Why? Was it the way the boundaries had been drawn up? Was it the way the colonial "masters" had not prepared their "charges" for independence? Was it the climate or the lack of education? Was it disease, poor health, low life expectancy? Poverty? Has excessive population growth occurred? The answer to every question is, obviously, at least partly yes.

> ### ! Geographically Speaking
>
> When you think of Africa, you probably think first of animals; their preservation has been a hot issue in Africa in recent years. As humans encroached, the big cats were imperiled (one lion pride needs at least 15 square miles of living space). Elephants were shot for their tusks, and rhinoceroses for their horns. All big animals were at risk from big-game hunters—something indeed had to be done.
>
> Tanzania and Kenya devoted substantial portions of their land to wildlife reserves to protect their animals. Western countries banned the import of ivory, and Kenya has outlawed hunting. For some species, the tide is now turning. Reserve names, such as Serengeti, Ngorongoro, Mara, and Tsavo, have become legend and annually draw almost 1.5 million tourists who spend more than a billion dollars.

That's enough of the past: The real question is whether these struggling countries will make it in the future. Perhaps even unfair elections are better than coups and civil wars, and they might even lead to fairer elections in the future. Perhaps these countries will be able to look to the south, to South Africa, to give them hope and maybe even a road map (blurred though it may be) to an improved future. So far, they have tasted little of the fruits of this world.

South Africa: Riches and Despair

GeoJargon
Veld is a Dutch word meaning "field." In South Africa, it refers to the tall grass, primarily on the central plateau. South Africa has its highveld, over 5,000 feet high and covering most of the plateau; its middleveld, the rest of the plateau; and its lowveld, which lies beneath the Great Escarpment (cliffs, sometimes thousands of feet high, surrounding the plateau).

This state of 44.5 million people on the southernmost shores of Africa—with more natural riches than almost anywhere in the world—should indeed be the cornerstone of the sub-Saharan region. It isn't, though. All the world looks at South Africa, hoping to laud it but waiting anxiously with fingers crossed.

South Africa, as shown on the map, is the only truly temperate country in sub-Saharan Africa. Dominated by a high plateau flatland veld in the center, it has mountains in the southeast and desert in the west. Although temperatures vary by season and altitude, hot summers and mild winters typically prevail. Rainfall is moderate in the east but sparse in the west.

Bantu-speaking tribes inhabited the area from early times. In 1652, the Dutch (later to be called Boers or Afrikaners) came and settled Cape Town. After protracted conflict with

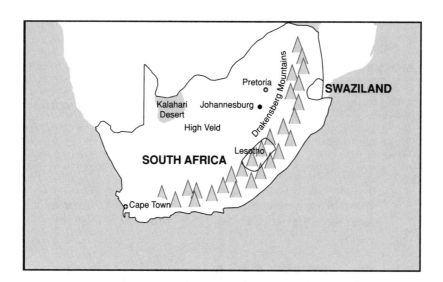

South Africa.

the British and local peoples, the Afrikaners established their government, which had a one-word solution to the problem of coexistence: *apartheid.*

Apartheid means "separation," and that's what it was. Blacks were separated from whites in every way—and resisted. Their chief spokesman, Nelson Mandela, was quickly imprisoned and was not released for 27 years, in 1989. Apartheid soon evolved into separate development, as blacks were denied citizenship in South Africa and set up their own homeland states.

The world was appalled. Foreign investment began to dry up. Neighboring African states infiltrated South Africa under the name of the African National Congress (ANC), and the Boers began to argue among themselves. Finally, in 1994, Mandela won the first general white–black election to become president of South Africa.

In 1996, a new constitution was signed. Expectations are high, and the problems the country faces are monumental:

➤ Of South Africa's 44.5 million people, three-fourths are black, and the rest are white, colored (mixed), and Asian. It's a truly enormous ethnic complexity.

➤ Urban crime has increased immensely to levels rivaling the world's worst.

➤ Unemployment is a staggering problem; of the country's 17 million workers, almost half are unemployed.

On the plus side:

➤ Annual economic growth in 1996 was a respectable 3 percent, and inflation was high but down to 9 percent.

243

➤ Many minerals are in this region in abundance. The world's needs for diamonds and gold are almost all met by South Africa.

➤ Literacy and life expectancy are high.

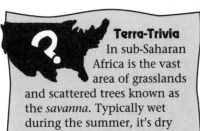

Terra-Trivia
In sub-Saharan Africa is the vast area of grasslands and scattered trees known as the *savanna*. Typically wet during the summer, it's dry during the rest of the year. Africa's large mammals roam on the savannas, and lions, elephants, giraffes, cheetahs, baboons, hyenas, rhinoceroses, and great herds of antelopes and wildebeests all live in this region.

➤ Although income levels are also high, at about $2,500, the brackets broken down by color probably would show blacks at about $400 and whites at about $12,000—a huge disparity.

Will this new nation move forward or falter and stumble, as so many of its northern neighbors have? Its people have been through much turmoil already, and most leaders on all sides have the necessary will. Only time will tell in South Africa.

The Rest of Sub-Saharan Africa

Sub-Saharan Africa has so many countries that a description of each one is beyond the scope of this book. This section just describes the highlights of the region by geographic sections (or subregions).

Western Africa

This subregion, as shown on the map, is the southern part of the western African bulge, from Senegal to Nigeria. All the countries are coastal except for Burkina Faso. Because the area is tropical, it's hot year-round. Rainfall is heavy in the south and diminishes to the north. Tropical rain forests in the south give way to savanna in the northern half of the subregion.

Terra-Trivia
Western Africa is building two new capitals: Abuja in central Nigeria and Yamoussoukro (the president's hometown) in Côte d'Ivoire. The president has just built a tremendous multimillion-dollar Roman Catholic basilica, Notre Dame de la Paix, looking a little like and rivaling in size St. Peter's Cathedral in Rome.

The dominant physical feature of the area is the Niger River. The world's thirteenth longest river (2,600 miles long) works its way through much of the subregion. Ghana also has Lake Volta, one of the largest artificial lakes in the world.

Western Africa has about 190 million people. Its largest country is Nigeria, with about 105 million. The coastal strip, including Abidjan, Côte d'Ivoire (also known as the Ivory Coast); Accra, Ghana; Lomé, Togo; Porto-Novo, Benin; and Lagos, Nigeria, comprises Africa's megalopolis, with more than 40 million people (Lagos alone has more than 10 million inhabitants). To the north, Dakar is also a large city in Senegal. Despite its large cities, western Africa is still 70 percent rural. Most of its people are subsistence farmers or herders or both.

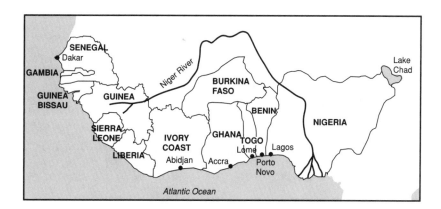

Western Africa.

Nigeria by itself has more than 250 ethnic groups. The region is quite illiterate; in 7 of its 13 countries, fewer than 20 percent of the people can read. Tribal and ethnic roots are extremely complex throughout the subregion.

Nigeria, in size and population, is the cornerstone of western Africa. The discovery of significant oil reserves in Nigeria added to its potential. Quickly tapped, the reserves soon accounted for 90 percent of all exports. This type of major dependency can be a severe problem if the world oil price drops, which it did—in one decade, Nigeria's gross domestic product was halved.

Eastern Africa

Eastern Africa, as shown on the map, is astride the equator on Africa's eastern shore; Kenya and Tanzania are on the coast; and Uganda, Rwanda, and Burundi are inland alongside them. The climate is tropical year-round but cooler in higher elevations. Although Uganda, Rwanda, and Burundi have ample rain, Tanzania is generally drier and Kenya considerably drier still. From west to east, this highland region ranges from light tropical forest vegetation to savanna lands.

Although the Nile starts in Burundi and flows north, it's small and not dominant in this region. A string of large lakes from north to south forms the entire western edge of the subregion. In the north, Lake Albert has a maximum depth of only 55 feet; Lakes Edward and Kivu are the smallest; Tanganyika is the seventh-largest lake in the world, and Malawi is number ten.

The region has two other disconnected major lakes. Lake Victoria, the third-largest lake in the world, lies on the equator where the boundaries of Uganda, Kenya, and Tanzania meet. The other is Kenya's Lake Turkana (formerly Lake Rudolf). Although eastern Africa is somewhat mountainous (it's no Ethiopia), in northern Tanzania is Africa's highest mountain, the snow-capped Mount Kilimanjaro, at 19,340 feet.

Eastern Africa.

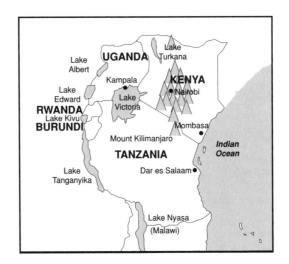

Centered around Lake Victoria is one of Africa's most populated areas (Uganda, Rwanda, Burundi, southwest Kenya, and northwest Tanzania). Almost half of all eastern Africa (which has 100 million people) live there! The population density exceeds 500 people per square mile, compared to 65 for all of sub-Saharan Africa. Dar es Salaam, Tanzania, and Nairobi, Kenya, each with about 2 million people, are its largest cities. Next in size are Kampala, Uganda, and Mombasa, Kenya.

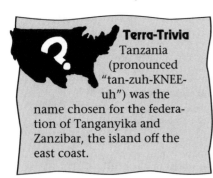

Terra-Trivia

Tanzania (pronounced "tan-zuh-KNEE-uh") was the name chosen for the federation of Tanganyika and Zanzibar, the island off the east coast.

Despite its big cities, the population of eastern Africa is highly dispersed, with 87 percent rural people. Population growth, which is high in the entire region, is alarmingly high in Kenya, where it's almost 4 percent—which means that the country's population will double in just 17 years. Literacy and life expectancy are about par for the sub-Saharan course. Income levels also parallel sub-Saharan Africa as a whole except in Tanzania, where its average income level of about $120 makes it one of the poorest countries in the world.

A major tribal dispute is taking place now in this part of the world: The implacable Tutsi–Hutu tribal warfare situation has been afire since anyone alive can remember. Although the dispute is centered in Rwanda and Burundi, massive numbers of refugees have fled to neighboring Democratic Republic of Congo and Tanzania. Figuring out how to resolve this crisis of massacres, civil war, and government takeovers is mind-boggling.

Eastern Africa also has a capital switch planned for the future. Tanzania, already poor and staying alive only because 40 percent of its budget is paid for by foreign aid, is planning to move its capital inland from Dar es Salaam (population two million) on the Indian Ocean coast to the middle of nowhere to Dodoma (population less than 50,000).

Central Africa

Central Africa.

Central Africa lies astride the equator in the center of the continent and on the western shore. On the coast heading south are Cameroon, Equatorial Guinea, Gabon, Congo, and the western tip and corridor of Republic of Congo. Inland, north of the Democratic Republic of Congo, is the Central African Republic. Off the western shore is the island country of São Tomé and Príncipe.

Rainfall is ample (as you might expect) in this tropical clime but heavier on the coast and mountain slopes. West Cameroon's slopes get as much as 400 inches of annual rainfall! Two of the major rivers (the region has hundreds) separate the Central African Republic from the Democratic Republic of Congo: the 2,720-mile-long Congo (the world's eighth longest) and its major tributary, the 660-mile-long Ubangi.

Except for northern Cameroon and the Central African Republic, which have savanna and shrub growth, the entire region is tropical rain forest. The subregion is centered around the Congo Basin Depression in the Democratic Republic of Congo, which is surrounded by mountains—in the east, the Mibumba Range rises to 16,000 feet.

The subregion is relatively thinly peopled. Only 40 percent of the population is urban, and the Democratic Republic of Congo's capital, Kinshasa, is its largest city. The country is the giant in the subregion, with about five-eighths of the land and people.

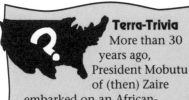

Terra-Trivia
More than 30 years ago, President Mobutu of (then) Zaire embarked on an African-ization program and changed the name of the country from Belgian Congo to Zaire and changed the Congo River to the Zaire River. The new names stuck until recently, when the country's name was changed *again*, to the Demo-cratic Republic of Congo. The Congo is a mighty river, ten miles wide in places, with more than 4,000 islands.

Although Bantu-based languages prevail, some smatterings of English, French, and Belgian remain, of course. Portu-guese is spoken in São Tomé and Príncipe, and Equatorial Guinea is the only country in Africa where Spanish is the official language. Literacy levels range from Congo's 35 percent to Gabon's 63 percent, and life expectancy is slightly higher than average for sub-Saharan Africa.

Most people farm, fish, herd, log, or mine for a living. Countries are gradually depleting their natural resources through exporting to stay fiscally alive. Gabon has the highest income levels in the subregion (per capita income of more than $4,000), and the Democratic Republic of Congo has the lowest (per capita income of less than $200). Eco-nomics almost always parallels politics: Although Gabon started off on the right foot with democratic leadership and stayed stable, most of the countries in this region have flirted with Marxism or suffered from unrest, revolutions, or civil or neighbor wars—and are just hanging on.

The Democratic Republic of Congo first stumbled and began to disintegrate under 30 years of Mobutu rule. Its infrastructure has collapsed, it imports one-fourth of its food, one-third of all its children die before they reach age 15, and it has starvingly low income levels. The country is emerging from a revolution that recently overthrew President Mobutu. Geographers can only hope that the political change will improve the sorry lot of the country's people.

Southern Africa and Madagascar

Southern Africa, as shown on the map, includes everything south of the Democratic Republic of Congo and Tanzania; heading south on the west are Angola, Namibia, South Africa (which surrounds Lesotho and Swaziland), and then Mozambique on the east—with Malawi, Zambia, Zimbabwe, and Botswana landlocked between them. Madagascar, the world's fourth-largest island, is off Mozambique's east coast, along with the island countries of Comoros and Mauritius.

Woodland forests dominate Angola, Malawi, and eastern Madagascar. The Kalahari Desert dominates Botswana, western South Africa, and southern Namibia, where it meets the Namib Desert. Almost everywhere else, grazing and arable land prevail.

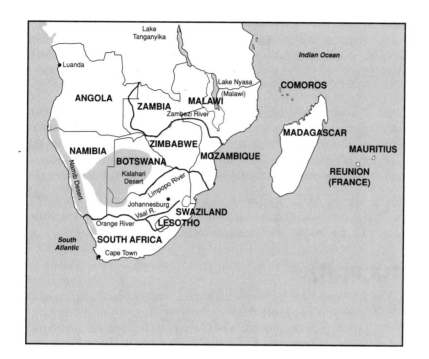

Southern Africa

Although Northern Angola, Zambia, and Malawi are tropical, the rest of the region is temperate with mild winters and hot summers. The deserts are hotter and dry. Northward, the forest areas get moderate amounts of rain; totals are lower elsewhere in the subregion.

The largest lake in the subregion is Lake Malawi, between Malawi and Mozambique. The region has several sizable rivers, including the Zambezi (Africa's fourth longest), the Orange–Vaal, and the Limpopo. South Africa's highveld is in the southeast with the Drakensberg Mountains; Angola has the high Bihe Plateau.

This subregion has some big cities, the largest of which are Angola's Luanda and South Africa's Johannesburg and Cape Town. Other than South Africa, the countries in this area are thinly populated: They have 25 percent of sub-Saharan Africa's land but only 13 percent of its people.

Terra-Trivia

Most people in sub-Saharan Africa speak languages belonging to the Bantu subfamily of the Niger–Congo language family, with two interesting exceptions. On the island of Madagascar, people speak Malagasy, a Malayo-Polynesian language betraying South Pacific roots. The bushmen and Hottentots of the Kalahari and Namib deserts speak a unique Khoisan "click" language that is spoken nowhere else on earth.

249

In numbers of languages, growth, literacy rates, life expectancy, and urbanization, this region is typical of the sub-Saharan region. Excluding South Africa and the small island countries, income levels are also typical but varied: Mozambique's per capita income hovers around an unbelievably poor level of $100; Botswana is relatively rich, at about $2,000.

In addition to the plague of unstable governments, the subregion has more minuses than pluses: Madagascar's damaged ecosystem is a top conservation priority; Angola's guerrilla and civil wars have left it as a world "leader" in infant mortality; after living off its great copper belt, Zambia's copper is running out; Zimbabwe's socialist government faces merciless unemployment, and the country still aches from its recent ten-year drought.

The subregion has a few positives: Mozambique sees increases in its new free markets; Malawi is leaning toward stability; and Botswana, stable and democratic, has an outstanding human-rights record and is even getting a taste of tourism.

Health (or a Lack of It)

Sub-Saharan Africa suffers from many ills, mostly man-made, as you have read. Nature hasn't been kind either. The region has more than its share of droughts and resultant famines. The tropical vegetation that covers much of the region is a natural breeding ground for insects that cause or carry disease.

Malaria (from mosquitoes) kills more than a million African children yearly and an untold number of adults. Yellow fever (also carried by mosquitoes) attacks hundreds of thousands of children annually. Tsetse flies transmit the African sleeping sickness that has ravaged people and livestock for hundreds of years. Flies also cause dreaded river blindness. Even snails infest waters, and the humans that swim in them, with schistosomiasis.

Possibly worst of all, the region now has its share of AIDS (acquired immune deficiency syndrome). This deadly disease is believed to have started in Africa and now ravages as much as 30 percent of the urban population. Because members of the rural population rarely see a doctor and Africa is two-thirds rural, the true degree of onslaught isn't known. The growth rate of this disease, if not checked, seems, unfortunately, almost exponential. What AIDS will do to Africa's already diseased population is alarming to even contemplate.

The Least You Need to Know

➤ Humankind started in sub-Saharan Africa.

➤ Europeans colonized the African continent in a domination that ended only in the twentieth century.

➤ Although sub-Saharan Africa is now all independent, only a few of the more than 40 countries have attained political stability.

➤ Slavery took a toll of 30 million people in central Africa.

➤ Wildlife reserves in Kenya and Tanzania are a step in the right direction that have saved some African animal species from probable extinction.

➤ The vast majority of sub-Saharan people are poor, and civil wars still exist. Now AIDS is one of the array of diseases endemic to the area. Life for this region's people is difficult.

The Middle East: Seat of Turmoil

When you think of the Middle East, as shown on the map, several geographic images might come to mind: desert sands, Muslims, Jews, Christian Holy Lands, Arabs, political turbulence—and don't forget oil! Although all are in the Middle East, no one country is characterized by them all. Many similarities exist across the region, and a host of differences too.

This region has been the birthplace of three of the world's most influential religions: Judaism, Christianity, and Islam. Religious and political conflicts have, unfortunately, marked the area in recent years (and throughout history). Although the people of the Middle East all claim to be "people of the book," the books don't seem to be similar enough to promote harmony. (Perhaps it's not the book but rather the readers?)

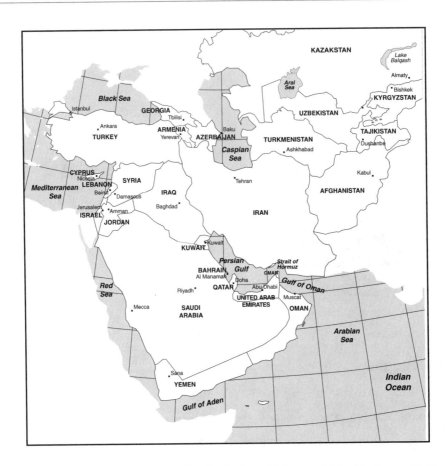

The Middle East has also become a hotbed of world interest because beneath its sands lie the earth's largest repositories of petroleum. Oil fuels the modern economy, and the Middle East has it! Not all the countries in the region have it, however. Some of the income ranges in this region are unmatched anywhere else in the world. It's a region of haves and have-nots.

Crossroads of Three Continents

Although the Middle East is located primarily on the continent of Asia, it includes a little of Europe and abuts Africa. The term Middle East is appropriate because these lands were, for the Western world, traditionally a transition zone, a place midway to the peoples of the east. The countries have been important throughout history, and many empires have come and gone. To control this strategic geographic spot on the earth is to control much of the world's trade—and to hold a key to power on three continents.

Land of Conquerors, Religion, and Great Seas

A look at the conquerors who have passed this way is a look at history's heavy hitters: Sumerians, Egyptians, Babylonians, Romans, Mongols, Crusaders, Alexander the Great's Macedonians, the Arab Empire, the Ottoman Empire, and the British Empire. Although many more existed, of course, these are just a few who realized the strategic importance of the Middle East. Not all the historic figures associated with the Middle East came to wrest temporal power, though—some had higher callings.

The Middle East was the crucible for three of the world's great religions. From the teachings and followers of Abraham, Jesus, and Mohammed sprang Judaism, Christianity, and Islam, respectively. Despite their teachings of peace, compassion, and a universal God, the message was often twisted, and religious conflict is as old as the region itself. Although rivers aren't in great supply, the region is ringed with sizable bodies of water. Starting in the north and moving clockwise are the Caspian Sea, the Aral Sea, and Lake Balkhash. In the southeast, the region is penetrated by the Persian Gulf and the Gulf of Oman. The Arabian Peninsula is bounded by the Arabian Sea, the Gulf of Aden, the Red Sea, and the Gulf of Aqaba. To the west are the Mediterranean and Aegean seas that are connected to the Black Sea by the Bosporus and Dardanelles straits (which divide Turkey into the continents of Europe and Asia).

> **Terra-Trivia**
> Within this region is the country of Iran. Although that's its modern name, for centuries it was known as Persia (its name changed in 1935). In the sixth century B.C., under Cyrus the Great, the Persian empire reached from the Indus River in Pakistan west to Egypt's Nile River. This ancient empire eventually fell to Alexander the Great in the fourth century B.C.

Generally Speaking

The geography of the Middle East region is much too diverse to cover in one broad sweep. Although commonalities exist between the countries in this region, the exceptions are many. Five generalizations are commonly associated with the Middle East:

➤ It's a region of deserts.

➤ It's an Arab world.

➤ It's rich in oil.

➤ It's a Muslim domain.

➤ It's in ethnic, religious, and political turmoil.

As you will see, some of these generalizations have only a limited application in this region. Because 24 countries in the Middle East region are too many to consider in detail, this chapter discusses the common generalizations by breaking this complicated region into five subregions. You can then think in terms of groupings of countries.

Sorting Through the Subregions

In this chapter, the Middle East region is grouped into these subregions:

➤ **The Arabian peninsula:** Bahrain, Kuwait, Oman, Qatar, Saudi Arabia, United Arab Emirates, and Yemen

➤ **The Middle Eastern core:** Cyprus, Israel, Jordan, Lebanon, and Syria

➤ **The mountainous population giants:** Afghanistan, Iran, Iraq, and Turkey

➤ **Former Soviet republics—the Caucusus:** Armenia, Azerbaijan, and Georgia

➤ **Former Soviet republics—the 'stans:** Kazakstan, Kyrgyzstan, Tajikistan, Turkmenistan, and Uzbekistan

By considering the generalizations associated with the region, you also get some insight into individual countries—but just the basics. If you try to take in too much information about each country, you run the risk of confusing yourself in an area that's already so complicated that neither conqueror nor diplomat has been able to figure it out. Let's start your Middle Eastern sojourn with the Arabian Peninsula.

The Arabian Peninsula

Seven countries are in the first Middle Eastern subregion described in this chapter, including Bahrain (Al-Manamah is its capital), Kuwait (Kuwait), Oman (Muscat), Qatar (Doha), Saudi Arabia (Riyadh), the United Arab Emirates (Abu Dhabi), and Yemen (San'a). These countries have more in common than just geographic proximity. This section explains how the preceding list of generalizations about the Middle East apply to this subregion.

These countries *are* in the desert, and much of it is the kind of desert you dream about. You've read elsewhere in this book about deserts, usually rocky places with scrub brush—not the *Lawrence of Arabia* type of sand dune place you want in a desert. Parts of the Arabian desert are loaded with dunes, and it really is Lawrence's stomping ground. It has caravan routes, turbans—the whole deal—and not a real river in the place.

Don't get the impression that this area is inhospitable and completely covered with drifting sand dunes—it has some rocky mountainous areas! Along the coast of the Red Sea and the Gulf of Aden, a mountain chain rims the peninsula, and some mountains reach higher than 10,000 feet; it also has a mountainous rise along the coast of the Gulf of Aden. Although the Arabian Peninsula is an arid place, a few goats survive, and a gasp of moisture does fall in the mountainous southwest.

Because it's the Arabian peninsula, of course, you might have guessed that the people of this subregion are almost all ethnic Arabs. Arabic is the primary language spoken in every country along with a smattering of English (a remnant of its British Empire days). Some Indian languages are also spoken, primarily in areas with an abundance of imported labor.

The area has oil in spades. In just the Big Three oil producers (Saudi Arabia, Kuwait, and the United Arab Emirates), more than 40 percent of the world's known oil reserves exist. These countries all have oil, and some have it in a big way. Oil incomes are reflected in the area's annual gross domestic product per capita figures. Although Yemen's per capita figure is fairly modest (less than $2,000), most countries' in the region are quite high, and the United Arab Emirates boasts a figure higher than $22,400.

> **Terra-Trivia**
> Mohammed was born in A.D. 570 in the Saudi Arabian city of Mecca. There, he received revelations, recorded in the Koran, that were to form the basis of the Islamic faith. Muslims from across the globe make pilgrimages (*hajj*) to this sacred city. Also in Saudi Arabia is the revered city of Al Madinah (Medina), where Mohammed died and was buried in A.D. 632.

Despite its abundance of oil, the subregion lacks what might be considered a more important resource: water. Although desalinization plants are used and are being built, it's an expensive way to produce water. Although it would be great to strike oil in the back-yard, of course, the resource that much of the world's people take for granted is what the Arabian Peninsula values most.

Except for some imported labor and a few foreign consultants, everyone in this subregion is a Muslim. It's the heart of the Islamic world—and the birthplace of the prophet Mohammed. From these sands, Islam spread to become the dominant religion of north Africa and southwestern Asia. It is now one of the world's largest religions, with more than one billion adherents.

Although some domestic unrest surfaces periodically, ethnic and religious tensions are for the most part less significant than elsewhere in the region. (It's not much, by Middle East standards.) The giant Saudi Arabia shares undefined borders with the small nations to the south and east of the peninsula, and it's a potential sore spot.

The political scene has been dominated in recent years by a military exploit from outside the subregion. In a dispute over oil prices and ancient land claims, Iraq invaded Kuwait in 1990. Although an Arab–Western military coalition ousted the Iraqis, relations between neighbors have been tense ever since. Ongoing political tension exists in Yemen and has existed as long as anyone can remember.

The Middle Eastern Core

Five countries are in this subregion, including Cyprus (Nicosia is its capital), Israel (Jerusalem), Jordan (Amman), Lebanon (Beirut), and Syria (Damascus). Although this subregion has fewer countries than the Arabian peninsula, this one's a political and ethnic quagmire.

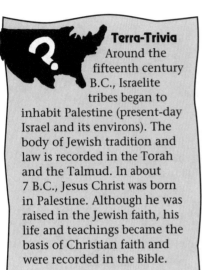

Terra-Trivia

Around the fifteenth century B.C., Israelite tribes began to inhabit Palestine (present-day Israel and its environs). The body of Jewish tradition and law is recorded in the Torah and the Talmud. In about 7 B.C., Jesus Christ was born in Palestine. Although he was raised in the Jewish faith, his life and teachings became the basis of Christian faith and were recorded in the Bible.

GeoJargon

The term *Semitic* refers to peoples of the Semitic language family, which includes the languages of Hebrew and Arabic. In ancient times, Aramaeans, Assyrians, Babylonians, Canaanites, and Phoenicians also belonged to this group.

Although Jerusalem is not one of the larger cities in this subregion, it would be inappropriate not to mention it. This city is sacred to Jews, Christians, and Muslims alike. To Jews, it's the City of David, the ancient capital of Israel, and the site of Solomon's Temple. To Christians, the Church of the Holy Sepulcher is the site of the crucifixion, burial, and resurrection of Jesus. Muslims revere the city as the location of the Dome of the Rock, the spot where Mohammed ascended to heaven.

This subregion does have deserts, the Negev Desert in southern Israel and the Syrian Desert in southeastern Syria that extends south through much of Jordan. Although much of the region is dry, it's more typically Mediterranean in climate, especially along the coast, where typical Mediterranean crops predominate.

Semites make up the bulk of the subregion's population but don't make up the majority in every country. Cyprus isn't an Arab country at all; most of the population is of Greek or Turkish origin. Israel is principally a Jewish state (with a sizable Arab minority—mostly Palestinian). Although Arabs are prevalent in Lebanon, it's a true ethnic mix and therefore isn't classified as a classic Arab country.

You can find some oil here, but not in the proportions you might expect from the Middle East. Syria has the bulk of the area's oil, and Jordan has some oil-based industries, including some refining.

Muslims live in every country and represent the majority in most of them. Israel is the most notable exception to the Muslim rule. Only one-fifth of Israel's population is Muslim, and more than 70 percent are Jewish. Cyprus also has a religious mix. On the northern third of the island, the population is almost entirely Muslim; on the southern two-thirds of the island, however, Muslims represent only 20 percent of the population. The ethnic mix in Lebanon is about 60 percent Muslim.

You would be hard-pressed to find an area more ethnically, religiously, or politically tumultuous in recent history. Here's a quick rundown of the worst of it:

➤ **Cyprus:** The northern third is Turkish, and the southern two-thirds are Greek. They have fought each other for centuries: Since 1993, the north has claimed separate-nation status.

➤ **Israel:** A Jewish nation surrounded by Arab nations hasn't proved to be a peaceful proposition. Every nation in the subregion has been in conflict with Israel, except

Cyprus, which has been too busy fighting with itself. Internally, struggles for Palestinian autonomy, terrorist attacks, and clashes between rival Jewish political factions have added to the ever-boiling pot of unrest. The world holds out high hopes for ongoing peace initiatives.

➤ **Lebanon:** Beirut, the capital of this shambles of a country, was long known as the "Paris of the Middle East." It's no longer true. Civil war between Maronite Christians, Shiite Muslims, Sunni Muslims, Druse militia, Syrians, Palestinian guerrillas, Christian phalangists, United Nations peacekeeping forces, Israelis, and just about any splinter group you can name have reduced the country and its capital to ruins.

The Mountainous Population Giants

Although only four countries are in this subregion, in terms of population, they're big. Afghanistan (its capital is Kabul), Iran (Tehran), Iraq (Baghdad), and Turkey (Ankara) are the Big Four that geographically occupy a band through the center of the Middle Eastern region. These countries are populous compared with the rest of the Middle East. Afghanistan and Iraq are both large, each with a population of more than 21 million people. Iran and Turkey are the giants of the region, with a population of more than 60 million apiece.

Unlike the other subregions described in this chapter, this one has several large cities, each with a population of more than two million people. Three of the capital cities are in this category: Ankara (population 2.7 million), Baghdad (4.6 million), and Tehran (6.5 million). The largest city in the entire region isn't a capital, however: the ancient city of Istanbul. The city, which sits at the juncture of Europe and Asia astride the Bosporus Strait, was founded by the Roman emperor Constantine the Great under its former name, Constantinople. The city served as the capital of the Eastern Roman empire and the Byzantine Empire until captured by the Turks in the fifteenth century.

This subregion is largely mountainous—except for Iraq. Iraq has largely low-lying areas of river basin and is mountainous in only the northern part of the country. The rest of the countries are covered with rugged mountains. Mount Ararat in Turkey is of particular interest because of its great height (almost 17,000 feet) and because it's the traditional resting place of Noah's ark (various parties claim to have located its archaeological remains).

Although much of the area is mountainous, the apex (highest point) in the subregion is the mighty Hindu Kush in Afghanistan. Its peaks top 24,000 feet. Also in Afghanistan is the famed Khyber Pass, the ancient

Terra-Trivia

Iraq's fertile lowlands are formed by the famed Tigris and Euphrates rivers. This area, called Mesopotamia, was the ancient cultural hearth of the Sumerians, Assyrians, and Babylonians. Dating back almost 12,000 years, some of the first human settlements and cities arose along these life-giving waters.

Sunni and Shiite are the two major Islamic sects. Worldwide, about 11 percent of Muslims are members of the Shiite sect, and 85 percent are Sunni. (The remaining 4 percent belong to smaller offshoots of Islam.) Shiite groups are located primarily in Iran, Iraq, and Bahrain, with minorities living in other countries throughout the region.

Terra-Trivia
Afghanistan is one of the poorest countries on earth. In a region where several countries have a gross domestic product per capita of more than $20,000 per year, Afghanistan's is less than $200. With no oil, few natural resources, and decades of war, it isn't an economic powerhouse. You know that a country is poor when its primary industrial product is rugs and its major export is dried fruit.

trade route through the Hindu Kush to Pakistan. All these mountains were formed by the collision of the Arabian and Eurasian plates, and the area is highly subject to earthquakes.

This subregion has just a few deserts. Southwestern Iraq is dominated by an extension of the Arabian Desert, and, despite irrigation from the Tigris and Euphrates rivers, this area receives less than ten inches of rainfall per year. A high central plateau dominates central Iran, and the Dasht-e Kavir and Dasht-e Lut deserts extend into southern Afghanistan. The rest of the subregion isn't exactly lush, and only northeastern and western Turkey receive heavier precipitation.

Iraq is inhabited mostly by Arabic peoples, and some minority Arab populations also live in Iran. The vast majority of peoples in this subregion, however, aren't Arabs. Turkish peoples are predominant in Turkey, and Afghans and some Turkic Uzbek peoples represent the primary population of Afghanistan. In addition to the Arab minority, Iran is made up largely of Iranians, with Persians dominating.

In northern Iraq and Iran and in eastern Turkey are the Kurds. They're similar to the Palestinians in that they're a large ethnic group that has occupied an area for long periods but that has no official political nation (a stateless nation). Turkey recently has used military force to capture Kurdish rebel leaders. Iraq's Saddam Hussein used chemical weapons on Kurdish villages in Iraqi territory. For now at least, most Kurds live the life of permanent refugees.

Although Iraq and Iran are loaded with oil and Turkey has some too, Afghanistan is out of the loop on oil (and just about any other resource). Despite being loaded with oil, Iraq's per capita income remains well behind that of Turkey and Iran. Massive military spending and decades of corrupt leadership have kept the average Iraqi well behind economically. In addition to oil, Iran also has some of the world's largest natural gas reserves just waiting to be tapped.

Almost everyone in this subregion is Muslim. The question isn't whether people in this subregion are of the Islamic faith, but rather to which sect they belong. In Turkey, the vast majority of Muslims are of the Sunni sect (as are most Muslims worldwide). In Afghanistan, Sunnis represent about 85 percent of all Muslims, and Shiites only about 15 percent. The percentage is reversed in Iraq, where Shiites outnumber Sunnis by about 2-to-1. Iran has the bulk of the world's Shiite Muslim population, with about 95 percent of its 66 million people belonging to the Shiite sect.

Unfortunately, this subregion is in tremendous turmoil. Turkey has long been involved in the Cyprus dispute and more recently has been militarily involved with Kurdish rebels. Iraq embarked on a costly war with Iran in 1980, invaded Kuwait in 1990, massacred its own Kurdish peoples before and after, and has been on a United Nations embargo list ever since the end of the Kuwait War (Desert Storm).

Iran ousted its leader, Shah Mohammed Reza Pahlavi, and established Iran as a Muslim state in 1979. Under the Ayatollah Khomeini, Iran took hostages from the United States embassy that same year. The Iran–Iraq war raged until 1988. Afghanistan has fared even worse than its neighbors in the turmoil department. Civil war brought on a Soviet invasion in 1978 that rained bloodshed and destruction on the country for ten years. Even after the Soviet withdrawal, Afghan rebel factions have proved themselves entirely able to continue the misery and devastation on their own.

Former Soviet Republics: The Caucasus

This region consists of only three countries: Armenia (its capital is Yerevan), Azerbaijan (Baku), and Georgia (Tbilisi). All three are former Soviet republics that spun off and formed separate nations when the Soviet Union fractured.

Mountain glaciers in the Caucasus drain to the Kur River. Although the Kur flows through Azerbaijan into the Caspian Sea, the area isn't known for its rivers. As with its neighbors, the Caucasus aren't always stable and are prone to occasional earthquakes.

This subregion is not a place of deserts. Although Georgia in the west gets more precipitation than Azerbaijan in the east, no place in this subregion is classified as a desert.

Although this subregion has many Georgians, Armenians, Kurds, Russians, Ukrainians, Azerbaijanis, and many smaller groups, it's almost devoid of Arabs.

Azerbaijan is just now developing its rich oil reserves along the Caspian Sea, which should prove to be a major shot in the arm for its economy. Unfortunately for Armenia and Georgia, they're both out of luck when it comes to major oil deposits.

This subregion isn't really Muslim. Although Azerbaijan is almost 90 percent Muslim, the other two countries have only a small Muslim minority. Georgia's people are Georgian Orthodox, Russian

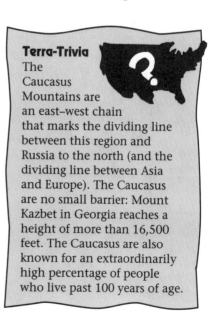

Terra-Trivia
The Caucasus Mountains are an east–west chain that marks the dividing line between this region and Russia to the north (and the dividing line between Asia and Europe). The Caucasus are no small barrier: Mount Kazbet in Georgia reaches a height of more than 16,500 feet. The Caucasus are also known for an extraordinarily high percentage of people who live past 100 years of age.

Orthodox, and Armenian Orthodox, with only a 10 percent Muslim minority. Because Armenia is well over 90 percent Armenian Orthodox, neither Georgia nor Armenia could be called Muslim countries.

This subregion, unfortunately, fits at least one of the generalizations because it's fraught with ethnic, religious, and political turmoil. Azerbaijan and Armenia have been at each other's throats in recent years over a small parcel of land along their southern border. Georgia has weathered a collapsing economy and civil war as portions of the country have attempted to secede. Don't be surprised to see more of the same in this subregion because the mountainous terrain has led to isolated ethnic pockets (much the same as in the Balkans).

Former Soviet Republics: The 'stans

The last subregion described in this chapter includes five countries that occupy one of the most remote and inaccessible locations on earth: Kazakhstan (its capital is Almaty), Kyrgyzstan (Bishkek), Tajikistan (Dushanbe), Turkmenistan (Ashkhabad), and Uzbekistan (Tashkent). These new countries (recently spun off from the former Soviet Union) are all landlocked and remote, and it's entirely possible that you don't recognize a single one of their names.

This subregion, at the extreme north end of the Middle East region, abuts two giant countries: Russia to the north and China to the east. The most famous city (and perhaps the only name you might recognize) is the oasis city of Samarkand. It was a key stop on the historic Chinese Silk Road trade route.

You may have noticed that to the south are two more 'stans: Afghanistan and Pakistan. They weren't included in this subregion because neither is a former Soviet republic and because Pakistan has historic and cultural ties to South Asia.

Geographically Speaking

Along the border of Kazakhstan and Uzbekistan is the Aral Sea. Although it's the world's sixth-largest natural lake, just a few decades ago it was the world's fourth-largest. Irrigation demands for agriculture (primarily cotton) caused the diversion of water from the Syr and Amu rivers. Because they serve as the primary feeders to the Aral Sea, with reduced flows, evaporation has taken its toll on the great lake.

Salt flats surround the once bountiful Aral, and its waters are now four to five times saltier than in the past. Salinization and agricultural pollutants have devastated fish life and have bankrupted the fishing industry. Steps are in progress to rectify the situation before the environmental catastrophe worsens.

The Kara Kum and Kyzyl Kum deserts (*kum* is a Turkish word meaning "sand") dominate this subregion, covering much of its land surface. Where the area is not composed of desert, it's composed primarily of mountains and steppes (dry, level, treeless plains). The high, glacier-clad Tian Shan range consumes Kyrgyzstan and Tajikistan before continuing into China. At 24,590 feet, Pik Kommunizma in Tajikistan is one of the world's highest peaks.

Ethnicities abound in this subregion. It has Kazaks, Russians, Kyrgyz, Uzbeks, Turkmen, Ukrainians, and Tartars—but nary an Arab.

The subregion has oil, but not everywhere. Although Kazakhstan is well stocked with oil, the rest have less, especially Kyrgyzstan. Kyrgyzstan is poor in fossil fuels but rich in uranium, as is Uzbekistan. Turkmenistan is loaded with natural gas.

Kazakstan is only about half Muslim, and all the other 'stans are heavily Muslim. Most countries in the subregion have a sizable minority of Russian Orthodox practitioners.

The subregion experiences slight turmoil in some places. Tajikistan has weathered sporadic outbursts of violence that pit Communist and anti-Communist ideologies. Authoritarian rule is causing some friction in Turkmenistan. This subregion generally has seen less violence than other parts of the Middle East.

The Least You Need to Know

➤ The Middle East is at the crossroads of three continents: Europe, Asia, and Africa.

➤ This region is the birthplace of three of the world's major religions: Judaism, Christianity, and Islam.

➤ Deserts, Arabs, oil, Muslims, and turmoil dominate the area.

➤ This region is the site of the famous and historically important cities of Jerusalem and Istanbul.

➤ Eight countries in the region are "new" because they were spun off from the former Soviet Union in the early 1980s.

Chapter 21

South Asia:
The Former Jewel
in the Crown

In This Chapter

➤ Looking back at South Asia's history

➤ Measuring the mountains and other natural features

➤ Climbing around Nepal and Bhutan

➤ Cruising through the island countries

➤ Packing off to Pakistan

➤ Wading through Bangladesh: Too much water and too little land

➤ Pioneering a passage to India

The region of South Asia, as shown on the map, includes the countries of Pakistan, India, Nepal, Bhutan, Bangladesh, Sri Lanka, and the Maldives (the subcontinent of India and its surrounding neighbors). South Asia is an appropriate name for this region because it represents the area of south–central Asia. The region is walled off on the north by the world's highest and mightiest mountain chain, the Himalayas.

To the east is Myanmar (formerly Burma) and the Bay of Bengal; to the south, the Pacific Ocean; and to the west, the Arabian Sea, Iran, and Afghanistan. The region results from the breakup of the ancient supercontinent Gondwanaland, when the Indian plate broke off from eastern Africa and Antarctica, drifted northeast, and crashed into the Eurasian plate. This process caused the Himalayan upsurge and created this region (it causes the Himalayas to continue to rise).

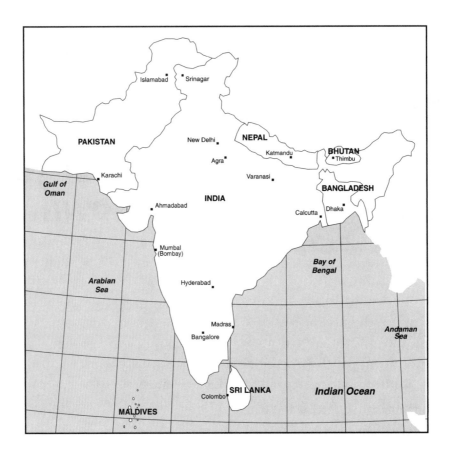

South Asia is small compared to other regions. Although it's not large in area, it more than makes up for this "shortcoming" in population. More people live in South Asia than anywhere else on earth. With more than 1.2 billion people, South Asia slightly edges out East Asia as the most populous region in the world. The higher rates of population growth in South Asian countries means that this difference will eventually increase.

The Road to the Present

One of the world's ancient Bronze Age cultures, the Indus Valley civilization, began in South Asia in about 3500 B.C. The fertile Indus River soils sustained this civilization until about 1700 B.C. At that time, Aryan tribes invaded from the west, overcame what was left of the Indus civilization, and settled in the *Punjab*.

The 33-mile-long Khyber Pass (elevation 3,500 feet) provides a passage through the Hindu Kush Mountains that separate Afghanistan and Pakistan. Through the centuries, the Khyber Pass has been a busy place. Many invaders have entered this realm: Cyrus the Great from Persia, Alexander the Great from Macedonia, the Huns, the Islamic Turks, and the Mongols under Genghis Khan and Tamerlane. India and Pakistan have experienced the rise and fall of many empires, including the Kushan, the Gupta, the Delhi Sultanate, and the Mogul.

> **GeoJargon**
> The *Punjab* is the "land of five rivers" where the Jhelum, Chenab, Ravi, Beas, and Sutlej rivers join the Indus and flow south to the Arabian Sea. Heavily populated, the Punjab is the core area of Pakistan.

In 1498, Europeans began to arrive. The Portuguese under Vasco da Gama came first. About a century later, the British East India Company was founded. The company was a better trader than it was a colonizer; in 1857, the Sepoy Mutiny occurred, and the Indians revolted and massacred British residents all over India. This event prompted the British government forces to step in, quell the revolt and take over the colonial reins to "the jewel in the crown."

Things did not go smoothly. The biggest problem was the religious rift between the majority (Hindus) and the minority (Muslims). The desire for independence was growing, and events such as the Amritsar Massacre, in which the British killed or wounded almost 2,000 Indians, didn't help. In the early 1920s, Mohandas (Mahatma) Gandhi and his nonviolent policies made independence a popular cause.

Geographically Speaking

The caste system, a hereditary social structure, dates back to Aryan priests in 200 B.C. Members of different castes could not intermingle, and marriage and occupation were restricted. The only way to a higher caste was by reincarnation (Hindu karma), and then only if you had adhered to your caste in your previous life. British rule did much to break the caste system, as did many leaders, such as Mahatma Gandhi, who referred to members of the Untouchable class as the "harijan" (children of God).

The castes are, in order: Brahmans (the priests), Kshatriyas (warriors, rulers, and large landowners), Vaisyas (farmers and merchants), and Shudras (laborers and artisans). At the bottom (not even a caste, but an outcaste, which is where the word originates) were the Untouchables, fit only for the most menial of tasks.

In 1947, the partition happened: Millions of Muslims were moved northwest and northeast into the new nation of (West and East) Pakistan, and millions of Hindus were moved from these areas into India. The relocation was not complete, however, and many millions were left as large minorities in their respective countries. Compounding the problem was the small but rabid minority of the Sikhs, heavily concentrated in the Punjab and particularly bitter toward the Muslims, who then prevailed in Pakistan.

Terra-Trivia

In 1984, Prime Minister Indira Gandhi (no relation to Mahatma, the daughter of Nehru) was assassinated by Sikh members of her own security guard. In 1991, her son, Rajiv, who was then prime minister, was killed by a bomb explosion at a Madras rally. India still continues, however, as the world's largest democratic state.

Mahatma Gandhi was assassinated in 1948 by a radical Hindu (one of his own) because Gandhi strove for tolerance of the Muslims and the unity of India regardless of religion. Under Jawaharlal Nehru, the new country's first prime minister, India began to tackle the many problems it faced, including religious differences, the Hindu caste system, feeding the populace, and developing an industrial base. Every problem was magnified, by the size and poverty of its population. Despite all its problems, however, India endured and has remained a democracy.

From the Bay of Bengal to the Top of the World

South Asia lines up latitudinally with Saharan Africa and Middle America but has little in common with either. A region of contrasts, South Asia has virtually every physical feature except major lakes. Beneath the Himalayan wall on the north and its foothills is the Asian subcontinent. It encompasses the fertile Indus Valley and the Great Indian (or Thar) Desert in the west, the Ganges basin in the center, and the Brahmaputra River in the east.

The great Deccan Plateau dominates the center of the subcontinent peninsula. The Deccan is flanked by the Western and Eastern Ghats (steep slopes) down to the narrow coastal plains. Mountains separate the Indus from Afghanistan, and rain forest hills separate the Brahmaputra from Myanmar.

Although temperature varies with the seasons, it's colder in the mountains and plateaus. The Himalaya Mountains average only 12 degrees F, whereas central India averages 79 degrees F. Rainfall also varies with topography. North of Bangladesh, Cherrapunji gets 450 inches per year (second only to Mount Waialeale in Kauai, Hawaii, with 460 inches). The Great Indian Desert gets only 4 inches; the rain forest in the east, 80 inches; the plain, 40 to 80 inches; and almost everywhere else, 20 to 40 inches.

The most famous of South Asia's many large rivers are the Indus (1,700 miles long) with its five river tributaries in the Punjab, the Hindu holy river the Ganges (almost 1,600 miles long), and the Brahmaputra (also almost 1,700 miles long) in the northeast. All these rivers originate in the mountains of Tibet to the north. Another, smaller river, the Jumna (only about 900 miles long), has religious significance where it meets the Ganges, but tourist significance in Agra, where it trickles by the Taj Mahal.

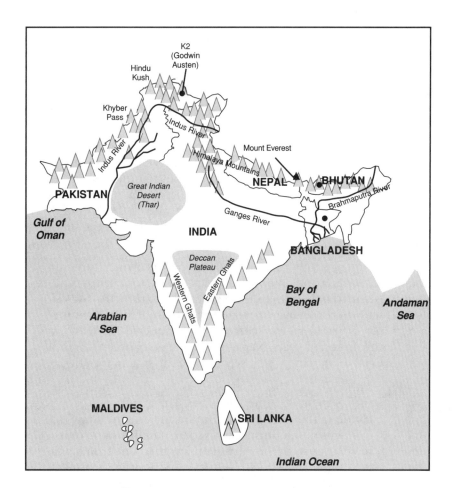

Physical features of South Asia.

Geographically Speaking

In the mid-seventeenth century, Shah Jahan, a Muslim ruler of the Mogul empire, built the Taj Mahal. It was designed as a mausoleum for his beloved wife, who had died in childbirth. It's all-white Makrana marble makes it perhaps the world's most beautiful building.

The building (186 feet square and 108 feet high) has a large onion dome rising 243 feet above garden level. At each corner of its base rises a 138-foot-tall minaret. The Taj Mahal has long reflecting pools, lawns, outbuildings, and surrounding walls. When Shah Jahan became ill, his son overthrew him and placed him under house arrest in an Agra fortress. For nine years, until he died, he could only view the distant Taj. In his death, he shares the mausoleum with his wife.

Mountain Monarchies

Nepal, a landlocked country in the Himalayas, is bounded on the north by China (the ancient area of Tibet) and on the east, south, and west by India. Its area measures about 57,000 square miles, and it has 22 million people who are Tibeto–Nepalese (the Sherpas, famed expedition guides and porters) or Indo–Nepalese (the Gurkhas, famed soldiers of British and Indian army days).

The country is dominated by the Himalayas: Nine of the ten highest mountain peaks in the world are in Nepal, including Mount Everest (29,028 feet), the world's highest on the Nepal–China border. South of the peaks are the foothills, and south of them is the Terai, a plain area of swamps and forests. In a small central valley is Nepal's capital, Katmandu, with more than a half-million people. The mountains are snowy and cold year-round, and the foothills are cool; in Katmandu and the Terai, it's hot and humid in the summer and cool in the winter.

Because Nepal was never a colony of another nation, it developed its own culture from its original settlers. They came from Tibet and, later, from India. (In the fourteenth century, Hindus fleeing from Muslims settled in Nepal.) In the eighteenth century, the Gurkas arrived while en route to Tibet, where they were defeated by the Chinese. In the nineteenth and twentieth centuries, Nepal lost a war with Britain, conquered Tibet, and was dominated by the Rama family. In recent years, Nepal has vacillated between Communism and a strong constitutional monarchy, which seems to be the system for now.

Terra-Trivia

Although Nepal is not a strictly Buddhist country, it holds an important place with Buddhist faithful. According to legend, Nepal's Kapilavastu was the birthplace of Siddhartha Gautama, the Buddha, in about 563 B.C. A four-year-old boy from Seattle, Washington, is now being educated in religious tenets outside Katmandu. Tibetan Buddhists believe him to be the reincarnation of the high lama who died in 1987.

Although half the Nepalese speak Nepali, the other half speaks about 30 other languages. The majority of people are Hindu (although Nepalese Hinduism contains many elements of Buddhism), with some Buddhists and only a few Muslims. Despite living on less than one-fifth arable land, nine of every ten Nepalese are subsistence farmers. Although in the 1950s only 1 percent were literate, now 28 percent are. The people are extremely poor—personal incomes per capita are among the lowest in the world.

Nepal suffers from high infant mortality and considerable disease, including cholera, malaria, tuberculosis, typhoid fever, and leprosy. Life expectancy is only 53 years. In a quest for arable land, people have denuded the hillsides of their forest cover, and deforestation presents a horrible environmental threat to the area. The country has only 32 miles of railroad but 3,500 miles of roads, half of which are paved. Although Nepal has air and road connections with India, footpaths are still the most common means of getting around. (And yes, most Nepalese believe that yeti, or abominable snowmen, inhabit the snowy Himalayan peaks.)

East of Nepal, and separated by a small strip of land (formerly the country of Sikkim) that's governed by India, lies the tiny mountain kingdom of Bhutan. Another monarchy in the Himalayas, it measures only 18,000 square miles (one-third of Nepal's area) and has less than two million people. It's bounded on the north by China and by India everywhere else.

Bhutan historically was influenced by Tibet and then later by Britain. Most people speak a Tibetan dialect, Dzongkha, and are ethnically Bhote and Nepalese. Three-quarters of the Bhutanese are Buddhist; the rest are Hindu. They're almost all subsistence farmers, and only 42 percent are literate. Health problems, low life expectancy, and poverty parallel Nepal's.

The terrain is almost all mountainous except for a small parcel of plain in the south called the Duars. Bhutan's capital, Thimbu, has only 30,000 residents. Although the country has no railroads, it has about 1,400 miles of roads (half are paved) and even has airplane service to neighboring countries. Slavery was not abolished in Bhutan until the 1960s, and foreigners were not even allowed in until the mid-1970s. Now it has some tourism, but less than in Nepal (and on a restricted basis).

The Island Countries

Sri Lanka, "India's tear," is a tear-shaped island off the southeast coast of India. Formerly Ceylon, this independent country is hilly in the center and surrounded by plateaus that fall off to beautiful beaches. Most of the northern island is lowland. Because Sri Lanka is only six degrees north of the equator, it has a tropical climate and rainfall.

Colombo, the capital of Sri Lanka, has more than a half-million people. Almost 20 million people live on the island: Three-quarters are Sinhalese, and one-fifth are Tamil. This ethnic division has proved to be a major stumbling block for Sri Lanka. In the past 15 years, the island has experienced near-devastation caused by civil war between the Sinhalese Buddhist majority and the Tamil Hindu minority. Almost 30,000 people have lost their lives, and their country is being destroyed.

Hindu legend says that the seventh reincarnation of Vishnu, the supreme deity, settled the island. In about 500 B.C., Hindus arrived and founded the kingdom of Sinhala, which lasted for eight centuries. Buddhism later swept the island. (Sri Lanka is now a stronghold of Hinayana, or Theravada, Buddhism, which exists only in Sri Lanka and Southeast Asia.) Then came the Tamil kings for nine centuries, and then the Portuguese, Dutch, and British. Sri Lanka emerged as an independent country in 1972.

> ! **Geographically Speaking**
>
> Although Hinduism's beginnings are obscure—it may be the world's oldest religion—it originated on the Indian subcontinent (the peninsular extension of Asia south of the Himalayas). An oddity among major religions, Hinduism has no founder. The Sanskrit Vedas are Hinduism's sacred text, and it has no specified creed. Encompassing all is Brahman, the source of the universe. Principal manifestations of Brahman include the deities Shiva and Vishnu.
>
> Reincarnation, a cycle of rebirth governed by the law of karma, is a central belief in Hinduism. The cycle is not broken until one enters the blissful state of nirvana. Also entwined with Hinduism is the caste system of social division. The Ganges River is sacred, as are holy cows and snakes (en route to Brahman). What you do, not what you think, is what counts.

The Maldives are an island republic in the north Indian Ocean, just south of the southern tip of India. The republic has about 2,000 islands in the chain, totaling only 115 square miles. Only 200 of the islands are inhabited by the Maldives' quarter of a million people. The people are Sinhalese, Arab, and African; all of them speak Divehi (a Sinhalese derivative), and all are Sunni Muslim. The capital, Male (that's its name, not its gender), has 55,000 citizens.

The Maldives have had Asian Buddhist, Arabian Muslim, Portuguese, Dutch, and British inhabitants. It has been a British protectorate, an independent sultanate, and is now a republic. Fortunately, Indian troops foiled a recently attempted Tamil coup.

The islands are tropical in rainfall and climate and, like all of southern Asia, are subject to annual monsoons. The islands are all coral atolls atop submerged defunct volcanoes; fortunately, they're not in a typhoon belt, because the highest elevation in the entire country is six feet. (Global warming and rising seas are not taken lightly in the Maldives.) Unfortunately, malaria is common, and food must be imported. In addition to a growing tourism business, the sale of postage stamps accounts for most of the island's income.

Pakistan: Land of Islam

Pakistan is bounded by India and China on the east, Afghanistan on the north and west, Iran on the southwest, and the Arabian Sea on the south. The eastern half of Pakistan is flat and dominated by the Indus River and its tributaries. The west and especially the north are quite mountainous. The southern parts of Pakistan are extremely dry (including its Great Indian desert, or *Thar*); the north, although not lush, receives slightly more precipitation.

Pakistan's 130 million people make it the seventh most populated country in the world. It's the second most populated Muslim country (Indonesia is first); the Muslim minority in India, however, is almost as numerous. Three-quarters of Pakistan's people are Sunni Muslim, and one in five are Shiite Muslim. Half of Pakistan's people speak Urdu, and half speak Punjabi.

The Sind is an area composed of the lower Indus Valley, which includes Pakistan's former capital, Karachi (its population is more than five million), and the large city of Hyderabad. The Arabs landed in the area in the eighth century and began spreading Islam throughout. Through irrigation, rice and wheat feed Pakistan. Commercially, cotton is important, and the Sind has quite a large textile industry. West of the Sind are the Baluchistan highlands, foreboding with their deserts, rocky plains, and mountains—and probably home to more camels than people.

North of the Sind is the Punjab, Pakistan's heart, with 60 percent of its people. In the Punjab is Lahore (it has about three million people), Pakistan's cultural, historic, and industrial center. The Northwest Frontier is just that—a frontier. It borders the Khyber Pass to Afghanistan, and Peshawar is its major city. With as many as four million refugees, this rugged, rather undeveloped area has recently been more attuned to the misfortunes of Afghanistan than to its own problems.

Northeastern Pakistan is the locale of Jammu and Kashmir, a trouble spot. Two wars between Pakistan and India have already been fought over this area, and another looms. (This possibility scares the world because both countries are known to have nuclear capability).

At the time of the partition of the British Indian subcontinent in 1947, the rulers of the Indian states had a choice: Align with Pakistan–Muslim or Indian–Hindu. Jammu and Kashmir had Hindu rulers and a majority (three-fourths) Muslim population. The choice was impossible. Although both countries claim the area, Pakistan controls it. Muslims don't like being under Hindu rulers, and Muslim Pakistan doesn't want Hindu India controlling the source waters and tributar-

Terra-Trivia

In 1856, T.G. Montgomerie charted all 35 peaks in the Karakorum Range on the Kashmir–China border. Each peak received a K designation. Pakistan's K-2 turned out to be the second-highest mountain in the world, at 28,250 feet (second to Mount Everest, at 29,028 feet). Five years later, it was unofficially named Godwin Austen, after a British topographer who was the second European to reach the area.

Terra-Trivia

Sikhism, founded by Nanak in A.D. 1500, is rooted in Hinduism and Islam. Centered around the sacred city of Amritsar on the India–Pakistan border, Sikhs are demanding an autonomous homeland. Their religion stresses active service, loyalty, justice, and the build-up of wealth; no castes, priests, pilgrimages, alcohol, or smoking are allowed. Sikhs are distinguished in appearance by wearing soldier shorts, a turban, an iron bangle, a steel dagger, and a comb.

273

ies of the Indus River.

In 1959, Pakistan announced a new capital, Islamabad, to be built just south of Kashmir. Both its name and location send a strong message about Pakistan's intent for Kashmir. The city, completed in the mid-1970s, now functions as Pakistan's capital city.

In its first 50 years of independence, Pakistan has come a long way but still has far to go. Irrigation has been expanded, land reform is in process, noted textiles are flourishing, some other manufacturing has begun, and rice is exported; the gross domestic product per capita is the highest in the mainland realm. Subsistence farming is the fate for most people, life expectancy is still in the 50s, only 62 percent can read, and population growth is still extremely high.

Politics have been dirty and unstable, and a military dictator now seems to be in control. In addition to the ethnic violence that plagues parts of the country, Sunni Muslim radicalism and nuclear capability alarm India and the world.

Troubles in Bangladesh

Before 1971, the country of Pakistan was a nation of two parts (east and west), with India in the middle. In 1971, after a bitter war that almost devastated East Pakistan, Bangladesh was born. East Pakistan no longer existed, and West Pakistan became Pakistan. Bangladesh is composed of almost entirely the double delta of the Ganges and Brahmaputra rivers. The country is all low-lying, except for wee bits of hills and forests around its perimeter.

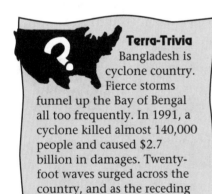

Terra-Trivia
Bangladesh is cyclone country. Fierce storms funnel up the Bay of Bengal all too frequently. In 1991, a cyclone killed almost 140,000 people and caused $2.7 billion in damages. Twenty-foot waves surged across the country, and as the receding water gushed out, it destroyed everything it missed on the way in. Eight of the world's worst recent natural disasters have been in Bangladesh.

Crammed into every inch of livable space are 123 million people, making Bangladesh the most crowded large country on earth and the ninth largest overall. It lies north of the Bay of Bengal, surrounded by India except for a small border with Myanmar (Burma) in the southeast.

The Bangladesh people are almost all Bengali and speak Bengali (this linguistic unity is about the only thing the country has going for it). Most people are Sunni Muslim, although a 16 percent Hindu minority exists. Only four in ten people are literate, and a life span of 50 years is about average. The climate is temperate; during the Bangladesh monsoon season, however, Bangladesh gets 95 percent of its rain—sometimes too much. In 1988, monsoon rains flooded the country and left 30 million people homeless.

Dhaka, the capital of Bangladesh, has almost 7 million people. Chittagong had almost 3 million people before a storm in 1991 and now has slightly more than 1.5 million; most of Chittagong's people were forced to relocate when the city was washed away. Although it has ample roads, only one in ten is paved. No road bridges exist across either the Ganges or the Brahmaputra rivers. Most transportation is by boat—350,000 of them.

Bangladesh has one major commercial crop: jute, used for producing coarse fabrics (such as burlap) and twine and rope. It grows 40 percent of the world's jute, its largest export. Because half its population is younger than 15 years old, rapid future growth is virtually ensured. The extremely fertile land, all silt, yields three rice crops per year that attempt to support the national diet of three daily rice meals. When floods are extensive, the country's people don't eat. Bangladesh is one of the poorest countries in the world. The political situation has been quite stable, probably because everyone is too busy just trying to stay alive.

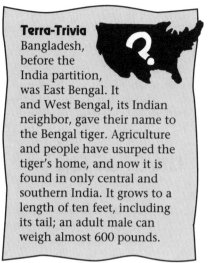

Terra-Trivia

Bangladesh, before the India partition, was East Bengal. It and West Bengal, its Indian neighbor, gave their name to the Bengal tiger. Agriculture and people have usurped the tiger's home, and now it is found in only central and southern India. It grows to a length of ten feet, including its tail; an adult male can weigh almost 600 pounds.

The Region's Titan: India (Bharat)

India is a large country with a land area equal to about one-third of the United States. Its population of 952 million, however, is almost four times the population of the United States. India's population represents about one-sixth of the world's population.

India's races have so intermingled over the centuries that its ethnic breakdown is generalized at best as Indo-Aryan, 72 percent; Dravidian, 25 percent; and "other," 3 percent. The language confusion is worse. Although Hindi is the official language, only slightly more than four in ten Indians speak it. Of world languages spoken by more than 40 million people, eight are in India. The largest, other than Hindi, are Bengali, Urdu, and Punjabi; Telugu, Tamil, Marathi, and Kannada are also common. Literally scores of other languages are spoken in India. Five language families are represented, and many or most of India's many languages are mutually unintelligible.

Sheer population size and diversity complicate every problem and aspect of Indian life. Education is a classic example: The British tried to organize a universal Indian school system, as Indian government leaders have been trying to do, yet only half of India's people are literate. Three-quarters of its people are rural and live in small villages, many without even a school. As this section examines Indian life more closely, keep in mind that Indians are generally poor and that almost half the entire nation exists below the poverty level.

Life in the Villages

The average Indian man is a farmer. He supports himself and his family and maybe has a few crops left over to sell at the market or fair. He usually lives in a one-room mud home with no windows, little or no furniture, and no indoor plumbing. He does have a veranda, which is the social center for his family (or families, because two or more related families often live together, sometimes with another room or two).

He might own a bike and probably an ox or two, and his plow is wooden with iron tips (he doesn't own a tractor, of course). He and his wife hope for male children because they're more helpful and tend to stay with the family throughout their lives. Girls are too expensive (because of their dowries) and aren't around to take care of elderly parents. Although the family doesn't have electricity, the town has perhaps one store where people can listen to the radio.

Life in the City

Although a quarter of Indians are urban, about half of them live in cities with populations of less than 100,000. The trend, however, is to move from a small village to a large city. Most large urban centers are growing at 5 to 7 percent a year. People living in smaller cities probably are in retail, small manufacturing, education, or government. Their residence is usually rented and crowded, and it's probably on a narrow street and is two stories high. Indians generally have plumbing, electricity, a radio, and maybe a TV. They own a bike and possibly a motorcycle. Maybe they know someone who owns a car.

Terra-Trivia
The Indian elephant has been tamed and put to work for more than 4,000 years. Smaller than the African elephant, it still weighs almost 11,000 pounds, stands more than 10 feet high, eats 500 pounds of forage a day, and drinks 50 gallons of water a day. Elephants have poor vision but have excellent hearing and smell (I don't know about their famous memories).

If Indians live in a big city, they may have just arrived from the farm. If that's so, they don't have a regular job yet. They're lucky if they have a shack in a squatter settlement on the outskirts of town; if they're not lucky, they sleep in the streets. Indians hire out as day laborers if they can, and beg if they can't. They try to find a real job, any job, just to get by. When they do get a job, they move into a real house not unlike their friends in the smaller city. Job opportunities are much greater, however, and the caste system is much less binding. Upward mobility is possible, likely if they're literate, and almost certain if they can speak English.

In new factories out of town, almost everything is being made: machine tools, autos, trucks and buses, bikes, motorcycles, railway cars, diesel engines, pumps, and sewing machines, for example. Jobs are available in town at small shops that make leather goods, carpets, or handicrafts. Department stores, banks, restaurants, and the government all seem available to an honest, hardworking, and educated person (the key is educated).

India already has the big-city blues, as described in Chapter 16, and is rapidly feeling more of them. The population of Bombay (or its new name, Mumbai) is reaching 13 million, and Calcutta, one of India's poorest and dirtiest cities, has more than 11 million people. Delhi and New Delhi (the capital, with attractive, British-built government buildings and wide streets) have almost 9 million people; Madras is nearing 6 million; and both Hyderabad and Bangalore have almost 5 million. Cities with more than a million people are too numerous to mention in this chapter.

Other Observations

Thanks to the British, India was left with and has added to an amazing railroad system; with more than 38,000 miles of track, it's the world's fourth-largest. With almost 1.5 million miles of roads (half are paved), every state is accessible, and now most villages are too. Even the paved roads can be uncomfortable, however, and some are axle-breakers. (The road from Jaipur to Agra, obviously well traveled because it's en route to the Taj Mahal, still troubles my kidneys.)

Forests make up a surprising quarter of India. They're state-owned and have little commercial value. Also state-owned but of great commercial value is the country's airline system. All major cities in India and most major international cities are connected by frequently scheduled flights. Coal is an important industry, and petroleum can meet only one-third the national needs. Many other minerals are mined, although none overwhelmingly. Hydroelectric power is important, as is iron ore, which is even exported.

The motion-picture industry is huge in India; even the smallest villages are likely to have a movie house. Although much of the movie content is escapist, some films are serious and have won international acclaim. Communications are slow and not as readily available as the expanding economy might need. Newspapers, despite the language diversity, are surprisingly extensive and extremely popular. Clothing, especially for women, is amazingly colorful. All of India, in fact, seems to be a sea of color.

The Least You Need to Know

➤ Nepal's 29,028-foot Mount Everest is the world's highest mountain.

➤ The devastating Tamil–Sinhalese civil war has killed almost 30,000 people in Sri Lanka.

➤ The nuclear aspect of the Pakistan–Indian Kashmir territorial dispute is world-threatening.

➤ Subject to devastating monsoon rains and cyclones, Bangladesh seems to be on nature's hit list.

➤ The world's largest democracy, India is poor and largely Hindu and has one-sixth of the world's people.

➤ Rapid population growth in this region means that the world's most populous region can soon expect to get even more populous.

Chapter 22

Southeast Asia: Shining Stupas

In This Chapter

➤ Understanding the terminology: Indochina or Southeast Asia?

➤ Describing the countries: Peninsulas and archipelagoes

➤ Hacking through jungles and deltas

➤ Spending one night in Bangkok—a look at the region's cities

➤ Getting to know the people of Southeast Asia

Southeast Asia, shown on the following map, has long been thought of as a remote and exotic region. Beautiful people, fabulous spices, and spectacular temples dot a region filled with rare wildlife and tropical forests. If not for Southeast Asia's tortured political history, this region would be truly magical.

The past several decades have not been altogether kind to Southeast Asia. Squeezed between the two most populous countries on earth, this little-known region is often overlooked, although warfare, genocide, and poverty have put it on the map. Despite these problems, Southeast Asia is a wondrous and intriguing place.

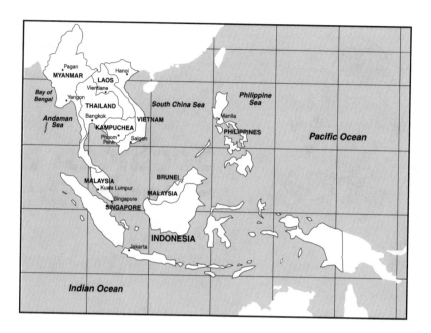

The Origins of Indochina

Southeast Asia is a region of peninsulas and islands that occupies the far southeastern corner of Asia. During the colonial period, much of this region was called Indochina or, specifically, French Indochina. Although this chapter does not use that name because it doesn't refer to the entire region, it does highlight two major characteristics of the area.

Geographically Speaking

The shape of a country is important because it can influence politics, administration, and infrastructure. Politically speaking, the ideal shape is compact and circular with a central capital. Other shapes have inherent disadvantages and can lead to fracturing and autonomy movements.

Compact countries are roughly circular—Cambodia resembles this model. *Elongated* countries are typically long and thin (Vietnam, for example). *Prorupt* countries are basically compact with a narrow elongation extending outward (Myanmar and Thailand, for example). *Fragmented* countries are part mainland and part island, such as Malaysia, or made up entirely of islands, such as the Philippines and Indonesia.

The first part of the term Indochina, *Indo,* refers to the region's western neighbor, India. Throughout time, it has been one of the world's great cultural hearths. From India, many elements of Southeast Asian culture have been drawn. Perhaps most important, the country has served a conduit for the region's primary religions.

Both Hinduism and Buddhism originated in India and have left a profound mark on Southeast Asia. Islam, via India's Mogul empire, also found its way to the region and is still an important influence. Indeed, India has had a profound cultural influence on much of the area.

The second half of the term *Indochina* refers to the China connection. To the north of Southeast Asia lies the country of China—the most populous on earth. For centuries, people from China streamed southward into Southeast Asia. Throughout most of the area, people of Chinese ancestry figure heavily in the population composition of the countries—especially in urban areas.

Pagan and Kublai Khan

In ancient Burma (modern Myanmar), on the western edge of Southeast Asia, stood one of the most fabled cities of history, the temple city of Pagan. In his thirteenth-century travels, the Venetian Marco Polo described this magical place as a city with thousands of pagodas sheathed in gold. It wouldn't last.

Kublai Khan's Mongol forces swept south and ravaged the city. The people were slaughtered and dispersed and the city burned. Pagan was never to rise again, its proud history buried in the ashes. Only the remnants of Pagan's thousands of pagodas remain, lonely reminders of past greatness.

Local villagers and farmers still worship at the many statues of the Buddha that remain in the temple niches. You can climb these ancient temples, some of which reach more than 300 feet high. From their pinnacles, you can gaze across the dry plains, watch the sun glisten off the mighty Irrawaddy River, and let your mind drift back to the glories of ancient Pagan.

GeoJargon
The pagodas of Southeast Asia are principally of two types: temples and stupas. *Temples* can be entered or ascended and often contain statues of the Buddha in large niches. *Stupas* are the characteristic "Hershey's kiss"-shaped structures often sheathed in gold; they usually house some revered relic, such as a hair from the Buddha.

Terra-Trivia
Pagan isn't the region's only archaeological wonder. In Cambodia, Angkor Wat (the massive temple complex built in the Hindu style) is one of the world's most spectacular architectural monuments. On the island of Java sits the world's largest Buddhist temple, the mountainlike structure of Borobodur. Each tier of the temple is filled with small stupas containing images of the Buddha.

Colonization ⚡

The Mongols weren't the only foreign power to make their presence felt in Southeast Asia. From the sixteenth century onward, Europeans made their mark on the region. The British, French, Dutch, and Spanish all made claims in Southeast Asia. Even the United States became a colonizing power in the late nineteenth century. The following map shows who used to control what in Southeast Asia.

Colonial realms of Southeast Asia.

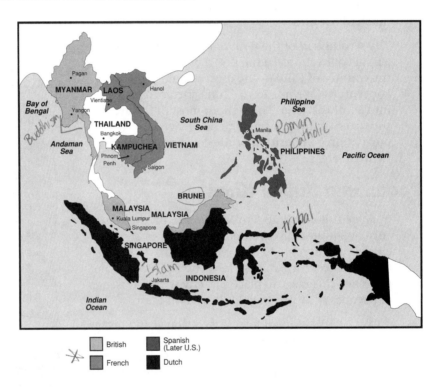

Before World War II, Japanese forces displaced the Europeans and became colonial masters of much of the region. The colonial yoke was not to be shaken until the years following World War II. Centuries of colonization left lasting imprints on the countries of the region. European administrative divisions, the infrastructure, and even religions remain in most of Southeast Asia's countries.

The Countries: Peninsulas and Archipelagoes

Mainland Southeast Asia consists of a peninsular outcropping that extends from the southeastern corner of the continent of Asia. From the peninsular mainland, a second peninsula (the long, thin Malay Peninsula that's often compared to a kite tail) winds its way southward. Beyond the Malay Peninsula is a vast archipelago consisting of thousands of islands. To make sense of all these peninsulas and archipelagoes, take a look at the preceding map of Southeast Asia.

Every country in Southeast Asia except Laos has frontage on the sea. Throughout history, this ocean access has been extremely important to the region. To describe the countries of the region, this chapter works from the mainland southward to the islands. Although many of these countries are little-known, you have to keep track of only nine.

The Peninsular Mainland

The large mainland peninsula of Southeast Asia contains five countries. (In this book, Malaysia is considered part of the island grouping even though part of it is on the mainland.) Although the peninsular mainland contains most of the region's countries, it contains less than half its population. All of Southeast Asia's large rivers are on the peninsular mainlaind.

Myanmar

You get many strange looks when you mention Myanmar—even the geographically informed balk at this one. In addition to being off the beaten track, Myanmar adopted a new name in 1989. You might be more familiar with its old name, Burma.

Diamond-shaped Myanmar, the westernmost and northernmost country in Southeast Asia, is roughly the size of Texas. Jutting southward is an elongated stretch of territory that extends down the Malay peninsula.

The country was colonized by the British and incorporated into British holdings in India. Although in the 1930s it became a separate colony named Burma, this existence was short-lived. Japan took over the area in the 1940s, and Burma was the scene of some of the most horrific fighting of World War II. Many battles occurred along the famous Burma Road, the Allies' lifeline to China.

Although Burma gained independence in 1948, it has survived a tortured history ever since. Socialist and military regimes have run Burma and isolated it from much of the global community while it languished largely undeveloped. Burma (under its new name, Myanmar) is now ruled by a heavy-handed military government that severely limits personal freedom.

> **Terra-Trivia**
>
> Myanmar, northern Thailand, and Laos are home to the "golden triangle," an untamed area where drug lords and their personal armies control much of the countryside and where heroin is its principal export.

Myanmar was once considered the rice bowl of Southeast Asia—the major rice exporter in the region. Although the country remains heavily agricultural, it now barely feeds itself. Despite agricultural potential and undeveloped mineral wealth, its 50 million people remain among the world's poorest.

Thailand

As with Myanmar, this country's former name sounds more romantic. Thailand, once known as Siam (as in Anna and the King of Siam), is the only country in Southeast Asia to not be colonized by a European power: It served as a buffer state between British and French holdings in the region.

Terra-Trivia
If you like spicy food, Thailand is the place for you. Thai chilis, a staple of Thai cooking, are hot enough to bring sweat to your face and smoke from your ears. Cilantro and other Thai seasonings give Thai cooking its exotic and steamy flavor.

Thailand, located in the heart of mainland Southeast Asia, is similar in shape to Myanmar. As with Myanmar, the bulk of Thailand lies in the north, and it has a thin, peninsular outcropping that reaches southward on the Malay Peninsula. Because of this unique configuration, Thailand accesses the Adaman Sea and Indian Ocean on the west and the South China Sea and Pacific Ocean on the east.

As in many countries in the region, Thailand's 60 million residents are largely farmers. Although the country has remained independent for much of its history, Thailand's government has also been somewhat shaky. Nonetheless, the country remains one of the more affluent in the region, with an annual $6,000 per capita income. Its economy has surged during the past two decades, and increasing tourism has provided a major boost.

Laos

Landlocked and mountainous, Laos was one of three countries (with Cambodia and Vietnam) that comprised French Indochina. Laos borders on all the mainland Southeast Asian countries (Myanmar, Thailand, Cambodia, and Vietnam) as well as on China in the north. Laos became free of French control in 1950, and, like its neighbors, has weathered political storms ever since.

In the past several decades, Laos has experienced an ongoing struggle between royalists, Communists, and pro-Western revolutionary forces. The country is now a "democratic republic." Because it has only one political party, the Communist Pathet Lao, you can probably surmise just how "democratic" it really is.

Economically, Laos is the poorest country in a poor region. The country is heavily agricultural: Its largest city and capital, Vientiane, has fewer than half a million people. This country of five million people has not a single mile of railroad track and little in the way of hard-surfaced roads. Remote, undeveloped, and tumultuous, Laos remains an enigma.

Cambodia

Cambodia is another country with a multiple-name complex. As part of French Indochina, it was long known as Cambodia. Political turmoil in the mid-1970s resulted in a name change, and Cambodia became Kampuchea. Its name has changed back to Cambodia, but don't hold your breath that it will stick.

Bordered by Thailand to the west, Laos to the north, and Vietnam to the east, Cambodia is sort of a lowland bowl facing south to the Gulf of Thailand. Ringed by highlands, its interior is hot and steamy. With a population of 10.5 million, the country has twice the population of Laos, despite being smaller. Like the other Southeast Asian countries discussed in this chapter, Cambodia is a poor and heavily agricultural country. Political turmoil and isolationism have hindered development in recent years, and its people struggle to feed themselves in this subsistence-based economy.

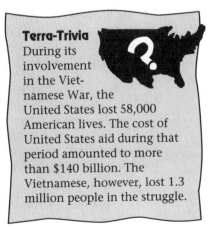

Terra-Trivia

Although political turmoil is the norm in Southeast Asia, few places have witnessed more violence and horror than Cambodia. In the late 1970s, Communist Khmer Rouge forces under the notorious (and recently captured) Pol Pot overthrew the government. Between executions, starvation, and forced labor, almost two million people died in the "killing fields" of Kampuchea.

Vietnam

Vietnam, the easternmost of the Southeast Asian peninsular countries, borders Laos and Cambodia to the west, China to the north, and the South China Sea to the east and south. Vietnam is another country that formed colonial French Indochina; its population of more than 75 million people makes it the largest on peninsular Southeast Asia.

In the same-old-story department, Vietnam is a poor, agriculturally based country with a turbulent political past. Vietnam shook off French colonial rule in 1954, which led to a divided country. In the north, Communist North Vietnam was supported by China and the Soviet Union. In the south, South Vietnam's authoritarian regime was supported by the United States.

Full-scale war between the north and south raged throughout the 1960s and into the 1970s. The United States had more than half a million troops stationed in Vietnam during the high point of the war, in an effort to thwart North Vietnamese penetrations into the south. Embroiled in an "unwinnable" war, the United States finally pulled out of Vietnam, and North Vietnam quickly overran the south, uniting the country under Communism.

Terra-Trivia

During its involvement in the Vietnamese War, the United States lost 58,000 American lives. The cost of United States aid during that period amounted to more than $140 billion. The Vietnamese, however, lost 1.3 million people in the struggle.

Vietnam has adopted a free-market economy despite Communist political leanings. Years of war and destruction have decimated the country, and it struggles to get itself back on track economically. In 1995, Vietnam and the United States again restored full diplomatic relations.

Island Countries

Southeast Asia has thousands of islands, including some of the world's largest. In this chapter, four countries are in the island group: Malaysia, Indonesia, Brunei, and the Philippines. Except for Brunei, all these countries are considered geographically fragmented.

Malaysia

Geographically, it's difficult to decide whether Malaysia should be considered part of the peninsular mainland or the islands. It's both, in fact. At the southern tip of the Malay Peninsula is Malaysia's core, the former British colony of Malaya. To the east, on the island of Borneo, is the rest of the country, the two former British colonies of Sarawak and Sabah. Although Malaya is the heart of the country, Sarawak and Sabah represent more than half of Malaysia's area.

Although Malaysia also included Singapore at one time, Singapore is now a separate country and is described as part of the Pacific Rim region (refer to Chapter 15). Although agriculture is important in Malaysia, the economy is more diverse than most of the economies in Southeast Asia. Technology and manufacturing hold an increasingly important niche in the economic scene.

Though more politically stable than other Southeast Asian countries, Malaysia struggles with ethnic divisions. Although the majority of its 20 million people are Malay, the country has large numbers of ethnic Chinese (about one-third of the population) and Indians. Although the Malay majority wields primary political power, the ethnic Chinese hold considerable economic power, so trouble brews.

Indonesia

Although some question may have existed about whether Malaysia should rightfully have been considered an island nation, no question exists regarding Indonesia. Indonesia's archipelago is the largest in the world, by number of islands, land area, and population. Between 13,000 and 17,000 islands comprise Indonesia (estimates vary); its 210 million people make it the fourth-largest population in the world.

Indonesia was rigidly colonized by the Dutch after initial contact in the late sixteenth century and remained a Dutch colony until World War II. Known as the fabled Spice Islands, the "Netherlands East Indies" became a valuable Dutch possession. During the war, Indonesia was harshly colonized by the Japanese and was returned briefly to the Dutch until Indonesian independence began in 1949. The Netherlands colony in New Guinea joined the Indonesian realm in 1973.

In 1965, a coup attempt against "president for life" Suharto prompted an anti-Communist purge in which 300,000 sympathizers were murdered. Although restrictions are still imposed on the press and the government wields a heavy hand, softening policies are in the wind. Recently, elections have been permitted and Indonesia has moved more strongly toward true democracy.

In addition to a developing manufacturing base and a wealth of minerals, Indonesia has timber, petroleum, and natural gas. Economically, it has shown the beginnings of strong growth. The primary problem facing Indonesia is keeping its many islands, peoples, languages, and religions from spinning apart.

Brunei

Brunei is a little different from any other country you've read about in this book. Located on the northwest coast of the island of Borneo, it's a peanut of a country, measuring a little more than 2,200 square miles. The population is mite-size also (about 300,000 people). Those facts by themselves set this tiny country apart from the rest of the region, and the story gets stranger.

Terra-Trivia
Of Indonesia's thousands of islands, three are among the six largest in the world. (New Guinea and Borneo are only partially owned by Indonesia.) In this list of the world's largest islands, square mileage is shown in parentheses, and an asterisk (*) signifies an Indonesian island:

Greenland (840,004), *New Guinea (309,000), *Borneo (287,300), Madagascar (226,658), Baffin Island (195,928), *Sumatra (182,860).

Long a British protectorate, Brunei became an independent country in 1994. Before British domination, Brunei had controlled much of northern Borneo and wielded a great deal of influence in the area as a powerful Islamic sultanate. Despite its diminutive size, Brunei again wields enormous influence. It has petroleum and gas—and lots of it.

Ruled by Haji Hassanal Bolkiah, the sultan of Brunei, the country is swimming in oil money. The revenues, which flow through the sultan, are dispersed to the populace in the form of free education, health care, and social assistance. Its per capita income of more than $16,000 far exceeds the rest of the region. (The sultan's personal income far exceeds just about everyone else's on earth!)

The sultan of Brunei is reported to be the richest man on earth. Fabulous stories abound regarding his wealth. Did he really fly in Joe Montana for a multi-million-dollar weekend at the palace to teach his son how to throw a football? Does the sultan's interest income alone exceed $1 billion every year? Does an international hotel chain run and staff his palace? Although the conjectures are endless, Brunei is undoubtedly not your typical Southeast Asian country.

Terra-Trivia
The world's largest Islamic country is not in the Middle East or in northern Africa. It's Indonesia, with more than 170 million Muslims.

The Philippines

The Philippines are another archipelago, an island group of about 7,000 islands clustered in the northeast portion of Southeast Asia. This group of islands became a Spanish colony in the sixteenth century and remained that way until 1899. As a result of the Spanish–American war, the country was ceded to the United States.

> **Terra-Trivia**
> In the South China Sea are hundreds of islands that include the Spratly Islands and Parcel Islands. Although most are little more than reef outcroppings and are uninhabited, China, Taiwan, Vietnam, Malaysia, the Philippines, and Brunei all make claims on them, and tensions are high. The reason? Whoever owns the islands of the South China Sea also owns the oil that's beneath it.

The Philippines emerged as a separate country in 1935, only to be invaded by the Japanese in 1941. They were the site of General Douglas MacArthur's famous line, "I shall return." Return he did—in 1945, the Philippines were free again. Strongman Ferdinand Marcos ruled the land from 1965 to 1986. (Perhaps equally well known is the former First Lady Imelda Marcos, infamous for her enormous shoe collection.)

Almost half the Philippines' population of 75 million is involved in agriculture. The steep slopes of the volcanic islands are heavily terraced with rice paddies throughout the chain. Timber, plantation crops, fishing, and a growing industrial base also figure heavily in the Philippines' economic picture. Recent U.S. pullouts from Clark Air Force Base and Subic Bay Naval Base have left an economic void that the Philippines hope to fill with industry. As in Indonesia, maintaining central control over many islands and peoples has proved difficult.

Tropical Rain Forest and Deltas

> **Terra-Trivia**
> In 1991, Mount Pinatubo violently erupted on the Philippine island of Luzon. The blast buried Clark Air Force Base and forced the evacuation of thousands of Filipinos. The eruption, the third-largest of this century, affected weather around the world.

Physically, the region of Southeast Asia straddles the equator and is located almost entirely within the tropics (only the northern portion of Myanmar peeks above the Tropic of Cancer).

As you might expect, most of the region is hot and wet. Rainfall is heavy over the entire region between May and October. Northern areas receive considerably less rainfall between November and April, resulting from the wet and dry monsoons. The area is dominated by tropical rain forest, although drier forests and even some savanna lands are in the northern interior.

The region abuts the Pacific Ring of Fire, and volcanic activity and earthquakes are particularly frequent in the Philippines and Indonesia. Because the volcanic soils are quite fertile, however, and despite the dangers, people have braved these islands in great numbers. The region typically has mountainous interiors with fertile coastal plains and river lowlands. The populations tend to be clustered in these lowland areas.

The Waters

Southeast Asia is home to several large and regionally important rivers, as shown on the following map. Myanmar is home to two of Southeast Asia's largest rivers. Dominating the entire central portion of the country is the Irrawaddy River. Along its shores are Myanmar's two largest cities and largest population concentrations. In eastern Myanmar, the Salween River runs roughly along the Thai border and originates deep in the heart of Chinese Tibet.

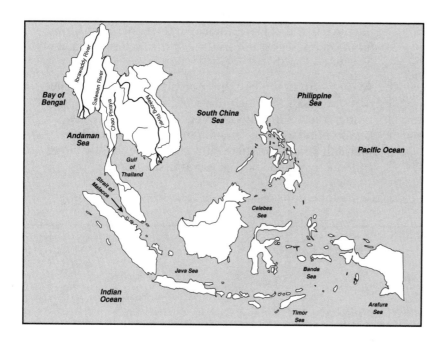

Southeast Asia's bodies of water.

Thailand is home to the region's third-largest river, the Chao Phraya. Known for the golden pagodas that line its shores, the river runs through the Thai capital. The Mekong River is one of the world's largest; from its origins in Tibet, it flows southward through China, along the Thai–Laotian border, through Cambodia, and to its huge delta in southern Vietnam.

As a region of peninsulas and archipelagoes, the sea is important to Southeast Asia. In the west, the Indian Ocean abuts the region with its Bay of Bengal and Andaman Sea to the north. The mainland is separated from the islands by the famous spice trade route, the Strait of Malacca. Amid the Indonesian islands are the Java Sea and the Banda Sea; the Timor Sea and the Arafura Sea separate Indonesia from Australia. Between Indonesia and the Philippines is the Celebes Sea. East of Vietnam lies the South China Sea; east of the

Philippines, the Philippine Sea finally gives way to the vast Pacific Ocean. (Are you getting seasick?)

The Islands

As noted, literally thousands of islands are in Southeast Asia—too many to describe in this book. The largest and most important are shown on the following map. The large island, Borneo, contains three countries: Brunei and portions of Malaysia (Sabah and Sarawak) and Indonesia.

Indonesia has several notable islands. In addition to Borneo, the large islands of Sumatra and Celebes (Sulawesi) form the Greater Sunda Islands. Indonesia also has the western half of the huge island of New Guinea. In the Lesser Sunda Islands, Java—Indonesia's core—has one of the highest population densities on earth.

To the north in the Philippine archipelago is a vast collection of islands. Although many of the islands are familiar to World War II buffs, this chapter describes only the two largest: Mindanao, the less-populous island toward the southern end of the chain, and Luzon, the slightly larger and much more populated island to the north.

Terra-Trivia
In the Lesser Sunda chain are the tiny islands of Komodo and Bali. The feared Komodo dragon is a rare species of huge, meat-eating lizards that walk the island of Komodo. Bali is known to travelers worldwide for its spectacular scenery, beautiful people, and the cultural traditions and architecture that accompany its Hindu faith.

Islands of Southeast Asia.

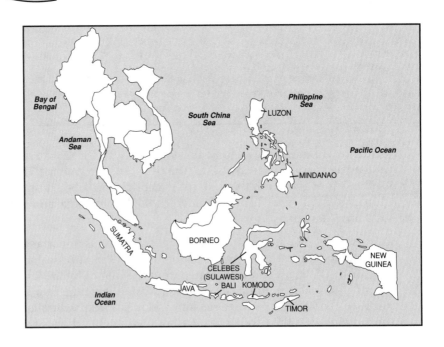

From Manila to Rangoon: The Cities

Because most of the countries in Southeast Asia are not heavily urbanized, this region does not have the abundance of large cities a more developed realm of its size might have. Here's a list of the region's most important cities:

➤ **Yangon (Rangoon):** The capital and largest city in Myanmar, Yangon is at the mouth of the Irrawaddy River. This city of 2.5 million people is known for its famous Buddhist shrine, the huge, gold-plated Swedagon Pagoda.

➤ **Bangkok:** The "Venice of the East," Thailand's capital is at the mouth of the Chao Phraya River. This city of almost six million people is known for its many canals, spectacular architecture, crowded streets, and exotic nightlife.

➤ **Phnom Penh:** Although the capital of Cambodia has fewer than one million people, it was severely depopulated during the harsh Khmer Rouge regime.

➤ **Ho Chi Minh City (Saigon):** This bustling center in southern Vietnam is home to four million people. It was the former capital of South Vietnam and French Indochina. Although Hanoi is half its size, it's the Vietnamese capital.

➤ **Kuala Lumpur:** This Malaysian capital of more than one million people is somewhat of a boom town that combines classical Islamic architecture with modern skyscrapers.

➤ **Jakarta:** Located on the island of Java, the Indonesian capital of more than eight million people is the largest city in Southeast Asia.

➤ **Manila:** Although Manila, the capital of the Philippines, has a little more than 1.5 million people, it pairs with nearby Quezon City to form a conurbation of almost 10 million people.

The Mix of People

Partly because of Southeast Asia's mountainous character, it's home to many ethnic groups. On the mainland, the major groups roughly coincide with national borders: Burman peoples in Myanmar, Thai in Thailand, Lao in Laos, Khmer in Cambodia, Malay and Chinese in Malaysia, and Vietnamese in Vietnam. A considerable overlap, however, is a source of dispute.

The islands of Indonesia and the Philippines contain hundreds of smaller ethnic groups; the primary groups are Indonesian, Malay, and Filipino. On Indonesian New Guinea, people of Papuan ancestry are the prevalent group. The individual languages of the region parallel the ethnic patterns.

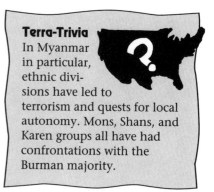

Terra-Trivia
In Myanmar in particular, ethnic divisions have led to terrorism and quests for local autonomy. Mons, Shans, and Karen groups all have had confrontations with the Burman majority.

Southeast Asia is also a land of many religions. Although it has several of the world's largest, their basic layout is not all that complex. On the mainland, Buddhism is dominant. Unlike Buddhism throughout most of the world, Southeast Asian Buddhism is primarily of the Theravada sect. Much more individualized and meditative than the more predominant Mahayana sect, Theravada Buddhism manifests itself on the region's landscape in its thousands of shining pagodas.

Exceptions to the Buddhist rule on the mainland show up in Vietnam and Malaysia. In Vietnam, Chinese religions are practiced (combinations of Confucianism, Taoism, and Mahayana Buddhism). Although Malaysia has pockets of Chinese religion, most of its people are Muslims. Islam also prevails in Brunei and Indonesia. Although Hinduism once existed throughout the region, now only remnants of it survive—especially on the Indonesian island of Bali. In the Philippines, as a legacy of its Spanish colonial heritage, Roman Catholicism is most common.

The Least You Need to Know

➤ Mainland Southeast Asia is a peninsular extension of Asia; the region also contains thousands of large and small islands.

➤ Southeast Asia is heavily influenced by its two large neighbors, India and China.

➤ The region has nine countries: Myanmar, Thailand, Cambodia, Laos, Vietnam, Malaysia, Brunei, Indonesia, and the Philippines.

➤ Southeast Asia is hot and wet, with rain forests and mountainous interiors.

➤ Jakarta, Bangkok, Ho Chi Minh City (Saigon), and Manila are the region's largest cities.

➤ Southeast Asia has many ethnic groups, numerous languages, and the world's largest concentration of Theravada Buddhists.

East Asia: The Far East

The East Asian region includes, as shown on the map, three countries and one wanna-be. The countries are China, Mongolia, and North Korea; the wanna-be is Tibet. Although the region has almost 4.5 million square miles, astoundingly, it has more than 1.2 billion people, more than one-fifth the earth's total population.

Most of the region's area and 98 percent of its people are in the huge country of China, whose name in Chinese means "central land." China is central in this realm and certainly dominates it. Before describing it in detail, this chapter looks at the region's other two countries and the wanna-be.

The Forgotten Republic of Mongolia

Large in area, Mongolia occupies an isolated and desolate spot in central Asia. (The last time Mongolia made front-page news was in the thirteenth century.) This lonely place is bounded on the north by Russia and on the east, south, and west by China. Though Mongolia is a large country, it has a population of only 2.5 million people. With a density of only 4 people per square mile, it ties French Guiana as the world's most sparsely populated country. Although its capital, Ulan Bator, has only 600,000 people, it's a quarter of the entire country.

To the west are the Altai Mountains; to the north are a plateau and more mountains. In the south is Mongolia's most prominent feature, the Gobi Desert. The Gobi—the coldest, northernmost desert in the world—covers more than 500,000 square miles. This dry, rocky, foreboding place does not conform to the classic desert image: Only 5 percent is covered by sand dunes.

Mongolia's climate is extreme—extremely cold in winter and extremely hot in summer. (Because it's in central Asia, no nearby large water body moderates climate extremes.)

Most Mongolian people are rural nomadic herders living in yurts (felt tents that protect against the intense cold and heat). Mongolia is poor: Its annual gross domestic product per capita is only $112, the world's second lowest (only Mozambique, with $80, is lower).

The country's history, however, is fascinating. In 1992, it became a democracy. It has been a Communist state, a part of China, a Buddhist–Mongol (religious–lay) dually controlled state, and initially the heart of the Mongol empire. What still fascinates the world (and Mongolians, for that matter) is Mongolia's more distant past.

In the thirteenth century, Genghis Khan and his sons and grandsons (most notably Kublai) forged the Mongol empire, history's largest land empire, consisting of China, India, Eastern Europe, Russia, most of the Middle East, and all of central Asia. The Mongols, fierce and devastating in war, were known as superb horsemen and excellent archers. Strong discipline and leaders with brilliant military strategies led to unrivaled military successes.

GeoJargon
The Mongol empire was also known by other names that have lived on in infamy. At various times and in different places, the Mongol armies were also called the Tatar, the Golden Horde, and the Horde from hell.

The Mongols left conquered rulers in place but took one-tenth of everything: women and children, men for troops, wealth, food, and animals. At the height of their rule, they lost only once on land—to Burmese elephants that sent the Mongol horses fleeing in fear. They always lost at sea, twice while trying to invade Japan and once in trying to conquer Java. In the fourteenth century, the empire fragmented, and its weakened armies were driven back to their Mongolian plateau. The great Mongol empire was history.

The Democratic People's Republic of North Korea

North Korea occupies the upper half of the Asian peninsula just west of Japan. It has almost 24 million people, all of whom speak Korean. North Korea's climate is temperate and its terrain mountainous. It's a Marxist, government-controlled state that makes China look liberal! Until 108 B.C., it was (old) Chosen, then part of China, then Mongol, and then independent again as (new) Chosen.

In 1910, Japan took North Korea over as a colony and mistreated it until the end of World War II, when Korea was split at the 38th parallel. The northern part went to the U.S.S.R., and the southern part went to the United States. In 1948, the two Koreas became North Korea and South Korea, respectively. Two years later, North Korea invaded South Korea and attempted a military unification. With help from the United States, North Korea was pushed back almost to the Chinese border; with Chinese help, South Korea was pushed back to the 38th parallel, where a no-man's land was set up in the 1953 truce.

While South Korea was forging a modern-day economic miracle, North Korea was going nowhere (or perhaps downhill, economically). Although the country supports one of the largest standing armies in the world, its people are getting poorer. They cannot feed themselves and are now receiving emergency grain from China. Starvation has recently forced the government to accept international aid.

North Korea's per capita income is less than $1/12$ of South Korea's. Continuous tension has existed between the two countries, and it may be getting worse. The world's fear is that North Korea, under the pretense of unification, will try to cure its economic ills with the stimulus of another war.

Ancient Tibet: The Wanna-be

Terra-Trivia
The Dalai Lama, the spiritual and temporal leader of Tibet and Tibetan Buddhists across the world, fled Tibet in 1959 in his mid-20s. He now lives in exile in India but lectures all over the world in nonviolent opposition to Chinese rule in Tibet (for which he received the 1989 Nobel prize for peace).

Tibet (or Xizang, its Chinese name) is the largest and least populated (2.3 million people) province of China. Its capital is Lhasa, renowned for the Potala Palace, the traditional home of the Dalai Lama. Tibet is in southwestern China, just northeast of India. The highest region on earth, it has an average elevation of more than 16,000 feet and is called the "roof of the world." The heart of Tibet is the Himalaya Mountains to the south (including Mount Everest on the India–China border, the world's highest mountain, at 29,028 feet), mountains to the west and north, and the world's highest plateau (an average elevation of 15,000 feet) in the middle.

Tibet's major rivers.

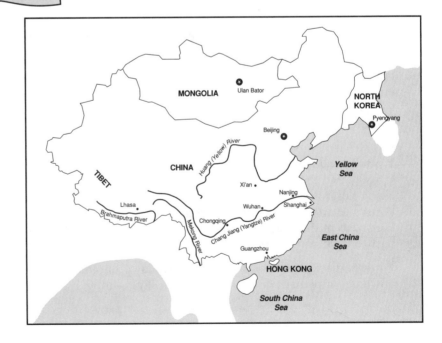

Tibet's plateau is Asia's principal watershed. It's the source of the Huang (Yellow), Chang Jiang (Yangtze), Mekong, Ganges, Indus, and Brahmaputra rivers, the largest of which are shown on the map. Tibet is semiarid (it receives only 15 inches of precipitation annually) and is generally cold and windy. Its people are mostly Tibetan (although a sizable Chinese minority lives there) and speak Tibetan. They are strongly Lamaists (a Buddhist sect), although China encourages atheism. Most Tibetans survive through herding and subsistence farming.

Tibet's history is obscure until the seventh century, when Buddhists began to penetrate. Then Lamaists evolved; then the Mongols in the thirteenth century; then Lamaists for four centuries; and then the Chinese for almost three centuries. Although the British invaded in 1904, Tibet went back to China two years later. In 1913, the Tibetans threw out the Chinese and set up their own state. In 1950, Mao Zedong took back Tibet, and that's where it now stands. Although China is content, the Tibetans certainly aren't.

The Big One: The People's Republic of China

This huge country in eastern Asia, which encompasses 3.7 million square miles, is the world's third-largest in area (behind only Russia and Canada). China has more than 3,400 offshore islands; Hainan is the largest. With more than 1.2 billion people, it's the world's most populous country. China has 50 cities with more than a million people apiece; 1 of every 5 people on earth is Chinese.

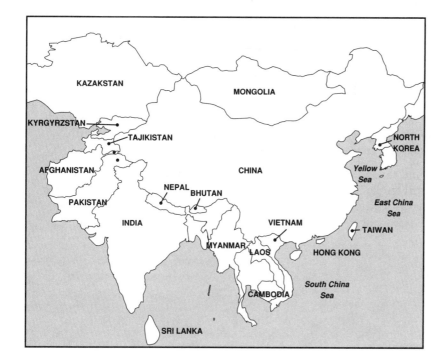

China's neighbors.

Although everyone knows where China is, few could name its neighboring countries and water features, at least not without looking at the map. To the north are Mongolia and Russia; to the northeast is North Korea; to the east are the Yellow and East China seas; to the south is the South China Sea, Vietnam, Laos, Myanmar, India, Bhutan, and Nepal; to the west, China touches Pakistan, Afghanistan, and Tajikistan; to the northwest are Kyrgyzstan and Kazakstan. Wow!

China is so extensive that it has almost every type of climate, every degree of rainfall, and every topographical feature. It's composed of 43 percent mountains, 26 percent plateaus, 19 percent basins, and 12 percent plains. The only way to get a handle on such diversity is to break it up into smaller areas and describe each one in detail. To make that process simple, this chapter refers to the areas as central, northeastern, southern, southwestern, and northwestern. (The southwestern area is composed primarily of Tibet.) The following section starts with the central section because that's where it all began for China.

In the Beginning: Central China

Legend has it that, in about 2000 B.C., China had its first dynasty: the Hsia. The Shang Dynasty overthrew the Hsia and ruled for almost 750 years. It was followed by the Chou Dynasty for another 800 years, which primarily consolidated the gains, culture, and rule of the Shang. The land these dynasties ruled over was dominated by the vast wheat and rice growing on the North China Plain. This area claims the 2,900-mile Huang (Yellow) River. It's now a major population center, and in its northern area lies China's capital, Beijing (with a massive population of 13 million), and Tianjin, its port (with another 11 million people). The entire region surrounding these cities is now a major industrial area.

In 221 B.C., the Ch'in Dynasty began a short, 15-year rule of China. The dynasty is famous for two momentous things: It gave China its name and consolidated and extended existing walls into a 1,200- to 1,500-mile-long complex now called the Great Wall of China. This wall provided central China with at least psychological protection from the Mongolian hordes to the north. Although the wall didn't prove to be militarily effective, it did produce an economic drain and sped the takeover by the Han Dynasty.

The Hans, who were militarists, set about conquering all the lands surrounding central China. They also ended feudalism and started the Silk Road westward to other empires. They so shaped Chinese life, in fact, that the Chinese now refer to themselves as the "people of Han." The Hans ruled for four centuries and left a strong mark on Chinese history.

> ## ! Geographically Speaking
>
> Midway through the Chou Dynasty, in about 500 B.C., China's greatest philosopher, Confucius, began to work as an itinerant teacher and soon became known for his learning and character. He deplored the disorder of Chou feudalism and urged a return to ancient concepts of morality.
>
> Confucius preached that rulers must set the example. At age 50, he became a magistrate and got a chance to demonstrate his credo. He was so successful that a fearful neighboring ruler had him ousted. His principles of morality were passed on only verbally by his disciples. His teachings were practical and ethical but not religious in nature. Now, $2^1/_2$ millennia later, they still lead the way.

China's Furnace: The Southern Section

South of the Northern Plain are mountains, plateaus, hills, and many basins. Through these basins flows the 3,400-mile Chiang Jiang (Yangtze) River. This fertile and heavily populated area is China's major rice- and tea-growing region. The Chiang Jiang finally drains into the East China Sea near Shanghai, China's largest city. Southern China is known for its hot temperatures and stifling humidity, from which it gets its "furnace" label.

Shanghai is also China's largest port and industrial center, although the recent addition of Hong Kong presents Shanghai with a formidable rival. Although the city itself has 8 million inhabitants, greater Shanghai has possibly 20 million. It's interesting that this great city began as a small fishing village some 5,000 years ago. In the rolling hills farther south are China's famed limestone pinnacles that rise to as much as 600 feet high—it's no wonder that they're so popular with Chinese painters and poets.

Farther south, the area becomes tropical in rainfall, temperature, and forestation, with its Xi (Pearl) River dominating. The area includes the metropolis of Guangzhou (Canton), with a population of five million people. It's also home to China's modern-day economic miracle, the Guandong Province, whose industrial and trade center, Shenzhen, adjoins Hong Kong. This area harvests two, and sometimes even three, rice crops a year. It also has China's largest island, Hainan, and abuts the major Asian peninsula of Indochina.

> **GeoJargon**
> DICTIONARY
>
> On a tributary of the Xi River is the favorite tourist destination of Guilin. It has limestone pinnacles that are part of a geologic formation known as *kharst* topography, formed when flowing water dissolves existing limestone and forms underground caverns and sinkholes. As caverns collapse and more sinkholes are formed, the higher original limestone rock that remains forms the spectacular pinnacles.

The Top Half: Northeastern and Northwestern China

Northeastern China is half of historic Manchuria (which used to include Mongolia). This extensive area, which has a large central plain surrounded on the east and west by mountain chains, is China's "Siberia"; it's cold and has a short growing season. The great Manchu Dynasty was rooted in this area.

Northwestern China has three large desert basins, each surrounded by great mountain chains. The entire area is sparsely settled; other than the Silk Road that passed through this area, not much else has passed.

Terra-Trivia

During the Yuan Dynasty, the Venetian adventurer Marco Polo visited China (or Cathay, as the Europeans called it). He made two visits, spent 26 years in Asia, crossed China six times (including once across the Gobi Desert), and was an administrative employee of Kublai Kahn for much of the time. Marco was even the governor of a Chinese city for three years!

Following the Han Dynasty, from 206 B.C. to A.D. 220, were three centuries of turmoil. The Sui Dynasty started reconsolidation, including the construction of the 1,000-mile Grand Canal linking Beijing with the south; the T'ang (A.D. 618–A.D. 907) and Song dynasties continued the cultural honing for three centuries apiece. The Great Wall turned to be not so great—it didn't keep out the Mongol conquest in the mid-thirteenth century. They set up the Yuan Dynasty (A.D. 1280–1368), which lasted fewer than 100 years before the Mings threw them out and restored greatness for another three centuries.

In the mid-1600s, the Manchus came out of Manchuria into China to establish the Manchu (or Qing) Dynasty. Unlike the Mongols, the Manchus were absorbed and became part of China. They were quite militaristic and conquered all the neighboring lands the Hans had missed. They also ruled for three centuries (China apparently runs on a three-century cycle) and proved to be China's last dynasty.

China's More Recent History

After years of isolation, the nineteenth century witnessed the opening of China to foreigners. The experience was not pleasant for China: It had two Opium Wars, the Boxer Rebellion, and the creation of foreign "spheres" directly on China's territory. China also lost possessions to foreign powers: Japan took the Ryukyu Islands and, later, Taiwan; France took Vietnam; Britain took Burma (now Myanmar); Russia took Manchuria and Northern Mongolia; and North Korea gained independence. By 1912, the Manchu Dynasty was gone, and the Republic of China began.

The emperors and their times left behind many awesome wonders. Many thousands of examples of the art are exhibited at the museum in Taiwan. China's rich culture extends to its poetry, music, dance, opera, and drama. No tourist lucky enough to have visited China can ever forget the Great Wall, the glories of the Forbidden City, or the beauty and grace of the Summer Palace. The wonders continue with the recent unearthing of more than 6,000 life-size terra cotta warriors, horses, and carriages at Xi'an, still only partially excavated.

The Republic of China

In 1912, the Chinese revolutionists had won. But which ones? Internal conflicts, first among themselves and then with the Communists and then with the Japanese, plagued the new republic for more than a decade. Not until 1928 did Chiang Kai-shek and the Nationalists join many of the warlords in an attempt to unify China. In the mid-1930s, after losing several battles with the Nationalists, Mao Zedong and what was left of his Communist army began the 6,000-mile Long March back to north China to regroup.

In 1937, Japan attacked the divided, strife-torn China and within one year had most of northeastern China under its cruel control. The Nationalists and Communists joined forces to fight the Japanese. In the mid-1940s, when World War II was over, the civil war resumed. In 1949, Mao was on a roll and officially announced the new People's Republic of China. Chiang was on the run, and the Nationalists withdrew to Taiwan.

Terra-Trivia

The emperor's fifteenth century palace, the Forbidden City (forbidden to the common people), sits by Beijing's Tiananmen Square. It sits on 250 acres and has 9,000 rooms, 35-foot-high walls, a 171-foot-wide moat, a ceremonial outer palace with five entry bridges and three halls, and a residential inner palace where hundreds of imperial concubines and their thousands of attendants lived. Now it's a museum.

The People's Republic of China: The First Half-Century

The new China then had to define itself. At first, it tried outside aggression—invading and seizing Tibet. China then took actions against the Nationalists in Taiwan, which it still considers its province and vows to take back.

By the 1980s, China witnessed a rise in capitalism and democratic thinking, only to be crushed by armored People's Liberation Army troops in the 1989 Tiananmen Square massacre (from 3,000 to 5,000 killed, 10,000 injured, and hundreds arrested).

Geographically Speaking

The People's Liberation Army, the world's largest standing army, is 3 million strong, including 2.3 million members in the army, 240,000 in the navy, and 470,000 in the air force. The People's Liberation Army is backed by a national militia of 12 million members and a security force of 2 million.

The Chinese navy has more than 1,700 vessels, including almost 100 submarines, at least one with nuclear arms. Although China has many nuclear weapons, it has relatively few compared to the United States and Russia.

China Today

It has been said that if you could give your child only one piece of advice, it should be to learn Chinese. China has the potential to outdo anything it has done in its remarkable past. Remember that because China is so big, whatever it does, it does in a big way!

On the Plus Side

China is extremely well endowed with energy resources, claiming almost 11 trillion metric tons of reserves. Even if its claim is exaggerated—wow! With recent discoveries of offshore oil, China now claims (accurately, perhaps) to be second only to Saudi Arabia in known oil reserves.

The country also has abundant mineral reserves—its tungsten and antimony are known to lead the world, followed by large supplies of aluminum and tin. The list continues: China has iron ore reserves of 40 billion metric tons: In the 1990s, the entire world's usage was only 920 million tons per year.

About a dozen Chinese dialects are spoken; Cantonese and Mandarin are the primary two. For 3,000 years, China has had only one written language, which not only eases communication but also contributes immeasurably to a common culture. Both internally and through the United Nations, China is pushing the use of Pinyin, a system of phonetic spelling, for personal and place-names (Beijing for Peking and Mao Zedong for Mao Tse-tung, for example).

Although the Cultural Revolution against elitism in education proved to be shortsighted and counterproductive, it did provoke debate and place new emphasis on education. It closed most schools for three years, and colleges for five. After the schools opened again, however, education took off: In 20 years, primary education increased 500 percent, and secondary education increased by an astounding 4,200 percent. By the late 1980s, almost two million students were in more than 1,000 colleges.

With a continuing emphasis on exporting, factory jobs and national gross domestic product have also taken off. In 1992, the per capita income was only $370; by 1996, it was $2,446.

Terra-Trivia
The world has only about a thousand giant pandas, mostly in the wild, living in the cool forests of Sichuan. They grow from a few inches at birth to about five feet and 350 pounds. Pandas have a second "thumb" to help them eat their main diet, bamboo. It's unfortunate that they're difficult to breed, because their numbers are small and they're hands-down zoo favorites.

On the Minus Side

Although China has concentrated on stressing late marriages, one-child families, and all kinds of birth control, its population increase rate is 1.5 percent. Multiplied by the country's population of 1.2 billion, the result is an increase of 18 million people per year—plenty of hands to do the work. Don't forget the flip side, though: That number means plenty of mouths to feed, bodies to house, and children to educate.

China cannot ever again become isolated. It must import the world's materials, goods, technology, and investments to create wealth and jobs for its citizens. It must not only avoid world disfavor, therefore, but also court it. For this reason, China may not always be able to have it its own way. China's recent relations with the United States have been dicey. China is under international scrutiny for its Tiananmen Square massacre, claims to Taiwan, nuclear sales to Pakistan, intellectual property rights issues, and most-favored-nation status.

Territorial disputes with Japan have worsened an already bad situation. Hong Kong, reclaimed by China in 1997, must be managed carefully. The world is watching to see whether—and how—China follows up on its promise to take back Taiwan. In addition, because China's territorial claims in the oil-rich South China Sea are disputed by several of its neighbors, a future feud may be looming.

China's economy has three main sectors: the private, the collective, and the state. All have major problems. The state sector is unprofitable (to say the least), hindered by flaws in the Communist system that have already torn apart the Soviet Union. The state sector is gigantic, producing almost 50 percent of China's gross domestic product, but using 70 percent of the workforce and two-thirds of capital investment. Although some state-owned factories have been closed, this "cure" only aggravates the problems in the other sectors.

Unemployment is officially only 3 percent, not including the 14 million people who have already been sent home (at reduced wages) from closed or downsized state factories. The official figure also doesn't include members of the "floating" migrant population who have left the private rural sector in search of either work or a better life.

Feeding its people has always been difficult for China—and it's getting worse. Despite improved fertilizers, more efficient equipment and methods, and more crops per year, available arable land has been shrinking because of advances in roads, railways, factories, and housing. The better quality of life

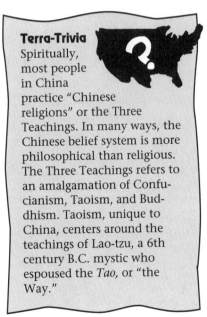

Terra-Trivia
An old Chinese proverb says, "Never waken a sleeping giant, unless you're a bigger giant." Is China the sleeping giant or the bigger giant? As the country emerges on the international scene, either answer will have profound implications for the rest of the world.

Terra-Trivia
Spiritually, most people in China practice "Chinese religions" or the Three Teachings. In many ways, the Chinese belief system is more philosophical than religious. The Three Teachings refers to an amalgamation of Confucianism, Taoism, and Buddhism. Taoism, unique to China, centers around the teachings of Lao-tzu, a 6th century B.C. mystic who espoused the *Tao,* or "the Way."

demands more than rice (wheat, meat, and cattle feed), and 18 million more people each year must be fed—which is China's biggest challenge.

In a recent interview, Chinese president Jiang Zemin cited these same problems (in less-scary terms) as his country's biggest challenges—and also expressed confidence in the Chinese people's ability to solve them.

The Least You Need to Know

➤ East Asia includes Mongolia, North Korea, and China.

➤ Mongolia is a democracy—one of the world's poorest.

➤ Staunchly Communist North Korea lags significantly behind South Korea economically, and tension between them continues.

➤ Although Tibet has a distinct people, language, and religion and wants to regain independence, China won't give up its control.

➤ With 1.2 billion people (one-fifth the world's population), China is the world's third-largest country in area.

➤ China is a Communist state with many assets but is struggling to feed its people and improve their standard of living.

The Vast Pacific

The Pacific is a region of islands and ocean. It has no continental landmass, and most of its islands are quite small. What dominates this region is not the sporadic scattering of land but rather the incredible expanse of water.

The planet earth is often called the "blue planet"; 71 percent of its surface is water, mostly in the oceans. Early mariners used to refer to the world's oceans as the Seven Seas, which included the North and South Atlantic, the North and South Pacific, the Indian, the Arctic, and the Antarctic oceans.

Because the oceans all flow into each other, most geographers have abandoned the notion of a separate Antarctic Ocean because it had no definite boundaries. Now the oceans border Antarctica. If you ignore the obscure north and south divisions of the Atlantic and Pacific, you see just four oceans. This chapter briefly examines the world's oceans and then looks more closely at the largest—the Pacific.

The Pacific.

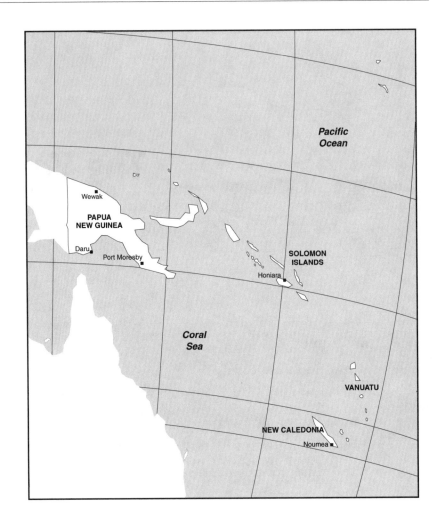

The World's Oceans

Although the earth's surface is mostly water, it's not evenly distributed. Sixty-one percent of the Northern Hemisphere is composed of water; 81 percent of the Southern Hemisphere is water. Ocean water has an average depth of 12,450 feet, $4^1/2$ times the average height of dry land above sea level. The depth of each ocean varies: The Pacific is the deepest ocean, at 12,925 feet, and the Indian and the Atlantic are just behind. The Arctic is quite shallow—it's only about one-fourth as deep as the Pacific.

> ## ! Geographically Speaking
>
> Eighty-three percent of the ocean floor is between 10,000 feet and 20,000 feet deep. Only 1.3 percent is deeper, in areas called, naturally, *deeps*. Extremely deep deeps are called *trenches* (they're narrow), *troughs* (wide), or *basins* (broad). A *bank* is a shallow area that's safe for navigation; a *shoal* is a shallow area with no rocks or coral that's unsafe for navigation; and a *reef* is a shallow area with rocks or coral that's unsafe for navigation.
>
> A *ridge*, or *rise*, is an elongated elevation or mountain chain; a *seamount* is an isolated elevation (a mountain whose summit is well below sea level), and a *plateau* is the upper part of an extensive flat-topped elevation.

Another indicator of ocean depth is the number of and depth of trenches. The Pacific leads again with 16 trenches more than 20,000 feet deep; the Atlantic has five, the Indian four, and the Arctic none, although its Eurasian Basin measures almost 18,000 feet. The Pacific's trenches are also much deeper than any others. The Pacific has three deeper than 30,000 feet: the Philippine Trench, just shy of 33,000 feet; the Tonga Trench, about 35,400 feet; and the deepest spot on earth, the Mariana Trench's Challenger Deep, at 35,840 feet. The Atlantic's deepest trench, the Puerto Rican Trench, measures about 28,200 feet.

The surface area of the oceans also varies, and more widely in the Pacific. The Pacific Ocean, the largest, measures more than 64 million square miles and is almost as large as the other three oceans combined. The Atlantic measures 33 million squares miles; the Indian, 28 million; and the Arctic, only 5 million.

The Not-So-Peaceful Pacific

Another way to get a grasp on the Pacific's size is to think of it in world terms. It covers one-third of the earth's surface and holds more than half its free water. (Remember that ocean boundaries are not always definite and are a matter of consensus.) Most geographers agree that the Pacific is bounded on the north by the Bering Sea and the Aleutian Islands. On the east, it's bounded by the Americas. Antarctica forms the southern limit of the Pacific; the ocean's western boundary is truly a matter of consensus (or lack of it).

Although much of the region called Southeast Asia in this book is physically within the Pacific region, this book considers it part of Southeast Asia for cultural reasons. For the same reasons, New Zealand could be in the Pacific region but is described in the same chapter as Australia. From north to south, the Pacific measures 9,600 miles; it measures 11,000 miles from west to east.

The dominant color of the Pacific on a map or in real life is blue. In the area mainly south of the Tropic of Cancer are hundreds of "dots"—the Pacific islands, perhaps as many as 30,000 of them. Yet all of them, even including New Guinea, make up only about two-thirds of 1 percent of the Pacific realm.

Terra-Trivia
The surface temperature of the Pacific Ocean varies from saltwater's freezing point (29.5 degrees F) in the polar regions to an average 79 degrees F in tropical waters. The temperature of the top 300 feet is near the seasonal air temperature. Between 300 and 3,000 feet down, the water temperature falls to 40 degrees F. Still deeper, the temperature decreases to 36 degrees F.

In 1520, the Pacific (meaning "peaceful") got its name from Ferdinand Magellan, who had just completed his rough and difficult passage through the strait he had discovered and that now bears his name (the Strait of Magellan, at the tip of South America). In contrast, the Pacific was indeed peaceful, which is how it got its name. If you look at the ocean floor, however, you can see that the Pacific isn't really all that peaceful.

From Antarctica, running north for about 6,000 miles, is the East Pacific Rise, which thrusts up about 7,000 feet. Along this rise, molten rock swells up from the earth's mantel and adds crust to the plates on both sides. The plates are forced apart, which causes them to collide with the continental plates on all sides, as described in Chapter 2. This process causes tremendous pressure because the continental plates are forced upward as mountains and the ocean plates are forced down to form deep trenches.

The Pacific Islands

Terra-Trivia
Coral reefs and islands form either in shallow water or on top of non-active volcanic sea-mounts. The typical Pacific coral island is an *atoll,* a large ring of coral above or below sea level, surrounding a calm lagoon. The coral above the surface supports tropical growth and palm trees. From the air, an atoll looks exactly like the rim and crater of a volcano, which in fact it grows on.

Pressures along the Pacific's perimeter result in earthquakes and volcanic eruptions that have created the Ring of Fire, the gigantic rim surrounding the Pacific. Hot spots in the earth's mantle can cause molten rock to break through its surface. When this rock is beneath the ocean, a seamount (volcanic island) is born, which is how thousands of the "dots" have appeared all over the Pacific, particularly in the southwest. Native inhabitants call them the "high" islands.

Thousands more of the Pacific dots are the "low," or coral, islands. Coral islands and reefs have been described as the earth's most complex ecosystems. They occur only in warm, sunlit water and result from thousands of live marine animals clustering in vast colonies. The skeletal deposits that remain after these creatures die eventually layer themselves and can reach a height of 300 or 400 feet.

Currents: Air and Sea

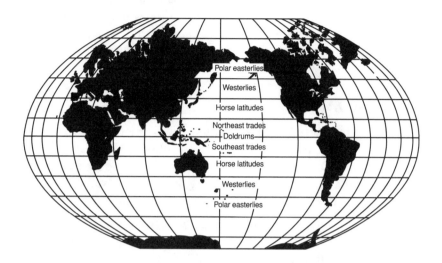

Pacific wind patterns.

To best understand the Pacific Ocean's wind patterns, start at the equator and work north and south from there (don't forget to look at the preceding map). Between 10 degrees north and 10 degrees south are the *doldrums,* a low-pressure area of light breezes. At 30 degrees north and 30 degrees south are the *horse latitudes,* belts of high pressure. The winds blowing from these high-pressure areas toward the low-pressure doldrums are deflected by the earth's rotation and become the tropical easterlies known as *trade winds.* Between 30 degrees north and 60 degrees north and between 30 degrees south and 60 degrees south, winds blow from the horse latitudes toward the poles and again are deflected by the earth's rotation to become the *westerlies.* The polar regions are high-pressure areas; the deflected winds blowing from them become the *polar easterlies.*

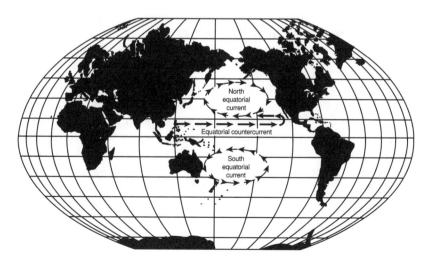

Pacific currents.

Although ocean currents in the Pacific are relatively complex, this chapter simplifies them with the usual broad brush. North of the equator, the ocean currents form a huge oval that moves clockwise. A similar oval current forms south of the equator, in which the current moves counterclockwise. Between the two ovals and running along the equator is a west-to-east movement of water, called the *equatorial countercurrent*.

> ## Geographically Speaking
>
> In the 1930s, while in Tahiti, Thor Heyerdahl noted many similarities between Polynesian and pre-Incan Peruvian art and culture. He and some associates used only native materials to build, according to old Peruvian designs, a 40-foot × 18-foot balsa raft. He wanted to prove that it was possible for pre-Incan Peruvians to sail west to settle Tahiti and its South Pacific neighbors.
>
> In 1947, Heyerdahl and five Norwegians sailed their boat, *Kon Tiki*, from Callao, Peru, to start their oceanic adventure. The ocean currents and southeast trade winds joined to carry them west. After sailing 4,300 miles, they arrived 101 days later at the Tuamotu Islands in French Polynesia, east of Tahiti. Heyerdahl proved that it could be done. But was it?

The Europeans Find the Pacific

Not until 1513 did Spanish eyes first glimpse the largest body of water in the world. The Spanish explorer Vasco Balboa crossed the Isthmus of Panama, stood on the shore, and claimed the ocean beyond and all the lands in it as the property of Spain. For almost a century, the Pacific Ocean became the "Spanish lake."

In 1520, Ferdinand Magellan, a Portuguese sailing for Spain, named the Pacific, discovered the Marianas and the Philippines, and then died there in a native skirmish. In the sixteenth century, the Spanish reached the Caroline, Marshall, Galápagos, Santa Cruz, and Marquesas islands.

The days of the Spanish lake were over, and the rest of Europe began to dive in. In the seventeenth and eighteenth centuries, Vanuatu was sighted by the Portuguese, later visited by the French, and again visited by the British explorer Captain James Cook. He named them New Hebrides, after the similarly rugged Scottish Hebrides Islands. The Dutch sighted the Tonga group, as did Cook. The Dutch also sighted the Fiji Islands, and again Cook later found these. The French discovered the Solomon Islands and Tahiti, claiming the latter for France. One year later, Captain Cook came to Tahiti.

History's Greatest Explorer

You might have noticed that Captain James Cook shows up in several places. This fantastic navigator, who sailed around the world twice, made three voyages to the Pacific Ocean and discovered Hawaii, eastern Australia, the Cook Islands, New Caledonia, Niue, and the Antarctic ice cap.

Cook laid British claim to both islands of New Zealand and charted the coastlines of eastern Australia, New Hebrides, the Marquesas Islands, Easter Island, some of Hawaii, and all 2,400 miles of New Zealand. Seeking what is now the Northwest Passage between the Atlantic and the Pacific, he charted North America's west coast from Oregon to the Bering Strait, where ice blocked additional exploration.

In addition to undertaking these prodigious efforts, Cook won fame for the diet and hygiene practices he implemented aboard ship to avoid *scurvy,* the dread disease resulting from vitamin deficiency. He was honored in England with the rank of commander, made a fellow in the Royal Society, and awarded the Copley Medal for scientific achievement. After all this, he was killed in Hawaii in a native skirmish over a small boat.

GeoJargon

Scurvy resulted from spending months at sea without vegetables and fruit, sources of vitamin C. Scurvy, which was usually fatal, caused weakness, sore gums, lost teeth, swollen joints, and anemia. As a deterrent, in 1795 lime juice was issued to all British naval vessels. The nickname "limeys" for British sailors is derived from this dietary supplement.

From Pearl Harbor to Nagasaki

No description of the Pacific Ocean would be complete without some mention of World War II. To consolidate its gains in Asia, Japan had to seize the Philippines, the East Indies, and all the Pacific islands that were near enough to threaten an attack on its homeland. Although the Japanese knew that the United States would not stand idly by and let this expansion happen, the United States could do nothing in the Pacific without its fleet, stationed at Pearl Harbor in Hawaii. On December 7, 1941, Japan made its infamous surprise attack on Pearl Harbor and its quick follow-up invasions of the Philippines, Singapore, the East Indies, and numerous Pacific islands.

The United States countered with surprising and devastating naval victories in the Coral Sea and off Midway. With Japan's navy seriously weakened, the push toward Japan on land was then supplemented with an island-hopping strategy through the Pacific. As the United States army moved forward on land, its navy, marines, air force, and army set to work on the islands: first, Guadalcanal, and then the recapture of Aleutian Attu, Bougainville, Tarawa, Kwajalein, Eniwetok, Saipan, Tinian, and Guam.

Gains in the Pacific ultimately brought the United States close enough for its B-29 Superfortresses to bomb Japan. The United States also began its historic take-back of the

Philippines, where Japan first effectively used kamikaze suicide planes. Finally, the bloody taking of Japan itself began with Iwo Jima and Okinawa.

With the costly lessons from the initial thrusts into Japan fresh in his mind, President Harry Truman made his decision to stave off an even more costly invasion of the Japanese heartland. He ordered the dropping of the world's first two atomic bombs on Hiroshima and Nagasaki. Five days later, on August 14, 1945, Japan surrendered.

Geographically Speaking

The first test explosion of a nuclear device occurred July 16, 1945, at Alamogordo, New Mexico. The next two were the bombs dropped on Hiroshima and Nagasaki. These 20-kiloton atomic bombs were equal to 20,000 tons of TNT. Since that time, much of nuclear history has occurred in the Pacific.

Because of its remote location and small population, Oceania gets the dubious distinction of the world's nuclear ground zero. The United States chose testing sites on the Marshall Islands' Bikini Atoll in 1946 and on Eniwetak from 1948 to 1958, and the French chose a site in Polynesia. A-bombs later became hydrogen bombs; the last United States test was a 15-megaton H-bomb equal to 15 million tons of TNT.

Countries in an Endless Ocean

New Guinea and Oceania combined have almost 12 million people and encompass 30,000 islands which can be divided into three major groupings—Melanesia, Micronesia, and Polynesia, along with and a few scattered outliers. The groupings are made up of 13 independent countries; one Indonesian province; one United States state; five other United States affiliations; one British, four French, one Chilean, and three New Zealand territories; and a part of Ecuador.

Melanesia

Melanesia has 80 percent of the region's population (9.3 million people) and 93 percent of its land area. The reason that it's so comparably large is that it includes the large, heavily populated island of New Guinea.

Papua New Guinea

The country of Papua New Guinea occupies the eastern half of the world's second-largest island, New Guinea. This turkey-shaped island's western half is the Irian Jaya province of

Indonesia; the eastern half forms the largest part of Papua New Guinea along with some 600 smaller islands. The country includes the Admiralty Islands, the Bismarck Archipelago, and Bougainville. Papua New Guinea has about 4.5 million people. Port Moresby is its capital; its almost 200,000 residents make it one of the largest cities in the region.

Papua New Guinea has a hot and humid tropical climate, dense forests, and some of the most isolated people on earth. Tribal people fish, hunt, and farm for subsistence wherever possible. In coastal areas, some plantations grow cocoa, coffee, and copra for export. Rainfall varies with topography, and some places receive more than 230 inches per year.

Before World War II, Papua New Guinea was controlled by Australia and Germany, then Japan during the war, and Australia afterward. Papua New Guinea did not become fully independent until 1975. Its more than 700 Papuan and Melanesian tribes speak more than 700 languages. Only 32 percent of the population is literate, with only a 54-year life expectancy and a low $730 per capita income. With its abundant minerals, Papua New Guinea hopes that its future will be bright.

Geographically Speaking

Running through the middle of the Pacific Ocean, at 180 degrees longitude, is the international date line. The line is always perplexing to travelers who get consumed in the intricacies of the construct of time. Here's the short version: When you cross the line while traveling west, you add a day (6 p.m. Monday becomes 6 p.m. Tuesday). When you cross the line while traveling east, you subtract a day (6 p.m. Tuesday becomes 6 p.m. Monday).

Fiji

A country with a British past, Fiji is now independent. It has 300 islands (only 100 are inhabited) and 782,000 people. Half the people are Fijian, and half are Indian (who were brought to work the plantations). This split has been a source of friction in recent years. Its capital is Suva, with a quarter of Fiji's population. Almost all the people speak one language, Fijian.

The Solomon Islands

The Solomon Islands have seen their share of colonizing powers, including the Spanish, German, British, Japanese, and Australian. As of 1978, the islands and their 413,000 people are independent. The country is composed of ten major volcanic, deeply forested islands (including Guadalcanal) and 30 small isles and countless atolls. Despite the country's small size, its language is far from universal, with many speaking pidgin (English-Melanesian) and an assortment of 80 local languages.

Vanuatu

Formerly New Hebrides, Vanuatu was jointly administered by the British and French for 74 years and is now independent. About 180,000 people inhabit Vanuatu's 70 islands. Although most of these islands are volcanic, some are also coral.

Palau

Formerly the Carolines, Palau was formerly run by Spain, Germany, Japan, and the United States and is now independent. More than 200 islands (half in Melanesia and half in Micronesia) make up the chain. Two-thirds of Palau's 17,000 people live in the capital, Koror. Island ruins indicate early Chinese visits.

New Caledonia

New Caledonia is not an independent country but is still a French colony. The population of 170,000 people is made up of two majority groups that are 43 percent Melanesian and 37 percent European (mostly French). Melanesians boycotted the last election, and residents voted to stay colonial; the 1998 election could decide New Caledonia's independence. New Caledonia is the world's second-largest producer of nickel.

Micronesia

Micronesia comes from a Greek word that means "small islands." With more than 2,000 islands, it has only 1,000 square miles and 300,000 people. Although tiny coral atolls are the norm, the larger high islands contain the bulk of the population.

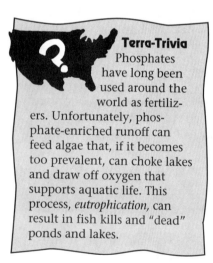

Terra-Trivia

Phosphates have long been used around the world as fertilizers. Unfortunately, phosphate-enriched runoff can feed algae that, if it becomes too prevalent, can choke lakes and draw off oxygen that supports aquatic life. This process, *eutrophication*, can result in fish kills and "dead" ponds and lakes.

Nauru

With more than 10,000 people packed on one 8-square-mile island, Nauru has a population density of more than 1,250 people per square mile. Nauru was, at different times, British, German, Australian, Japanese, and part of the U.N. Trust before it finally became independent in 1968. The country boasts one of the world's higher per-capita incomes because of its abundant phosphates that will run out by the year 2000.

Guam

Guam is large by Micronesian standards, with 209 square miles and 150,000 people. It's not a country, but rather an unincorporated territory of the United States. One-third of Guam's land is owned by the United States military.

The Federated States of Micronesia

The Federated States of Micronesia encompasses more than 600 atolls, many uninhabited. The independent country boasts 125,000 citizens and tropical conditions with heavy year-round rainfall. Economically, it depends primarily on United States aid.

The Marshall Islands

The Marshall Islands are two parallel chains of atolls and reefs with about 60,000 inhabitants. The 34 islands include Kwajalein (the world's largest atoll, with 90 islets surrounding a 650-square-mile lagoon), Bikini, Eniwetak (of nuclear testing fame), and the capital, Majuro, with one-third of the country's total population.   US

The Northern Mariana Islands

The Northern Mariana Islands are a United States Commonwealth (similar to Puerto Rico) composed of 16 coral and volcanic islands. Its major islands are Saipan, Tinian, and Rota. Only 20,000 people live there.

Polynesia

Polynesia stems from a Greek word that means "many islands." This wide array of many thousands of coral and volcanic islands is by far the most farflung island grouping. About two million people in diverse political units inhabit the chain called Polynesia.

Tonga

Tonga is made up of 170 coral and volcanic islands in a 289-mile-long north–south chain. A monarchy for 1,000 years and a British protectorate for 70 years, it's now an independent monarchy again.

Kiribati

Little-known Kiribati features about 80,000 people living on one coral island and 33 atolls. Its capital, Tarawa, is remembered as the site of a horrific battle between the United States Marines and the Japanese during World War II.

Tuvalu

Tiny Tuvalu has 10,000 people packed on 10 square miles of land on nine atolls in a 360-mile chain. "Global warming," which could cause rising seas, is a nasty phrase in Tuvalu because the country's highest elevation is only 15 feet.

New Zealand Territories

New Zealand has three territories: Tokelau, with three coral isles, 4 square miles and 2,000 people; Niue, composed of one coral island, 100 square miles, and 3,000 people; and the mostly volcanic (but some coral) Cook Islands, which measures 91 square miles and has 18,000 people.

Terra-Trivia

Paul Gauguin was a French stockbroker who was married with five children. At age 35, he left his family to paint full-time in Paris. At 43, he left France and went to Tahiti. His paintings of the lush tropical islands and their people eventually made him one of the art world's most noted impressionists. At 55, he died and was buried in the Marquesas Islands.

French Territories

The French territories include the islands of Wallis and Iles de Horne (Futuna), two island groups 125 miles apart, with 106 square miles and 20,000 inhabitants. A second French possession, French Polynesia, includes the Society, Tuamotu, Gambier, Austral, and Marquesas islands. This widely spaced collection of islands (regarded as some of the most beautiful islands on earth) is the home of about 210,000 people. The best-known islands are Tahiti (the site of the territory's capital, Papeete) and the islands of Moorea and Bora Bora. Tourism is extremely big business.

Western Samoa

The country of Western Samoa incorporates two large volcanic islands (Savai'i and Upolu) and seven smaller islands. Western Samoa, the western half of a 300-mile chain, has a population of more than 200,000 people. Its economy is boosted by New Zealand and Australian aid. Native legend claims that the island of Savai'i was the original home of Polynesians.

American Samoa

American Samoa is a United States territory made up of seven islands. Located in the eastern half of the chain containing Western Samoa, American Samoa has 50,000 people. Tutuila, the biggest island, is the site of its capital, Pago Pago.

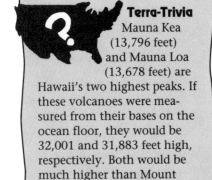

Terra-Trivia

Mauna Kea (13,796 feet) and Mauna Loa (13,678 feet) are Hawaii's two highest peaks. If these volcanoes were measured from their bases on the ocean floor, they would be 32,001 and 31,883 feet high, respectively. Both would be much higher than Mount Everest (29,028 feet) and therefore the two highest mountains in the world.

Hawaii

In 1959, the islands of Hawaii became the 50th U.S. state. Although *Hawaii* is a Polynesian word, now no one knows what it means. The arc of the Hawaiian islands is 1,490 miles long. The Hawaiian chain has eight main volcanic islands and 124 islets and reefs. Across its 10,932 square miles are more than one million diverse people (33 percent white, 22 percent Japanese, 15 percent Filipino, 13 percent Polynesian, 6 percent Chinese, 3 percent black, 2 percent Korean, and 2 percent Samoan).

Honolulu, on the island of Oahu, is the Hawaiian capital, with more than 350,000 residents. On the big island of Hawaii, Mauna Kea (13,796 feet) is the state's highest volcano. The air at the top of this mountain is so dry and clear that it's the world's best astronomical site. Mount Waialeale, on the island of Kauai, is the world's wettest place, receiving an annual 486 inches of rainfall. The island of Molokai earned renown as the home of Father Damien's famed leper colony.

Captain Cook discovered the Sandwich Islands (now Hawaii) in 1778, and Europeans began to arrive in 1790. From 1782, however, until 1810, King Kamehameha unified all of Hawaii and established his dynasty that ruled until 1872. United States interests began in 1819 with missionaries and then later with plantations for sugar cane and pineapples. Although the Hawaiians were decimated by disease, it fortunately leveled off in the late 1800s.

In 1898, Hawaii became a United States territory. Agriculture, in addition to military spending, made up most of its economy. As agriculture began to wane, the big boon of tourism accelerated. More than seven million tourists visit the islands annually, spending more than $11 billion and creating almost 250,000 jobs. The tourism industry is vital to the Hawaiian economy.

Miscellaneous Islands

A few outliers from these three main groupings of islands in this chapter include Wake Island (north of the Marshalls) and Midway (west of Hawaii), both of which are closed to the public and occupied solely by American troops. Other outliers are France's uninhabited Clipperton (west of Costa Rica) and the Galápagos Islands (west of and part of Ecuador). Easter Island is roughly halfway between Chile, its owner, and French Polynesia. Galápagos is famous for its gigantic tortoises and iguanas; Easter Island, for its hundreds of huge-headed, long-eared, long-nosed stone statues.

The People of New Guinea and the Oceania Dots

Melanesia comes from a Greek word meaning "black islands." Geographers believe that the black-skinned people of interior New Guinea mixed with the people of Papau and then spread out into the islands between the equator and 20 degrees south latitude. In Melanesia, more than 200 Malayo-Polynesian languages are now spoken, often several on the same island. Scattered throughout many of the southern Solomons and the northern Vanuatus are some of the small-statured tribes who to this day practice polygamy, infanticide, and perhaps even cannibalism.

The Micronesians descend from a historic mixture of Asian, Indonesian, Filipino, and Papuan peoples, and generally are slender with lighter skin and straight hair. Micronesians were skilled seamen who navigated their craft between the islands to supplement their diets. The high islanders farmed, and the low islanders fished.

The Polynesians probably came from Malay Peninsula via Indonesia and probably in two major waves. The first wave settled into Samoa, Tonga, and Fiji; the second, a few hundred years later, settled westward, northward, and southward (into New Zealand, whose Maori are thought to be of Polynesian descent). Polynesians have light brown skin and have long been famous for their gracefulness and beauty. The Polynesians were also renowned for their seamanship, both inter-island in small canoes and in large craft that carried them over vast distances to discover, explore, and settle new islands thousands of miles away.

> ## ! Geographically Speaking
>
> In 1789, in Tongan waters, Fletcher Christian led the famous mutiny on the *Bounty* against Captain William Bligh. The mutineers went back to Tahiti and then set sail with some adventurous Tahitians, ending up on the then-uninhabited Pitcairn Island. The descendants of the famous mutineers still live on the island.
>
> After the mutiny, Bligh was set adrift in a 23-foot longboat with 18 loyal men and few provisions. Although Bligh had no charts, he did possess extraordinary navigational skill and some knowledge of the local waters (he had been with Captain Cook on his second voyage around the world). For two months, the men lived through severe hardships and sailed the tiny craft 3,618 miles to Timor, off Java. Bligh's voyage ranks as one of the most truly remarkable feats of seamanship on the high seas.

The Least You Need to Know

➤ The size of the Pacific Ocean almost equals the Atlantic, Indian, and Arctic oceans combined. It covers one-third of the entire earth's surface.

➤ The Pacific's entire rim, the Ring of Fire, has active volcanoes and is prone to earthquakes.

➤ New Guinea is the world's second-largest island and the largest in the region.

➤ The Pacific has almost 30,000 islands (mostly atolls) that are volcanic and coral.

➤ Oceania is divided into subregions called Melanesia, Micronesia, and Polynesia.

➤ Many islands, many peoples, many languages, European explorers, World War II, independent countries, and foreign territories—all combine to tell the tale of the Pacific.

Chapter 25

The Frozen Poles

In This Chapter

➤ Identifying polar limits

➤ Racing to the poles with the early explorers

➤ Divvying up ownership

➤ Highlighting the region's similarities and differences

➤ Taking a closer look at the poles

➤ Keeping an eye on the ozone hole

At the far reaches of the earth, where ice and cold prevail and human beings are mere visitors, is the final region in this part of the book. Many geographers don't consider the poles a region because of its obvious dearth of human settlement. To help you in your quest for geographic knowledge, however, this chapter leaves no iceberg unturned, no snowdrift unprobed, and no region unexplored (no matter how dubious it may seem).

This book typically defines regions by both physical and cultural characteristics. Because the poles don't fit this mold, however, this chapter defines them by physical features.

The Poles.

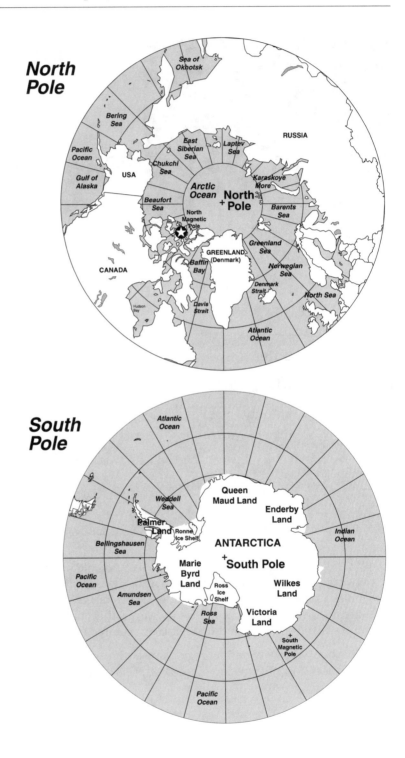

Defining the Polar Region

The process of defining the limits of the polar region is not as clear-cut as you might expect. Some definitions of this region are based on climatic conditions. Climatically, the Arctic and Antarctic regions are the areas in which the mean temperature for the warmest month never exceeds 50 degrees F. The polar region is also considered to be the area in which the warmest mean monthly temperature never rises above freezing (32 degrees F). Some geographers define the Arctic and Antarctic regions by vegetation: areas at a latitude above which trees no longer grow (the tree line).

You might think that the polar region's logical boundary lines would be the Arctic and Antarctic circles, at about 66$\frac{1}{2}$ degrees north and south latitude, respectively. As tidy as those lines would be, they don't quite work.

Although Antarctica is contained almost entirely within the Antarctic Circle, the continent peeks outside the circle in several locations. In the Eastern Hemisphere, much of Wilkes Land and Enderby Land flirt with the Antarctic Circle and nibbles of land poke out along the way. In the Western Hemisphere, Antarctica is contained entirely within the Antarctic Circle except for the northernmost portion of the Antarctic Peninsula.

The opposite situation exists in the Northern Hemisphere. In the north, the Arctic Circle is too large to consider everything inside it as region (and to avoid treading on other regions). A quick look at the map shows that Russia, Northern Europe, and North America all extend into the Arctic Circle. Because the polar region includes everything north of these other regions, the North Pole portion of the polar region contains no land.

Terra-Trivia

The magnetic poles don't coincide with the earth's true poles; rather, they represent the direction indicated by the arrow on a magnetic compass. The north magnetic pole is located off King Christian Island, in Canada's Northwest Territories. The South Magnetic Pole is off the coast of Wilkes Land, Antarctica. These locations change because the magnetic poles move around slightly from year to year.

The History of Discovery

The quest to be the first to reach the poles began in earnest only in the twentieth century. Earlier sailors had searched for a northwest passage between the Atlantic and Pacific oceans, and Captain Cook had circumnavigated the South Pole. When the race began, however, the situation got somewhat dicey. Countries issued claims and counterclaims, and scientific boards weighed the evidence. No satellite photos existed to verify explorers' stories about their conquests in the most isolated parts of the earth.

The Charge to the North

The United States explorers Robert Peary and Matthew Henson usually get credit for being the first to reach the North Pole. On April 6, 1909, the two explorers and their small party were the first humans to reach the North Pole (or close to it—maybe). After they returned home, they learned that another American explorer, Dr. Frederick Cook, claimed to have made it to the North Pole five days earlier.

After a review by scientific experts, Cook's claim was discredited based on astronomical observations that appeared incorrect or fabricated before his journey. Most experts have sided with Peary and Henson, although in recent years many have questioned whether that team ever even made it to the pole. Although that debate will undoubtedly rage on, one thing is certain: No one can check Peary's statement "Stars and Stripes nailed to the North Pole" because the pack ice long since has drifted into the Arctic night.

The Trek to the South

The quest to find the South Pole was not as controversial as the effort up north. Nonetheless, the adventure in the south was marked by tragedy. The Norwegian Roald Amundsen and his expedition became the first to reach the South Pole on December 11, 1911. A second party, also en route to the South Pole at the same time, was not as fortunate.

The second party, headed by the British adventurer Robert Scott, made it to the South Pole on January 17, 1912, about five weeks after Amundsen's party. The second party did not return alive, however. Buffeted by hellish winds and bone-numbing temperatures, the explorers ran short of supplies. Just a few miles from their base station, the entire group perished from starvation and frostbite.

Making National Claims

Answering the question of who claims the North Pole is easy: No nation claims possession because the pole is located in the heart of the Arctic Ocean. Even with the 200-mile exclusive economic zone provided by the Law of the Sea Convention, no country's jurisdiction reaches the pole. At the South Pole, however, Antarctica's situation is not so simple.

Antarctica is land, and although it has no permanent inhabitants and features no local government, it has not escaped political squabbling. Seven separate countries make claims to portions of this continent. No nation claims the part of Antarctica called Marie Byrd Land; it's the place to be if you're looking to claim a chunk of a continent.

The claims, all pie-shaped, originate at the South Pole (as shown on the map). Most of the claims have some logic behind them. Australia (two claims), New Zealand, Argentina, and Chile—the countries closest to Antarctica—all make claims representing southern extensions of their territories. Although Norway also makes a (somewhat dubious) claim, remember that a Norwegian, Amundsen, was the first person to make it to the South Pole.

Because Robert Scott, from the United Kingdom, was a close second, the British claim also has some rationale. The other claimant is France (if you don't get the logic on this one, join the club).

South Pole territorial claims.

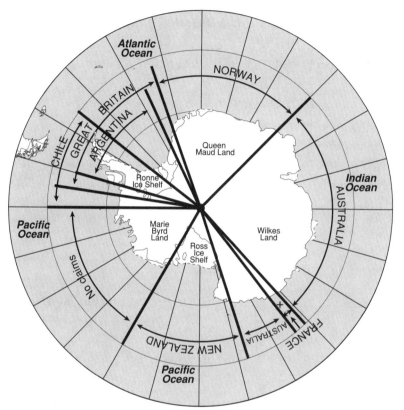

All this claim business is sort of a moot point anyway because of recent Antarctic treaties. In 1961, the signing of the Antarctic Treaty banned new territorial claims and prescribed peaceful, scientific use of the continent. A more recent agreement, signed in 1991, prohibits commercial mineral extraction on the continent. (If Antarctica had been left to politicians rather than to scientists, you just *know* that trouble would have ensued.) You don't have to look any further than the Antarctic Peninsula to see the potential for conflict: Three countries all claim the same area.

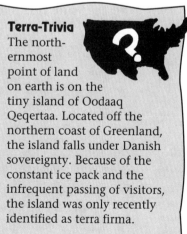

Terra-Trivia
The northernmost point of land on earth is on the tiny island of Oodaaq Qeqertaa. Located off the northern coast of Greenland, the island falls under Danish sovereignty. Because of the constant ice pack and the infrequent passing of visitors, the island was only recently identified as terra firma.

323

Looking at Polar Similarities and Differences

Even though the two areas of this region are "poles apart," they must have a few similarities or else they wouldn't be described in the same region, right? Sort of. Part of the reason the two have been joined to form one region is that they didn't fit anywhere else in this book. Nonetheless, they do have some similarities. Here's a list of the top nine similarities between the North and South poles:

➤ Both are located at 90 degrees latitude (ignore the fact that one's north latitude and one's south).

➤ During the winter, the sun never rises in either place.

➤ During the summer, the sun never sets in either place.

➤ Both have lots of ice and snow.

➤ If you stand on either pole, no matter which way you face, you can travel in only one direction (only south from the North Pole and only north from the South Pole).

➤ Seals live on both poles.

➤ Both are extremely cold.

➤ Neither pole has trees.

➤ You can forget about using your magnetic compass in either place.

Are the North and South poles similar enough to include in the same region? Maybe. Before you begin to think that the poles are almost identical, however, peruse the top ten differences between them:

➤ One is north, and the other is south. The poles are 180 degrees apart, on opposite sides of the globe—as far apart as you can get on earth.

➤ No polar bears live on Antarctica.

➤ The South Pole is on land; the North Pole is in the middle of an ocean.

➤ Santa lives only at the North Pole (apparently he's a seafaring sort of elf).

➤ Although they're both cold, the South Pole is considerably colder than the North Pole.

➤ The South Pole is almost entirely covered by glacial ice; the North Pole has nary a glacier.

➤ The South Pole has penguins; the North Pole doesn't.

➤ Antarctica has a mountain more than 16,000 feet high and a depression more than 8,000 feet deep. The North Pole is pretty much at sea level.

➤ No national claims to the North Pole have been made, although many countries have claimed pieces of Antarctica.

➤ Several "permanent" scientific stations exist on Antarctica, though only temporary floating stations exist at the North Pole.

Taking a Closer Look at the Poles

This section looks more closely at some of the similarities and differences between the North Pole and the South Pole. You might think that some of this information borders on trivia; when an area has no people, no countries, no soil or vegetation, and little history to talk about, however, trivia starts to sound pretty good.

Which One Is Land?

As mentioned, the North Pole is in an ocean, and the South Pole is in the middle of a continent. Of all the differences between the north and south, this one is the most important (except maybe for the Santa issue). This difference affects accessibility, geology, climate, and virtually everything else about the two locations.

The North Pole is in the Arctic Ocean. The smallest of the world's four oceans, it's also the shallowest and, obviously, the world's northernmost ocean. Most of the Arctic Ocean, including the North Pole, is always covered with ice. This ice is not a solid sheet; it's called *pack ice*—chunks of ice that slowly circulate about the pole.

The South Pole is on the continent of Antarctica. Despite almost its entire surface being covered with continental glacial ice, the continent is solid land. Antarctica is the fifth-largest of the earth's seven continents. Asia, Africa, and North and South America are larger; Europe and Australia are smaller.

For centuries, geographers speculated about whether the great southern continent even existed. Captain Cook circumnavigated Antartica in 1772–75; because the continent was protected by huge drifting icebergs and shifting pack ice, however, he never was able to get close enough to sight land. Charting (mapping) didn't begin until the 1820s, and not until the 1840s was the continent finally conclusively identified.

Terra-Trivia
At one point on the Antarctica ice plateau, the ice thickness has been measured at 15,670 feet. To put this measurement in perspective, consider that the tallest building in the United States is the Sears Tower in Chicago, at 1,454 feet. The depth of Antarctic ice is, therefore, more than ten times the height of the Sears tower.

Which One Is Colder?

No contest! The South Pole wins this contest hands-down. Plateau Station on Antarctica is the coldest place on earth, with an annual average temperature of –70 degrees F. The single coldest temperature ever recorded was also on Antarctica, at Vostok Station, where the thermometer dipped to –129 degrees F. These temperatures would be hideously cold even if the wind were calm on Antarctica, but it's not. Unfortunately for the Antarctic tourism bureau, Antarctica is also one of the world's windiest places. At Australia's Commonwealth Bay Station, cold air constantly rushes down from the ice cap, and wind speeds have been recorded at more than 180 miles per hour.

Terra-Trivia
Because extremely cold air can hold little moisture, snowfall totals in the Arctic and Antarctic are moderate—less than you might expect to receive in New York City. If the poles retained the same average annual precipitation but were warmed to lower latitude temperatures, in fact, you might find cactus, sand dunes, and camel caravans crossing their expanse.

Terra-Trivia
The highest point on Antarctica is the Vinson Massif, at 16,863 feet above sea level. The lowest surface point on earth is also in Antarctica, where a point on the continent drops to 8,327 feet below sea level. This low point on Antarctica's bedrock surface was caused by the weight of thousands of feet of ice pressing down on the continent.

Gauging Arctic temperatures and winds is a little more difficult than in the Antartic because no fixed, permanent stations exist. (In contrast, the United States Pole Station on Antarctica has been continuously occupied since 1956.) Although you would think that the coldest temperatures in the north would be at the North Pole, they're really farther south along the Arctic Circle. This phenomenon occurs because the North Pole is located over water, and the cold temperatures recorded along the Arctic Circle have been recorded over land.

Just outside the region, at 76$\frac{1}{2}$ degrees north latitude, is Thule Air Base in northern Greenland. Average annual temperatures average about –11 degrees F—cold, but not approaching the –70 degrees F recorded at Antarctica. Verkhoyansk Station in northern Siberia has recorded January temperatures as low as –90 degrees F—again, it's cold, but not as bad as at Vostok Station in Antarctica.

Several reasons exist for the temperature differences between the poles. At the height of the South Pole winter, around July 1, the earth is at its *aphelion,* or farthest point in its orbit from the sun. Although that's not a huge factor in temperature, it does mean that the South Pole is about three million miles farther from the sun during its winter than the North Pole is during its winter.

The windier conditions at the South Pole make the windchill a factor to be reckoned with. To demonstrate the impact of the windchill effect: If the recorded temperature were –45 degrees F (a warm day at Plateau Station, Antarctica), winds of 45 m.p.h. would make the temperature feel like –125 degrees F. If you're considering a vacation in Antarctica, bring your long woollies.

Where Are the Polar Bears, Seals, and Penguins?

Aside from an occasional mite, terrestrial life does not inhabit the poles—all life is dependent on the seas. Under the ocean ice, algae and diatoms support small shrimplike creatures, called *krill,* that provide food for fish and larger aquatic life at the poles.

Penguins live only at the South Pole. Antarctica sports five varieties of penguins, in fact. The largest and best-known of the penguins is the emperor penguin, which can reach four feet in stature. Penguins are the only large animals that stick it out through the nasty Antarctic winter.

Both poles have seals. In the south, the best known are the fur seal and the leopard seal. Up north, ringed seals and bearded seals are the best known (walruses also live in the north). Seals feed on krill, fish, and, in the south, penguins. Seals are not at the top of the food chain, however. Killer whales and polar bears love seal meat and will go to great lengths to get a tasty bite of seal blubber.

In the far north, polar bears as well as an occasional Inuit hunter have been known to venture out on the pack ice in search of seal meat. No such indigenous land predators exist on Antarctica to plague the penguins.

Where Are the Continental Glaciers?

When you think of the Ice Age, continental glaciers usually come to mind. The last ice age occurred during the Pleistocene era, when vast ice sheets spread to cover about one-fifth of the earth's continental surface. Only two remnants of the continental glaciers remain: One covers the island of Greenland, and the other covers Antarctica.

Continental glaciers form when climatic factors combine to allow a gradual buildup of snow and ice. When the accumulation of ice builds to a critical mass, the weight of the ice mass causes it to flow outward. As the glaciers advance, they grind and scrape the landscape with dramatic results.

Scientists do not agree on what climatic conditions must exist to cause glacial advance. Although you would suspect that colder temperatures would be a prerequisite, that's not necessarily the case. To build and flow, glaciers require snowfall in heavy amounts. Because cold air holds less moisture than warmer air, the coldest temperatures are often associated with lighter snowfall totals. Following this argument to its logical conclusion, warmer temperatures might be required to produce a new ice age.

Either way, glaciers have not historically formed over the North Pole. The continental glaciers of the Northern Hemisphere formed in subpolar areas and then spread northward toward the pole and southward toward more temperate climes. The ice cap on Antarctica is not advancing because of the scant amount of snowfall the continent receives. Nonetheless, if you want to trek across a continental glacier (even a semidormant one), head for the South Pole, not the North Pole.

Terra-Trivia
Ocean ice around the perimeter of Antarctica and on the Ronne and Ross ice shelves averages about three feet thick. Sea ice thickness averages around ten feet at the North Pole, despite its warmer temperatures. This difference in sea ice thickness is caused primarily by the rough, choppy seas at the South Pole that inhibit thicker ice buildup.

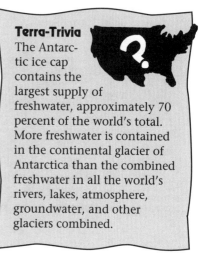

Terra-Trivia
The Antarctic ice cap contains the largest supply of freshwater, approximately 70 percent of the world's total. More freshwater is contained in the continental glacier of Antarctica than the combined freshwater in all the world's rivers, lakes, atmosphere, groundwater, and other glaciers combined.

Night and Day

If you're standing on either the North Pole or the South Pole, you're in for a boring diurnal (daylight) pattern. When the sun rises at either pole, it stays daylight for six months. When the sun finally sets, nighttime settles in for six months (as explained in Chapter 9, in the section "The Land of the North").

The cause of it all is the $23^1/_2$ degree tilt of the earth on its axis. While it's summer in the Northern Hemisphere (March 21–September 23), the North Pole remains in the sunlight and the South Pole stays in the darkness. During the Southern Hemisphere's summer (September 23–March 21), the South Pole experiences constant daylight and the North Pole languishes beneath dark winter skies.

The Ozone Hole

One of the most disturbing manifestations of the diminishing ozone layer is directly associated with this region. This layer in the earth's upper atmosphere shields people from the harmful effects of ultraviolet radiation from the sun. The ozone layer exists across the entire earth, but recently not so much over Antarctica. (Chapter 27 discusses the problems associated with the depletion of atmospheric ozone.)

Terra-Trivia
The land area of Antarctica is approximately six million square miles. The permanent population of the continent is zero, which yields a population density of zero people per square mile. Antarctica's birth rate is zero, as is its rate of natural increase. It has no official language, and the per capita income level remains a steady $0.00.

In 1985, scientists noticed that a perceptible hole appeared in the ozone layer directly over Antarctica. Since that time, observations have confirmed a 15 to 70 percent depletion in ozone over the South Pole every spring. Although ozone depletion has been observed over the North Pole also, the ozone loss there has been much less severe.

Ozone layer depletion is unfortunately a result of human activities. The release of certain chemicals (discussed in detail in Chapter 27) have broken down ozone concentrations and will continue to do so. Thanks to the circulation and concentration of cold air called the Antarctic *polar vortex,* the human impact on the ozone layer is most acute over the South Pole.

Because it's believed that excessive ultraviolet radiation can hinder plant development, Antarctica's basic food chain may be at risk. Remember that leopard seals eat penguins and penguins eat krill and krill eat algae. If algae are affected, so too are the higher vertebrates on the chain. Because algae also absorb carbon dioxide and produce oxygen, the reduction in algae may also have atmospheric effects. Each year since the ozone hole was first observed, its duration has increased—watch this story as the poles pop up in the news.

The Least You Need to Know

➤ The poles are over the earth's rotational axis at 90 degrees north and 90 degrees south latitude.

➤ Although both poles are extremely cold, the South Pole is colder.

➤ The North Pole is over the Arctic Ocean, and the South Pole is over the continent of Antarctica.

➤ The poles have no permanent human inhabitants.

➤ The South Pole is covered with a continental glacier.

➤ Each pole experiences six months of continuous daylight and six months of continuous night.

➤ A troubling ozone hole forms over the South Pole every spring.

Part 4
A Global Overview

Now that you have completed your tour of the world's developed and developing regions, it's time to stand back and look at the big picture. Although you can easily get caught up in the lines that define the boundaries of regions and countries, remember that if you take a stroll on the physical landscape, those lines don't really exist (unless someone decides to violate the good-neighbor policy and puts up a fence). And the lines of human construct certainly mean nothing to the forces of nature!

Part 4 of this book looks at the issues that don't fit neatly into a discussion of a single country or even a single region. These global issues affect us all. More importantly, it will take international cooperation to solve some of the problems we have created for ourselves. Too many people live in the world and our problems have become too large for any of us to adopt an island mentality. We all have to be informed and take responsibility for our planet—the ultimate geographic task!

Chapter 26

Coping with Population

In This Chapter

➤ Understanding population rates

➤ Pondering the prospect of people, people everywhere

➤ Finding where the people are—and where they're not

➤ Coming to grips with the food dilemma

➤ Taking a step in the right direction: demographic transition

In your whirlwind tour through this book, you have visited every region on earth. Now that you know a little about the geography of each place, it's time to stand back and look at the entire earth. What problems are faced by every region on earth? What solutions can humankind turn to?

In virtually every region explored in this book (except possibly the poles), population pressures are straining the environment. Although some regions have relatively small populations, their land is typically unable to support large numbers of people. In the fertile areas of the earth, millions of people pack into the usable land, creating massive clusters that dot the landscape.

At one time in human history, it was possible for overcrowded populations simply to move elsewhere. For centuries, crowded South Asian and East Asian populations emigrated to Southeast Asia. Crowded Europeans moved to distant lands in the New World, Africa, Australia, and New Zealand. Although emigration was once a solution to overpopulation, the earth is running out of places to put people.

As the human population continues to expand, every resource on earth is taxed. Many environmentalists consider the world's population problem to be the greatest environmental threat—because the sheer volume of the world's people lies behind virtually every other environmental problem humans face. Although the earth has shown remarkable resiliency, are humans stressing even the earth's ability to heal itself?

A Little About Demographics

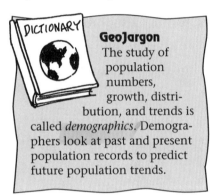

GeoJargon
The study of population numbers, growth, distribution, and trends is called *demographics*. Demographers look at past and present population records to predict future population trends.

Geographers look closely at *demographics,* the study of population, because it has so great an influence on the workings of the earth. They're particularly concerned with the distribution and movement of populations. As the twenty-first century begins, it's important to address the world's burgeoning population problem and to consider potential solutions.

Malthusian Predictions

In 1798, the famous demographer Thomas Malthus made some startling predictions. Among them, he noted that food production on earth would grow arithmetically (2, 4, 6, 8, 10, and so on), an encouraging prediction. He also predicted, however, that human *population* would grow geometrically (2, 4, 8, 16, 32, and so on). This prediction is discouraging because, according to his predictions, the ever-doubling human population will soon outstrip its ability to feed itself unless the population is checked by some powerful controls.

These controls fall into two categories: destructive and private. *Destructive* checks on human population growth include war, famine, and disease—not cheery prospects. *Private* controls include celibacy and chastity—or general reproductive restraint. Either way, without these curbs, humankind will outstrip the earth's ability to provide.

Although many believe that Malthus' predictions are coming to fruition, his theory has not been entirely correct. The ability to produce food on earth has grown at a rate that has exceeded the arithmetic rate Malthus postulated. New farming techniques and the introduction of chemical fertilizers, insecticides, herbicides, and bioengineering all have contributed to a dramatic rise in food production over the past decades. (Later, this chapter describes the downside to this technology.) Although Malthus was off the mark in his food-production rates, he was on target in his population-growth predictions.

The Doubling Dilemma

The world's population has doubled—and doubled, and doubled, and doubled. In Malthus' time (A.D. 1766-1834), the earth's human population was fewer than one billion people. Almost six billion people now live on the earth—and we're still cooking. How long does it take for the earth's population to double?

In the year A.D. 1, between 200 and 250 million people were living on the earth. This figure did not double to 500 million until the year 1650, which means that it took 1,650 years for the world's population to double. The population took only 200 years to double again, and then again in 80 years, and most recently in just 45 years.

Birthrates and Death Rates

Demographics are filled with rates: Birthrates and death rates are used to determine rates of population increase. The birthrate gauges the annual number of live births per thousand people; the death rates indicate the number of annual deaths per thousand people. To determine the natural growth rate, you simply subtract the death rate from the birthrate. The remainder is the increase (or, in rare instances, the decrease).

Population growth occurs when the birthrate exceeds the death rate. The greater the difference, the larger the growth. Population growth can occur because of an increase in the birthrate, a decrease in the death rate, or both. The only way to decrease population growth is to increase the death rate (not a people-friendly option) or decrease the birthrate.

Terra-Trivia
A quick way to determine doubling time for a population is to divide the number 70 by the growth rate. If a country's growth rate is 2, you divide 70 by two to get 35. That country's population will double in just 35 years! The calculations get a little scary when you realize that countries such as Kenya have a growth rate approaching 3.5, which means that its entire population will double in just 20 years.

How Bad Is It, Really?

You might say, "So almost six billion people live on the earth—what's the big deal? What are the consequences of having that many people, and should people be worried?" For comparison, a stack of $100 bills equaling $1 million would stand about 3½ feet tall. A stack of $100 bills equaling $1 billion would measure about 3,583 feet tall—more than twice the height of the Sears Tower in Chicago.

Some people might say that although six billion people is a large number of people, the earth is a large place and can handle that number. Although the earth *is* large, according to the Population Education Training Project, the earth doesn't have that much space, after all. Here's the math:

➤ About three-fourths of the earth's surface is water; the remaining one-fourth represents the earth's land surface.

➤ About one-half of that land surface is uninhabitable, such as the poles, rocky and high mountains, swamps, and deserts. That puts the total at about one-eighth of the earth's total surface area.

➤ About three-fourths of the remaining one-eighth section can't be used to grow food because the land is too dry, too cold, too steep, too rocky, or too infertile, or it has been built over with urban areas, factories, parking lots, roads, houses, schools, and parks. All that remains to produce food is $1/32$ of the earth's surface.

➤ The top five feet of topsoil on this $1/32$ remnant of the earth's surface represents the entire amount of the earth's food-producing land for six billion people—and the topsoil is eroding and washing out to sea at an alarming rate.

Feeling cramped? You should, because conservative estimates place the earth's population at ten billion within 50 to 75 years. Will humans colonize another planet? Maybe. Will technology save the world? Hopefully. Without that technology in hand, however, humans have reason to be extremely concerned.

The Scattered Masses: Where the People Are

Geographers look at not only the numbers of people on earth but also their distribution. People are not distributed evenly across the face of the earth. The earth's largest population clusters are in South Asia and East Asia. The population of Asia represents more than half the world's total population. Although other, smaller clusters exist around the world, vast areas of the earth are essentially uninhabitable.

Terra-Trivia

Industrialized countries shouldn't blame the world's woes on the rapidly expanding population of developing countries. Not so long ago, industrialized countries were the global leaders in population growth. In terms of an effect on the world's resources, the developed world gobbles up much more than its share. Each American's energy use equals the energy used by 422 Ethiopians.

Just as population distribution varies dramatically around the world, so too does food production. The ability to grow food can rely heavily on the length of the growing season, rainfall, soil fertility, and available technology. Droughts, floods, disease, war, and pest infestation can decimate food production in an entire region. Although international aid can do much to help these situations, it's often too little and too late.

In this century, horrible food shortages have caused starvation and suffering at previously unheard of levels. Each day, hundreds of millions of the earth's people (perhaps one-eighth of the world's population) go to bed hungry. The world's resources are being stretched, and desperate people are encroaching on marginal land. For example, steep, wooded slopes in Nepal are being terraced for marginal farmland, resulting in serious erosion and flooding. Despite the environmental consequences, rain forests are being cut as farmers attempt to scratch out a living in the nutrient-poor soil.

Although some countries, such as the United States and Canada, produce huge amounts of food for export, the food race also has its losers. Many countries in Africa, for example, already have more people than their land is able to feed. Even in a good year, these countries are dependent on imported food. Even more disturbing is that these same countries exhibit some of the world's highest growth rates.

More People, Less Food

As you will see, changing the population-growth trends in a region takes years (except in places with a state-mandated control policy, such as China). Because of the time it takes to decreases growth rates, especially in developing countries, humans soon will face increasing food shortages unless they're able to produce more food on earth. Is this possible?

As mentioned, the problem can't be solved by bringing new land under cultivation. Most of the world's arable land is already being used, and the remaining land offers little potential for any sizable agricultural yields. Although some people hope for the building of Martian colonies, bushel baskets of corn aren't likely to be shipping in from the Mars colony for many more years.

With no more land to place under the plow, people must turn to methods of increasing production on lands already being used for farming, which is already being done in many areas. This section looks at some advantages and disadvantages of this effort.

Although this section focuses on agriculture in its discussion of food resources, keep in mind that another major source is available: fish. More than 90 percent of all fish consumed on earth come from the oceans and seas. Not so long ago, most people viewed the immense oceans as an unending resource—but not anymore. Since the mid-1970s, overfishing of the world's oceans has led to a steady decline in the average catch per person. Despite the increase in *aquaculture,* or fish farming, the bounty of the seas is no longer seen as a limitless resource for the world's hungry.

GeoJargon
Aquaculture, or fish farming, occurs in shallow coastal waters or inland ponds around the world. If you drive through the flatlands of Mississippi, you see acres of catfish farms. In these artificial ponds, catfish are fed highly nutritious diets to maximize yields. When you order that spicy blackened catfish, you can bet that your catch is from a farm, not from the Mississippi River bottom.

The Green Revolution

During the past three decades, the world has experienced a phenomenon known as the *green revolution,* a term that describes many scientific advances applied to agriculture to increase crop yields. Some technologies include bioengineered hybrid high-yield seeds, organic and chemical fertilizers, chemical pesticides, chemical herbicides, mechanization, high-tech irrigation systems, and multiple-cropping.

The application of green revolution techniques initially produced significant agricultural increases. Eventually, however, problems with the technology also became evident. One problem is expense: Hybrid seeds, irrigation systems, machinery, and fertilizers and other chemicals are all expensive—and the energy they require is intensive. Although the developing countries stand to gain the most from adopting green revolution techniques, those countries have the least amount of money to spend on expensive technologies.

Another problem is that large portions of the populace in developing countries typically are involved in agriculture, primarily small subsistence family plots. Small fields do not handle the same kind of mechanization as do the great plains of North America. Green revolution techniques produce the highest yields when they're applied to large expanses of a single crop.

Geographically Speaking

One possible alternative to the widespread use of chemical spraying is *integrated pest management* (IPM). This system addresses the entire ecosystem and is sustainable over long periods while remaining environmentally friendly. In addition to its selective and limited use of chemicals, IPM relies on the use of natural pest predators, low-till agriculture (minimal soil disruption), and crop diversity to control pests.

A farmer using the IPM system accepts a small crop loss to pests in return for long-term sustainable yields. Many farmers would describe IPM as working with the ecosystem rather than against it. Although a farmer must be well educated about the subtleties of the local ecosystem, the lower costs of implementation are attractive to people in developing countries.

Disease, Rats, and Superbugs

Although hybrid seeds produce higher yields than indigenous crops, they're also more vulnerable to local blights. Large areas of a single hybrid crop are much more susceptible to drought, pests, and disease than are small plantings containing multiple varieties of indigenous crops. To protect these high-yielding but vulnerable hybrids, farmers have turned to ever-increasing applications of chemical herbicides, pesticides, and fungicides.

In addition to the pollution and contamination these chemicals cause, they have only a short-term effect on the pest problem. The extent of the problem cannot be minimized:

Each year, pests account for the destruction of an estimated half of the world's food production. To increase food production, the loss to insects, rodents, and fungi must be minimized—but are chemicals the answer?

Indiscriminate chemical spraying kills not only the targeted pest species but also the helpful natural enemies of that species. Although spraying typically produces a short-term benefit, the pests eventually return in increased numbers. After natural controls are destroyed, pests can decimate crops with abandon.

Although farmers can repeat the expensive spraying, of course, a higher dose is required the next time. Why? Because of the superbug phenomenon. Although most of the pests are killed during each spraying, some survive. The survivors have a natural immunity to the given level of toxin. These highly resistant survivors reproduce and breed a new generation of pests that are more resistant to spraying than the preceding generation. The only way to kill off these varmints is to hit them with yet a stronger dose of chemicals—and the cycle continues. The world eventually will become too toxic for human habitation—but will be just fine for the newest generation of superbugs.

A Glimpse of Hope: The Demographic Transition

This chapter has not painted a rosy picture of a world straining under the weight of a population doubling out of control and hungry people tapping marginal lands for minimal yields. You've read about food production that can't keep pace with an exploding populace and superbugs running rampant through the farmlands, laughing off all the poisons humankind can throw at them. Is there hope, or will Malthus be proved correct after all?

Yes, humans certainly have hope (I wouldn't leave you without at least a glimmer). Although the world's people might have to face difficult times for a few generations, if humankind can just survive the short term, a phenomenon called the *demographic transition* could prove to be the answer for the long term, as shown in the following graph. To understand the demographic transition, you have to look at its three stages. The concept of demographic transition was developed around the population of industrialized Europe that has passed through each stage.

GeoJargon
To depict age and sex characteristics of a population, geographers use a graphic device called a *population pyramid,* as shown in the following figure. In the pyramid, males and females are divided into age divisions that represent an age range. If the percentages remain fairly constant through the age cohorts, the population is stable. A pyramid with a wide base indicates lots of young children and an expanding population.

A typical population pyramid.

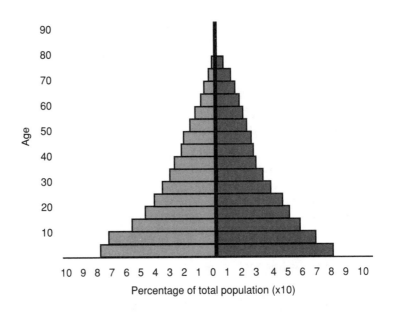

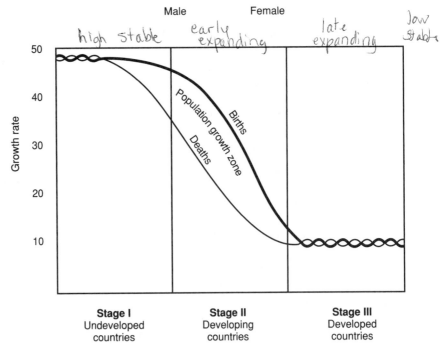

The demographic transition.

Stage I

Several hundred years ago, in preindustrial Europe, the population was relatively stable. The average woman in that society began having children early in life. If childbirth didn't kill her, she was likely to give birth to at least eight to ten children.

Why? One reason is that birth-control options were, at best, limited. Also, preindustrial Europe relied heavily on an agricultural-based economy, as do developing countries today. In this type of society, children are an economic asset: They don't eat much, and they can work in the fields and provide labor and income for their family. The more, the merrier.

The typical eighteenth-century peasant farmer did not have a chunk of his paycheck deducted each week for Social Security. When he reached his golden years (his 30s), he depended on his family to sustain him if he could no longer work the fields. If he had no family, his prospects for a happy retirement were dim. The conclusion: The bigger the family, the better.

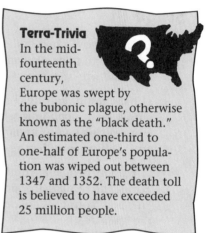

Terra-Trivia
In the mid-fourteenth century, Europe was swept by the bubonic plague, otherwise known as the "black death." An estimated one-third to one-half of Europe's population was wiped out between 1347 and 1352. The death toll is believed to have exceeded 25 million people.

Although you might think that the high birthrates of preindustrial Europe would give rise to high population growth rates, it wasn't necessarily so. Don't forget the death rates. Infant mortality was also extremely high. Perhaps half the children who were born never reached adulthood. People also died young by today's standards. Anyone who made it to age 40 was doing quite well. Poor hygiene, disease, famine, and war all took their toll. Death rates were high. With both high death rates and high birthrates, population growth was slow and fluctuated with war or famine.

Stage II

As the industrial revolution created new jobs and opportunities in Europe, the region entered demographic Stage II. New crops (such as potatoes and maize) were imported from the New World and proved to be a boon to agriculture. The situation was looking up as medical advances, improved nutrition, and hygiene increasingly enabled people to live longer lives. Although the death rate plunged, the birthrate did not. (Death rates can decrease dramatically within just a couple generations, although birthrates take much longer.)

Terra-Trivia
Between 1820 and 1940, Europe sent more than 32 million people to the United States. More than 6 million Germans and more than 4 million English, Irish, and Italians made the Atlantic passage. Virtually every country in Europe was represented, and now people of European ancestry represent the largest portion of the United States population.

The Europeans still had no Social Security and no birth control. Children still helped on the farm, and child labor entered the factories. People still wanted to have large families, and infant mortality had dropped. In a family with ten children, eight might have made it to adulthood. The result of high birthrates, lower infant mortality, and lower death rates were huge population growth rates. Europe's population exploded.

Europeans were lucky. Although the population explosion had made the region somewhat crowded, people could always take off for the New World, Africa, and Australia. Europe weathered its population explosion through emigration. Regions lagging 100 or 200 years behind Europe in the demographic cycle would not be as lucky.

Stage III

Europe is now a highly developed region. Few people farm the land, and most live in cities. Death rates are even lower now, with people living into their 70s. Social Security is available, and most people live in apartments, not on farms.

The net result: Raising children is expensive, and Europeans have stopped having large families. The economic tables have turned, and a large family is no longer an economic windfall; it's an economic liability, in fact. Although birth-control methods are effective and readily available, the choices are difficult: The BMW or another baby?

In Europe, people are opting for the BMW. Families are small: One or two children is the norm. The result is low death rates and low birthrates—and stable populations. Some European countries are even experiencing declining rates—which hasn't happened since the black death.

The Engine

What drives the entire process of the demographic transition is economics. The more economically advanced a society becomes, the lower its growth rate. Stable, and even declining growth rates, are possible under the proper economic conditions. Can this model be applied to today's developing countries? Perhaps.

Given time, education, and the proper economic conditions, the pattern followed by Europe's demographic transition certainly can be repeated in developing countries. No population on earth is still considered to be in Stage I. Basic hygiene and medical advances have reached just about every corner of the globe, and the extremely high death rates that were considered normal 300 years ago no longer exist. The last of the Stage I populations are now in Stage II.

The problem lies in the jump to Stage III. In many developing countries, the population is expanding so fast that economic growth simply can't keep pace. Until the shift is made from an agrarian rural economy to an industrial urban-based economy, the demographic transition will lag.

Geographers hope that education and improved birth-control techniques will accelerate the move to a Stage III population; without the economic stimulus, however, the jury is still out. However the population dilemma resolves itself, it's not just the problem of a handful of countries—the problem threatens the entire earth.

The Least You Need to Know

➤ Demographics is the study of population numbers, growth, distribution, and trends.

➤ In the late eighteenth century, Thomas Malthus predicted that population growth would outstrip food production.

➤ Almost six billion people live on the earth; within 50 to 75 years, it will likely have ten billion people.

➤ Although the Green Revolution has increased crop yields, it's expensive and can prove vulnerable.

➤ Demographic transition describes the lower growth rates that come with improved economic conditions in an urbanized society.

Chapter 27

Healing the Atmosphere

In This Chapter

➤ Sorting through the layers of the atmosphere

➤ Keeping an eye on global warming

➤ Dealing with the sorry state of the upper atmosphere

➤ Realizing that the acid rain in Spain might very well stain

➤ Reeling from Chernobyl

Surrounding the earth is a thin layer of gases that's essential to all life. Humans unfortunately have not always treated that layer, the *atmosphere,* kindly. This chapter looks at some problems the atmosphere is experiencing and what we can do about them.

Each region on earth has its own environmental problems, which have been described elsewhere in this book. When you're discussing the environment, it's easy to lapse into a lifeboat mentality (save yourself and disregard the plight of others). "It's a shame that the rhinoceros is becoming extinct in Africa—they really ought to do something." "You hate to see that deforestation in Nepal; I'm glad that we don't have the problem here." And so it goes with many environmental issues in the news. What about the problems you can't ignore because they affect you directly?

The atmosphere is one of those universalizing agents. Many atmospheric problems that trouble Eastern Europe will also eventually trouble North America, for example. What's bad for South America might also prove to be bad for South Asia. The air on which all humans depend is constantly mixing as it swirls around the earth. In the process, human problems are also distributed.

For that reason, this chapter looks at the atmosphere outside the regional context. Some problems are certainly more acute in a particular region; all in all, however, humans are in this one together. Likewise, responses to atmospheric problems cannot be addressed with a regional approach. Despite human differences, if people hope to solve the array of atmospheric ills they face, they must learn to work together.

What Does "the Atmosphere" Mean?

You can subdivide the layer generally called the atmosphere into several smaller zones, as shown in the following figure. If you put a "sphere" on the end of each one, things begin to look official. Closest to the earth is the *troposphere*. "Tropos" is a Greek word meaning "turn," which is exactly what happens in this layer, with its storms, temperature swings, winds, and general turbulence the earth's surface experiences. This zone encompasses the entire earth, although it's deeper at the equator and a little shallower at the poles. As you move higher in the troposphere, the temperature decreases.

Beyond the troposphere are several other atmospheric zones that are of less importance to geographers. About seven miles up is a transitional edge called the *tropopause*. Above the tropopause is a zone of fairly constant temperature, the *stratosphere*. This zone continues through the atmosphere to about 30 miles, where it turns into the *stratopause*.

The earth's atmospheric zones.

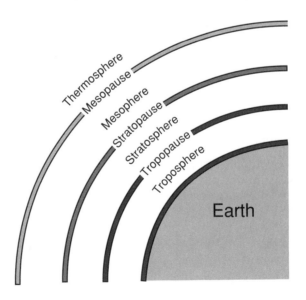

The stratopause is the dividing edge between the stratosphere and the *mesosphere*. Within the mesosphere, the temperature again decreases with height until you reach another transitional edge, the *mesopause*. Beyond the mesopause is the *thermosphere*, where temperatures remain constant for a while and then begin to increase with height.

Geographically Speaking

More than 35 miles above the earth's surface begins a layer known as the *ionosphere*, composed of electrically charged particles, or ions. The ionosphere permeates parts of the mesosphere and continues outward through the thermosphere.

The ionosphere is well known to ham radio operators as the layer that reflects radio waves and allows long-distance radio communication to take place. The ionosphere is also responsible for the famous aurora borealis, or northern lights, that grace the higher latitudes.

Turning Up the Heat: Global Warming

Global warming is almost constantly in the news now. This theory is hotly contested, and the intricacies of the process are not fully understood. The theory generally states that human activities have begun to produce warmer temperatures on earth. Unless the effects of these warmer temperatures are addressed, they could be catastrophic to human life. This section looks at the causes and some potential consequences of global warming.

The Greenhouse Effect

The engine behind global warming is a process known as the *greenhouse effect*. To understand it, think about the way a greenhouse works, as shown in the following figure. Light from the sun passes through the glass of a greenhouse and is absorbed by the earth and plants within. The ground surface reradiates the energy as infrared radiation, or heat energy.

Although the short-waved sunlight initially passed through the greenhouse glass, the reradiated long-wave infrared heat energy is not able to pass back out efficiently. This heat energy becomes trapped and causes the greenhouse to warm. (The same process causes your car to get unbearably hot while sitting in the mall parking lot on a summer day.) The earth's atmosphere acts in much the same manner as a greenhouse.

The atmosphere contains gases that enable sunlight to pass through to the earth's surface but hinder the escape of the reradiated heat energy. This process is helpful because planets and moons without much atmosphere are much too cold to sustain life. The problem arises when too much of the heat-trapping gases exist in the atmosphere. Then temperatures begin to rise, and the troubles begin.

The greenhouse effect.

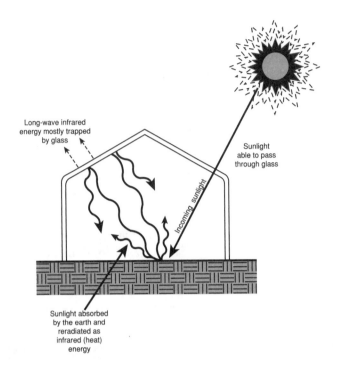

Long-wave infrared energy mostly trapped by glass

Sunlight able to pass through glass

Incoming sunlight

Sunlight absorbed by the earth and reradiated as infrared (heat) energy

What Causes the Greenhouse Effect?

Terra-Trivia
Other greenhouse gases include methane from landfills, livestock, and rice paddies; nitrous oxides from industrial processes; and chlorofluorocarbons (discussed in detail later in this chapter). In addition, several natural sources produce greenhouse gases, the most important of which is water vapor produced by evaporation and transpiration. (Water vapor accounts for as much as 85 percent of the greenhouse effect.)

Naturally occurring gases in the atmosphere account for the bulk of the greenhouse effect. Human activity has added to some of these gases, however, and is entirely responsible for others. Although the primary greenhouse gas is carbon dioxide, methane, nitrous oxides, and chlorofluorocarbons also contribute to the problem. The burning of fossil fuels, agriculture, mining, and industrial chemicals are all human-induced sources of these greenhouse gases. Because carbon dioxide is the gas that has caused more than 50 percent of the human-induced warming, this section focuses on that aspect of the complex equation.

All life is based on carbon. (Remember that favorite sci-fi phrase—earth creatures are "carbon-based life-forms"?) The coal, oil, and natural gas that humans value so highly once comprised primeval forests. Buried under ages of sediment, heat and pressure transformed the vegetable matter into the fossil fuels people use today. When fossil fuels are burned, they release into the atmosphere the carbon that was the original building block of the ancient forests.

Carbon is released directly when people simply breathe or when a living forest is burned. Plant life does much to moderate carbon dioxide buildup in the atmosphere, through the process of photosynthesis. Plants absorb carbon dioxide to build their carbon-based structure and produce oxygen. This process is good for humans because they do just the opposite—take in oxygen and exhale carbon dioxide. By burning fossil fuels, by cutting and burning the forests, and by just breathing, humans generate greenhouse gases.

Consequences of Global Warming

"So what if the temperature rises a few degrees; I hate shoveling snow anyway!" Why are people making a big deal about a slight rise in temperature? This question is a tricky one to answer because it's based to some degree on scientific conjecture. Although no one really knows what might happen, some disturbing possibilities exist. Here are the two most probable:

> **Terra-Trivia**
> Carbon dioxide in the atmosphere increases by .5 percent every year. Atmospheric carbon dioxide levels are now about 25 percent higher than they were in preindustrial times. By the year 2050, estimates are that the levels will rise to more than 50 percent higher than 1850 levels.

➤ **Rising sea levels:** Warmer temperatures might cause the massive ice caps (especially in Antarctica) to melt. The huge addition of water that's held in the ice caps would cause the level of the earth's oceans to rise. Ocean temperatures would warm, causing thermal expansion of the water molecules and additional rising. Some of the world's largest population clusters exist along coastal areas, and rising sea levels would obliterate these areas. If you live on the Maldives Islands or in the Netherlands, for example, rising sea levels are especially bad news.

➤ **Changing climates:** Most people could survive if their hometown became a little hotter or drier. Unfortunately, agriculture around your town might be altered dramatically. Because temperature and precipitation would not be uniformly affected by global warming, this situation is difficult to predict (supercomputers are modeling at this moment). If the Great Plains of the United States became too hot and dry to grow grain, for example, residents would be in trouble. Although other areas farther north in Canada and Siberia might benefit from the climatic shift, their soils are not necessarily conducive to agriculture. Considering the earth's burgeoning population, climatic threats to the food supply are worth worrying about.

Is Global Warming Fact or Fiction?

Most scientists agree that the earth is experiencing a warming and that greenhouse gases might be responsible. Scientists have reached no consensus, however, and they certainly haven't agreed about how the entire scenario will play out. Some scientists assert that the greenhouse gases will cause not a global temperature rise but rather an *icebox effect*, in which global temperatures will drop.

This theory speculates that heightened evaporation will cause an increase in cloud cover. Clouds will act to screen and reflect incoming sunlight and ultimately cause surface temperatures to decrease. In addition, global climatic systems are extremely complex and entail both positive- and negative-feedback systems. These systems could moderate climatic change or exacerbate it—no one knows.

While we're in global-warming limbo, is there anything we should be doing? Yes. Whether global warming occurs or not, humans should be reducing their production of greenhouse gases and conserving forests. These practices are sound, no matter what the outcome of supercomputer global-warming models. The conservation of fossil fuels will ensure an energy supply well into the future. Saving forests not only cleans the air but also helps to preserve biodiversity on the planet. Who knows? We might also just avoid the catastrophic consequences of global warming.

Ozone Depletion, or Pass the Sunblock!

Before looking closely at the atmospheric problem of ozone depletion, you should be able to distinguish between the ozone layer and ozone smog. In the lower layer of the earth's atmosphere (the troposphere), ozone is a pollutant. A by-product of auto exhaust and sunlight, ozone builds to produce ozone smog that often menaces cities during the sunny and warm summer months. Although this form of ozone is a health threat, don't confuse it with the ozone in the upper atmosphere.

The ozone in the stratosphere is essential to life on this planet because it protects humans from the harmful effects of the sun's ultraviolet radiation. "Ozone depletion" refers to the degradation of this protective shield in the stratosphere. Unfortunately, humankind has devised several chemicals that, when released to the atmosphere, attack and degrade the protective ozone shield. The most notorious chemical family, responsible for the bulk of the world's ozone destruction, is the chlorofluorocarbons, or CFCs.

CFCs are found in foam packaging, aerosol spray-can propellants, refrigerator and air conditioner coolants, and insulation and are used as an industrial cleaner. When CFCs are released, they break down the molecules of ozone in the stratosphere at an alarming rate. The reduction over the South Pole each spring is so severe, in fact, that it's called the "ozone hole" (see Chapter 25 for more details).

What are the consequences of ozone depletion? Without an ozone layer or with a reduced ozone layer, much more of the sun's ultraviolet radiation reaches the earth's surface. Ultraviolet radiation can be extremely harmful to life on earth in a number of ways. An excess of this radiation can damage plant life, which produces oxygen for humans. This radiation is also worrisome in its effect on the minute phytoplankton that abounds in the earth's oceans. Phytoplankton, which is at the base of the ocean's food chain, is a tremendous consumer of carbon dioxide and producer of oxygen. No one wants radiation messing with the world's life-giving phytoplankton.

The direct effect on humans has also become increasingly apparent in recent years. Increases in skin cancer (including the most severe, *melanoma*), eye cataracts, and damage to the human immune system are just a few of the consequences. Many of the world's industrialized nations have agreed to reduce and ultimately phase out the production of CFCs. Unfortunately, these long-lived chemicals continue to deplete ozone long after their initial release to the atmosphere. Even if all CFC production were to halt today, ozone destruction would continue for perhaps another 100 years.

> **Terra-Trivia**
> *Melanoma* is a type of skin cancer that's dramatically on the rise as a result of increased exposure to ultraviolet radiation. Melanoma first appears as a mole on the surface of the skin; if the cancer is not diagnosed early, it can spread quickly to other parts of the body. Each year, 32,000 people in the United States are diagnosed with melanoma.

That Burning Sensation: Acid Rain

Although people often refer to acid rain as "death from the sky," it's sometimes in the form of acid snow or acid fog. In the past several decades, acid rain has become a persistent and global problem. Its causes are believed to originate from industrial sulfur dioxide emissions and from nitrogen dioxide from automobile and truck exhausts. Placed into the atmosphere and mixed with water vapor, sulfur dioxide forms sulfuric acid, and nitrogen dioxide forms nitric acid.

The acidity that falls from the sky can result in a host of environmental problems. Ancient art and architecture exposed to the elements have suffered severely. The temples of the Athenian Acropolis and the statues of ancient Venice have been literally eaten away by the high concentrations of acid in the atmosphere.

When acidity levels rise in lakes and rivers, fish and aquatic life can no longer survive. The results are biologically dead lakes that are now prevalent in the northeastern United States and in eastern Canada. Acid rain, perhaps combined with other factors, is more apparently responsible for forest die-offs that are occurring worldwide. The eastern United States, and also Western and Northern Europe, have been hit especially hard.

> **Terra-Trivia**
> Acidity is measured on a *p*H scale that ranges from 0 to 14. The 7 on the scale is considered neutral; lower values indicate increased acidity. Although rainwater is normally about *p*H 5.6 (slightly acidic), acid rain has been measured as low as *p*H 1.5. In these instances, acidity levels exceed those of lemon juice, which is strong enough to etch a copper plate.

351

The solution, of course, is to reduce the emission of sulfur and nitrogen dioxides. This reduction is not easily accomplished, however. Because weather systems often move the problem far from its original source, identifying a specific polluter is difficult. It's also difficult to legislate mandates that would require one country to invest billions of dollars in a cleanup when the impact of its pollution occurs in its neighbor's backyard. In the meantime, Canadians are none too happy about their country's thousands of dead lakes that have resulted largely from pollution originating in the United States.

Nuclear Fallout

Nuclear energy is not generally regarded as an atmospheric problem. This section mentions it because of one specific event that made everyone aware of just how potentially dangerous the problem can be. That nuclear event now lives in infamy: Chernobyl.

On April 26, 1986, the number-four reactor at the Chernobyl nuclear power plant blew up. The roof of the containment building was blown off, and a deadly nuclear cloud rose into the atmosphere. The Soviets evacuated everyone within a 40-mile-diameter circle later known as the "zone of estrangement." The evacuation ultimately displaced about 116,000 residents of the area.

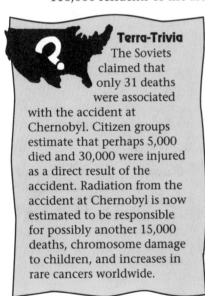

Terra-Trivia
The Soviets claimed that only 31 deaths were associated with the accident at Chernobyl. Citizen groups estimate that perhaps 5,000 died and 30,000 were injured as a direct result of the accident. Radiation from the accident at Chernobyl is now estimated to be responsible for possibly another 15,000 deaths, chromosome damage to children, and increases in rare cancers worldwide.

Although the accident occurred in the present-day country of the Ukraine, the bulk of the radioactive fallout fell on nearby Belarus. As it circulated in the atmosphere, the radiation did not just stop at national borders. Much of Western and Northern Europe received large doses of radiation from the accident. Even in the far Scandinavian north, the Lapp's reindeer milk was contaminated by radiation.

The number of people exposed to Chernobyl radiation in the Northern Hemisphere may have exceeded three billion. No longer can anyone think of nuclear accidents as a localized phenomenon. Everyone shares the same atmosphere, and you don't have to live next to ground zero to be affected.

By the way, you might not have heard the last of Chernobyl. Approximately 180 tons of uranium fuel and another 10 tons of radioactive dust remain sealed within a 24-story concrete-and-steel enclosure erected around the ruined reactor. The core of the reactor is still active, and the enclosure is believed to be structurally unsound. To make matters worse, two of the four original Chernobyl reactors continue to operate, and 13 other reactors of the same design are still on-line in Russia, Lithuania, and the Ukraine.

The Least You Need to Know

➤ Although what happens to the atmosphere is not a regional phenomenon, it affects the entire earth and its inhabitants.

➤ The most important part of the atmosphere to geographers is the lowest layer in which humans live, the troposphere.

➤ Global warming, resulting from the greenhouse effect, threatens to cause global climatic change and rising sea levels.

➤ Ozone depletion is caused primarily by chemicals called CFCs, and loss of the ozone shield poses a health threat to all life.

➤ Acid rain is caused by industrial and vehicular emissions; dying forests and lakes are the result.

➤ Chernobyl has opened people's eyes to the potential global consequences of a nuclear accident.

Chapter 28

Working to Save the Earth

In This Chapter

➤ Sailing an ocean of trouble

➤ Planting a tree and saving the earth

➤ Finding the energy

➤ Changing your diet to help the environment

➤ Understanding whether the earth is an organism: A look at the Gaia hypothesis

➤ Looking back at your journey through this book

During the past several decades, people have become more concerned about the environment. They have generally become better educated about the problems they face and more active in their efforts to help save the earth. In the process, people have begun to get a glimmer of the complex and amazing web of life that supports life on this planet.

The culmination of physical and human geographic studies is an understanding of how the two interrelate. How are human beings affected by the cycles of the physical world? How is the physical world, the environment, affected by human activities?

As these questions have been asked, the answer all too often has been that the human impact on the environment has had negative results for other life on this planet—as well as for our own.

This chapter looks at some of the problems humans face. Although Chapter 27 considers the atmosphere all humans share, this chapter focuses on the terrestrial realm. Despite the environmental problems that threaten humans' future, hope exists. This chapter describes some of the ways that people, drawing on the remarkable human spirit, are solving problems and how their efforts are holding out the prospect of a brighter future.

Taxing the World's Oceans

Chapter 26 describes how overfishing has affected the world's oceans. Unfortunately, that is not all humans have done to the earth's most precious resource. All life on earth sprang from the oceans. Because oceans cover 70 percent of the entire earth, it's difficult to imagine that something so huge could ever be threatened. But they are.

The tiny creeks, the streams, the rivers—all comprise a huge highway to the sea. When you fertilize your lawn, the rain washes the chemicals to the storm drains. The chemicals reach the streams and then reach the rivers, which eventually flow to the sea. And so it is with sewage, industrial waste, and leaching landfills. Sometimes no one bothers with the indirect path: They just take it out to sea and dump it directly.

The old adage "The solution to pollution is dilution" summed up the conventional wisdom regarding the seas. People used to think that the oceans could never be mucked up because they were simply too big. That time has passed, however. The oceans are mucked up already and, unless humans act soon, will quickly get worse.

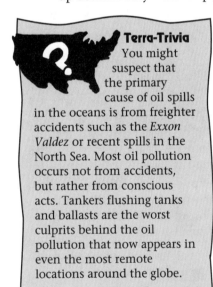

Terra-Trivia
You might suspect that the primary cause of oil spills in the oceans is from freighter accidents such as the *Exxon Valdez* or recent spills in the North Sea. Most oil pollution occurs not from accidents, but rather from conscious acts. Tankers flushing tanks and ballasts are the worst culprits behind the oil pollution that now appears in even the most remote locations around the globe.

Many countries and municipalities simply cart their wastes out to sea and dump them. Although the type of waste varies, it might include treated or untreated sewage, industrial waste, garbage, radioactive waste, construction materials, and even medical waste. Although most people want to forget about this refuse as it collects in Davy Jones' locker, the stuff has been coming back to haunt us.

As bad as these wastes are, they can seem almost benign next to the hideous materials humans spew into the oceans, such as toxic chemicals, heavy metals, oil, and PCBs (polychlorinated biphenyls, which have been linked to birth defects, cancers, and immune system damage).

Although solutions are possible, they require a will and a wallet. Only ten years ago, Boston Harbor was the cesspool of the east coast. Sewage from Boston and its surrounding communities was dumped directly into the harbor with little or no treatment. Political clout was mustered and a budget hammered out. After investing billions of dollars and years of labor, workers constructed a huge sewage-treatment facility on Deer Island at the mouth of the harbor. It's now possible again to find porpoises swimming in the clear blue waters of Boston Harbor.

The Dwindling Forests: Global Concerns

Deforestation has received increasing press attention in the past several decades. Although much of the attention has focused on the world's tropical rain forests, temperate forests have also been under the ax. Current estimates suggest that an area of forest the size of Pennsylvania is cut each year around the globe.

In the United States, the greatest concern has been voiced about the ongoing cutting of North America's last stands of virgin forest. Located primarily in the western states, these forests represent a unique wildlife habitat and spectacular beauty. Having never been cut, these forests are unique and by definition are not renewable resources.

The world's largest remaining concentrations of tropical rain forests exist primarily in three countries: Brazil, the Democratic Republic of Congo, and Indonesia. These forests are also being rapidly exploited. What are the causes of this rapid tree cutting, what are its consequences, and what can be done to save the vanishing forests?

Terra-Trivia
In the Asian mountain country of Nepal, forests are being cut at an alarming rate. Extreme poverty has forced people to clear steep hillsides in search of marginal farmland and firewood. The result has been a devastating loss of topsoil. Without tree cover to control runoff, flash flooding has increased, and Bangladesh, downstream from Nepal's rivers, has been subjected to yet another source of flooding.

The Causes of Deforestation

In different parts of the world, trees are cut for different reasons. As with most environmental ills, the reasons all stem from increasing pressures placed on the environment by the burgeoning human population. Here are four of the many causes:

➤ **Lumber and paper:** This industry is a major cause of forest cutting in developed countries. Unfortunately, the demand for lumber and paper by developed countries outstrips their own resources, and multinational companies have turned to the forests in developing countries to meet the demand.

➤ **Farming:** Rapidly growing populations are cutting and burning forests in marginal lands in a desperate search for subsistence plots. The demand for farmland has pushed people into tropical rain forests with poor soils and steep, sloped land where, without the forest cover, erosion washes away soils and denudes the land.

➤ **Cattle ranching:** This industry often follows on the heels of failed farming efforts. Poor rain forest soils that are too nutrient-poor to support farming are often scooped up by large ranching concerns to raise meat for export. Forests around the periphery of range lands are often cut to extend areas suitable for livestock grazing. These practices are especially prevalent in Middle and South America.

➤ **Fuel:** In the poorest countries, forest lands are being cut for firewood to heat homes and cook food. Because the impoverished people of these developing countries are typically unable to afford more expensive fuel alternatives, they're forced to denude the area forests for even the most basic essentials. This problem occurs in Haiti and is especially troublesome in Africa and Nepal.

The Consequences of Deforestation

Although other chapters have mentioned some of the consequences of deforestation, they don't even make a dent in the long list. The impact of deforestation is far-reaching:

Terra-Trivia
Tropical forest cutting and burning in the Brazilian state of Rondonia has been so severe that the smoke and heat have caused commercial air flights over the area to be rerouted. The Brazilian government's policy of settlement has been responsible for sending thousands of settlers into the forest to clear the land. Settlement has been enabled by international groups, such as the World Bank, which financed highway construction into the forests.

➤ **Atmospheric carbon dioxide buildup:** Trees remove carbon dioxide from the atmosphere and add oxygen. When you remove trees, you also remove a major atmospheric cleanser. The burning of trees releases carbon dioxide into the atmosphere, adding to the greenhouse gases that are responsible for global warming.

➤ **Topsoil loss:** Trees hold precious topsoil in place. Without forests, the land is subject to erosion and soil loss. The quick release of runoff rainwater is responsible for flooding.

➤ **Climate change:** Absorption and transpiration (taking in rainwater and giving off water vapor) of moisture by the forests is a critical cog in the hydrographic (water) cycle. The cutting of trees can result in altered patterns of precipitation.

➤ **Species decline:** Forests provide the habitat for an unrivaled diversity of plant and animal species. Cutting even small parcels of forest can result in species extinction. A loss of species results in diminished biological diversity on earth and a loss of potential sources for future foods and medicines.

The aesthetic component of a loss of trees cannot be minimized. For most people, a world without forests would be a much poorer place.

What Can Be Done?

As with many global problems, it's easy to feel that, because you're just one person, you probably can't do much to help. This belief is far from the truth. The way you choose to live your life can not only affect the environment but also influence people around you. Here are a few simple options that can affect the problem of deforestation:

➤ Support rain forest industries. Many small operations have begun to export products harvested from tropical forests, including nuts, fruits, natural rubber, and other products that help to make the forests economically viable without resorting to cutting.

➤ Avoid purchasing "rain forest meat" products that were raised on former forest lands.

➤ Reduce, recycle, and reuse wood products, especially newspaper—its sheer volume is staggering.

➤ Support groups that work to preserve forest lands and that litigate for forest conservation.

➤ Consider alternative building materials. For construction or furniture, avoid using tropical hardwoods or woods (such as redwood) from an old-growth forest.

➤ Plant a tree.

The Energy Dilemma

People need energy to run factories, drive cars, heat buildings, and operate just about every gizmo in the house. As society has become more industrialized, human energy needs have grown. Energy demands in the developed world far exceed those in the developing countries. Americans are the worst: Although the population of the United States comprises less than 5 percent of the world's total population, it consumes more than 25 percent of the world's annual energy production.

Nonrenewable Resources: Here Today, Gone Tomorrow

Most of the world's energy needs come from *nonrenewable* sources. Although these sources of energy are principally fossil fuels, nuclear energy is also nonrenewable. In addition to having only limited reserves of these fuels, these sources also present threats to the atmosphere, as described in Chapter 27.

> **GeoJargon**
> *Nonrenewable* resources take so long to form that they're essentially irreplaceable. They include fossil fuels, metal ores, and uranium used for generating nuclear power. Although wood is generally considered a renewable resource, if cleared forests are not replanted, wood can also be considered a nonrenewable resource.

Although the burning of fossil fuels releases greenhouse gases, the extraction of fossil fuels also takes a terrestrial toll. You've read about the possible consequences of a nuclear accident—how about the sticky problem of disposing of spent fuel and low-level radioactive waste? Considering the potential environmental costs and the limited amounts of the world's nonrenewable resources, what are the alternatives?

GeoJargon

The terms *old growth forest* and *virgin forest* are often used interchangeably. Although they generally mean the same thing, virgin forest refers more specifically to a forest that has never been cut. An old-growth forest generally includes extremely old forests, even those that might have been cut a long time ago.

Renewable Resources: Here Today, Here Again Tomorrow

Renewable resources can recoup or continuously regenerate within a person's lifetime. Trees stands harvested and replanted can regenerate within a lifetime (not virgin forests, obviously—"virgin" refers to forests that have never been cut). Other renewable resources include flowing water, wind, and sunlight. If any of these resources ceases to be renewable, humans are in big trouble. This section looks at how some of these renewable resources are being used to meet the world's energy needs.

Hydroelectric Power—Dam It!

Hydroelectric power, which uses the force of falling water to turn turbines and generate electricity, now generates about 25 percent of the world's electricity. Although this process would seem to be a safe, clean, and long-lasting solution to the world's energy problems, it presents its own problems.

First, potential sites for hydroelectric dams are limited and many have already been used. Second, huge reservoirs that build up behind hydroelectric dams can displace people and cause unforeseen ecological damage, as was the case in the Hydro-Quebec projects in northern Canada, which displaced the Cree people and contaminated one of their primary food sources with mercury poisoning.

Rivers typically flood during their annual cycle, cleansing the land within their floodplain and revitalizing it via nutrient-rich silt. Damming a river prevents the annual flooding, which has long-term consequences for floodplain fertility. Silt that no longer reaches farmlands downriver collects behind the dam, often clogging turbines and affecting future power generation. These problems have been evident on such large-scale projects as the Aswan High Dam on the Nile River in Egypt.

The Solar Solution

Every day the earth is bombarded with energy from the sun, so why not use it? The field of solar energy makes new technological advances every year that are making it possible to do exactly that. Energy from the sun is clean, unlimited, and affordable. At least it's clean and unlimited—"affordable" is still in the works.

Energy from the sun is used in two principal ways: to provide electrical energy and to provide heat energy. To convert solar energy directly into electricity, *photovoltaic cells* are used. Even if you have not seen these cells at work on a roof or power station, you have undoubtedly seen them on your pocket calculator. Just give the cells a little light, and you're ready to start balancing your checkbook.

As new techniques for manufacturing these photovoltaic cells are developed, the cost might soon be competitive with traditional power sources. These systems are becoming sophisticated enough to work well with even ambient light and will therefore be adaptable to most places on earth (except maybe at the North and South poles).

A second application for solar energy is as a source for heating. Many traditional building technologies have for centuries used passive solar-heating techniques, such as orienting buildings to take maximum advantage of the sun's warming rays. This list shows some of the traditional techniques that have been used in both temperate and cold climates:

➤ Heavy-mass walls that can absorb energy in the daytime and radiate it back into the living space during the cooler nights.

➤ The proper orientation of buildings (primary openings on the sunny side) to take maximum advantage of solar gain.

➤ Roof overhangs that admit sunlight in the winter months while the sun is lower on the horizon can then provide shade during the summer when the sun is higher in the sky.

➤ Deciduous plantings enable tree leaves to block sunlight in the summer months and let sunlight enter the living space in the winter months when the leaves have fallen.

These traditional building principles have been adapted and combined with new techniques and materials to create energy-efficient designs for a minimal cost. In addition to passive solar procedures, heating might be accomplished via *active solar systems*, in which some form of mechanical power is required to implement the systems.

These active systems typically employ solar panels that absorb sunlight and heat fluids circulated through panels in pipes. The heated fluid is then used to heat the living space or, more frequently, to heat water for the domestic hot-water supply. These systems work best, obviously, in areas that annually receive sufficient days of sunshine. Areas that are especially rainy or foggy, such as the Pacific Northwest in the United States or the British Isles, are worse candidates than drier areas.

Terra-Trivia
Active solar systems became popular in the United States following the 1973–1974 Arab oil embargo. Spurred by tax credits offered by the United States government, the installation of active solar systems increased steadily until the government withdrew tax credits in 1985. After the withdrawal of the tax credit, the installations markedly declined.

A Few More Natural and Renewable Resources

People are exploring other renewable resources that work in cooperation with the earth and reduce the environmental impact on the soils, oceans, and atmosphere. One such renewable resource is geothermal energy. By tapping the heat of the lower layers of the earth to provide an energy source, it can be used to create steam to turn turbines and generate electrical power.

Chapter 9 talks about how geothermal energy is an important source of power in Iceland. Unfortunately, geothermal energy is not easily accessible everywhere on earth (except via a different system that uses heat pumps to heat or cool spaces utilizing constant ground-water temperatures). The largest geothermal facilities are located at geologic hot spots on the earth's surface (generally, seismic areas). In addition to Iceland, California and Hawaii also have sizeable geothermal energy plants.

Another option is the wind. People have used wind power for centuries to run mechanical equipment and pumps. The most notable examples are the beautiful windmills of the Netherlands (Holland). The Dutch windmills were workhorses that constantly pumped water up and out of canals used to drain land reclaimed from the sea. Windmills (of higher-tech design than the old Dutch versions) are now used to turn generators for the production of electricity.

Terra-Trivia

You want dry? Move to the Atacama Desert in Chile. The place is so dry that virtually nothing grows. The Atacama Desert can go years without receiving a single drop of rain. Average rainfall in the desert is only $\frac{3}{100}$-inch per year. Although that's great news for your solar panels, it makes it difficult to maintain your front lawn.

Some places are windier than others, of course. Windmills work well in the Netherlands because of the steady wind from the North Sea. Many places are simply too calm to make windmills a feasible option. They also tend to be noisy, and the whirling blades take their toll on the local bird population.

California is the world's leader in wind-generated electricity; the European leader is no longer the Netherlands—now it's the North Sea neighbor, Denmark.

Another major source of energy worldwide is *biomass*, which includes wood, plant, and animal wastes. Although biomass can be burned directly for cooking or heating, it's a dirty fuel that can add significantly to greenhouse gases. Clever and simple techniques have been developed, however, for extracting energy in a much more environmentally friendly manner.

The Vegetarian Variation

In many parts of the world, especially Asia, diets are essentially vegetarian. The bulk of the diet consists of grains and vegetables with some fruit and little or no meat. Despite the damper this diet puts on backyard barbecues, it's part of an environmentally friendly lifestyle.

The other side of the coin is the typical American diet, which relies on meat and dairy products that are intensively packaged and heavily dependent on long-distance transportation. This diet is about as environmentally unfriendly as diets get. Here are a few ways that a person's diet has global environmental consequences:

➤ **Land use:** A plot of land can support 20 times as many people on a vegetarian diet as on a typical American diet (assuming fertile land; some less fertile lands can support livestock but not farming).

➤ **Soil loss:** In the United States, an estimated 80 percent of topsoil loss is associated directly with the raising of livestock. The Asian diet places much less demand on the soil than does a typical American diet.

➤ **Forest destruction:** Since 1960, Central America has experienced an 80 percent loss of tropical forests. The number-one cause for the destruction is cattle raising—almost all of which is exported to the United States.

➤ **Water conservation:** More than half of all water consumed in the United States is used in the livestock industry.

➤ **Waste:** Feedlots used by the livestock industry generate massive amounts of animal waste. (Animals generate almost 20 times as much waste as humans in the United States.) Methane gas given off by this untreated waste contributes to global greenhouse gases, and runoff is a major water pollutant.

➤ **Energy:** The typical American diet is tremendously energy-intensive. For each calorie of fossil fuel expended, you can cultivate 20 times more corn protein than feedlot beef.

Many environmentalists agree that the most environmentally friendly thing you can personally do is to alter your diet. If even a small percentage of the world's population moved toward a diet heavier in grains and vegetables and lighter in meat, many environmental ills in developed countries would be dramatically eased.

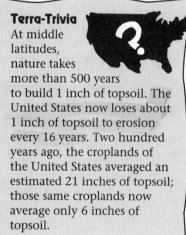

Terra-Trivia
Although people in North America, Europe, East Asia, and Southeast Asia share adequate nutrition, dietary differences are striking. Animal protein comprises more than 55 percent of the diet in North America and Europe but is less than 25 percent of the diet in East Asia and Southeast Asia.

Terra-Trivia
At middle latitudes, nature takes more than 500 years to build 1 inch of topsoil. The United States now loses about 1 inch of topsoil to erosion every 16 years. Two hundred years ago, the croplands of the United States averaged an estimated 21 inches of topsoil; those same croplands now average only 6 inches of topsoil.

Gaia: The Earth As an Organism

The more you learn about the complex systems of the earth, the more questions are raised. No ecological system is independent of another. The actions of individual organisms influence entire systems to an extent much greater than you might have believed. Is it possible that the earth is not just a collection of individual organisms and ecosystems but rather is itself a huge living entity?

The idea of the earth as a living organism of which people are but a small part is not new. This idea has been central to the beliefs of indigenous peoples across the globe. A key phrase in the spirituality of the Lakota people of the North American Great Plains is "Mitakuye Oyasin," or "We are all related." The Lakota understand the earth to be truly the mother of us all. The plants, animals, humans, minerals, and sky above are all interrelated parts of this great web. The Gaia hypothesis brings science back to this traditional belief.

Is the earth a "self-maintaining organism"? This question is now being debated by scientists. If the earth is a huge organism, if its rivers and streams are its circulatory system, if forests are the earth's lungs and the ocean is its heart, what then is humankind? Hopefully, humans will not prove to be analogous to spreading cancer cells.

A Final Word or Two

As you've toured the entire globe in this book, you have learned a thing or two about geography. You've investigated the physical world and the human world and looked at how the two interact and are dependent on one another. Perhaps as we humans increase our understanding, we all can work together to solve our mutual problems.

The best way to understand geography is to live it. Perhaps this guide has whetted your appetite and provided a spark. The entire world awaits you, filled with wonders to see, people to meet, and adventures to experience. Pack your bag, set your course, and, with a map in hand—set forth!

The Least You Need to Know

➤ The world's oceans are being contaminated at an unprecedented rate; unless people take massive and expensive efforts to clean the oceans, one of the earth's most valuable resources is in danger of faltering.

➤ Forest cutting has global consequences that will take a universal effort to abate.

➤ Nonrenewable resources are finite in quantity, and their use presents pollution and disposal problems.

➤ Renewable resources offer energy potential into the distant future, including hydroelectric, solar (passive and active systems), wind, and biomass systems.

➤ People can greatly influence many environmental concerns by adopting a diet that relies more heavily on vegetables, fruit, and grains as primary calorie sources.

Country Statistics

Population data generally are de facto figures for the country's territory as of 1996. Population estimates were derived from information available as of early 1996. A minus sign (–) indicates a decrease.

Country and Capital	1996 Population (1,000)	Annual Growth (%)	People per Square Mile
World	**5,771,939**	**1.4**	**114**
Afghanistan, Kabul	22,664	5.9	91
Albania, Tiranë	3,249	0.5	307
Algeria, Algiers	29,183	2.3	32
Andorra, Andora la Vella	73	4.2	418
Angola, Luanda	10,343	3.1	21
Antigua and Barbuda, St. John's	66	0.7	386
Argentina, Buenos Aires	34,673	1.1	33
Armenia, Yerevan	3,464	0.3	301
Australia, Canberra	18,261	1.1	6
Austria, Vienna	8,023	0.5	251
Azerbaijan, Baku	7,677	0.9	230
Bahamas, Nassau	259	1.1	67
Bahrain, Manama	590	2.5	2,469
Bangladesh, Dhaka	123,063	1.8	2,380

continues

continued

Country and Capital	1996 Population (1,000)	Annual Growth (%)	People per Square Mile
Barbados, Bridgetown	257	0.2	1,548
Belarus, Mensk	10,416	0.3	130
Belgium, Brussels	10,170	0.3	871
Belize, Belmopan	219	2.4	25
Benin, Porto–Novo	5,710	3.3	134
Bhutan, Thimphu	1,823	2.3	100
Bolivia, La Paz	7,165	1.8	17
Bosnia and Herzegovina, Sarajevo	2,656	–5.1	135
Botswana, Gaborone	1,478	1.8	7
Brazil, Brasília	162,661	1.2	50
Brunei, Bandar Seri Begawan	300	2.6	147
Bulgaria, Sofia	8,613	–0.2	202
Burkina Faso, Ouagadougou	10,623	2.6	100
Burundi, Bujumbura	5,943	1.4	600
Cambodia, Phnom Penh	10,861	3.3	159
Cameroon, Yaounde	14,262	2.9	79
Canada, Ottawa	28,821	1.2	8
Cape Verde, Praia	449	2.9	289
Central African Republic, Bangui	3,274	2.3	14
Chad, N'Djamena	6,977	2.8	14
Chile, Santiago	14,333	1.3	50
China, Beijing	1,210,005	1.0	336
Colombia, Santa Fe de Bogotá	36,813	1.7	92
Comoros, Moroni	569	3.5	679
Congo, Brazzaville	2,528	2.2	19
Costa Rica, San Jose	3,463	2.1	177
Côte d'Ivoire, Yamoussoukro	14,762	3.0	120
Croatia, Zagreb	5,004	0.6	229
Cuba, Havana	10,951	0.5	256
Cyprus, Nicosia	745	1.3	209
Czech Republic, Prague	10,321	0	213
Democratic Republic of Congo, Kinshasa	46,499	3.1	53
Denmark, Copenhagen	5,250	0.3	321
Djibouti, Djibouti	428	2.0	50

Country and Capital	1996 Population (1,000)	Annual Growth (%)	People per Square Mile
Dominica, Roseau	83	0.4	286
Dominican Republic, Santo Domingo	8,089	1.8	433
Ecuador, Quito	11,466	2.0	107
Egypt, Cairo	63,575	2.0	165
El Salvador, San Salvador	5,829	1.8	729
Equatorial Guinea, Malbo	431	2.6	40
Eritrea, Asmara	3,428	4.3	73
Estonia, Tallinn	1,459	−1.0	84
Ethiopia, Addis Ababa	57,172	2.8	132
Fiji, Suva	782	1.1	111
Finland, Helsinki	5,105	0.3	43
France, Paris	58,041	0.4	280
Gabon, Libreville	1,173	1.4	12
Gambia, Banjul	1,205	3.6	312
Georgia, Tbilisi	5,220	−0.6	194
Germany, Berlin	83,536	0.8	618
Ghana, Accra	17,698	2.4	199
Greece, Athens	10,539	0.6	209
Grenada, St. George's	95	0.4	726
Guatemala, Guatemala City	11,278	2.5	269
Guinea, Conakry	7,412	2.5	78
Guinea–Bissau, Bissau	1,151	2.4	106
Guyana, Georgetown	712	−0.7	9
Haiti, Port–au–Prince	6,732	1.8	633
Honduras, Tegucigalpa	5,605	2.7	130
Hungary, Budapest	10,003	−0.6	281
Iceland, Reykjavík	270	0.9	7
India, New Delhi	952,108	1.7	829
Indonesia, Jakarta	206,612	1.6	293
Iran, Tehran	66,094	2.3	105
Iraq, Baghdad	21,422	2.9	128
Ireland, Dublin	3,567	0	134
Israel, Jerusalem	5,422	2.6	691
Italy, Rome	57,460	0	506

continues

continued

Country and Capital	1996 Population (1,000)	Annual Growth (%)	People per Square Mile
Jamaica, Kingston	2,595	0.8	621
Japan, Tokyo	125,450	0.2	823
Jordan, Amman	4,212	3.6	119
Kazakstan, Almaty	16,916	0.1	16
Kenya, Nairobi	28,177	2.4	128
Kiribati, Tarawa	81	1.9	292
Korea, North, Pyongyang	23,904	1.7	514
Korea, South, Seoul	45,482	1.0	1,200
Kuwait, Kuwait	1,950	1.3	283
Kyrgyzstan, Bishkek	4,530	0.6	59
Laos, Vientiane	4,976	2.8	56
Latvia, Riga	2,469	–1.2	99
Lebanon, Beirut	3,776	2.0	956
Lesotho, Maseru	1,971	2.0	168
Liberia, Monrovia	2,110	3.0	57
Libya, Tripoli	5,445	3.7	8
Liechtenstein, Vaduz	31	1.3	501
Lithuania, Vilnius	3,646	–0.2	145
Luxembourg, Luxembourg	416	1.5	417
Macedonia, Skopje	2,104	0.6	212
Madagascar, Antananarivo	13,671	2.8	61
Malawi, Lilongwe	9,453	0.9	260
Malaysia, Kuala Lumpur	19,963	2.1	157
Maldives, Male	271	3.5	2,338
Mali, Bamako	9,653	2.8	20
Malta, Valletta	376	1.0	3,030
Marshall Islands, Majuro	58	3.9	835
Mauritania, Nouakchott	2,336	3.2	6
Mauritius, Port Louis	1,140	1.1	1,597
Mexico, Mexico City	95,772	1.9	129
Micronesia, Palikir	125	2.0	463
Moldova, Chisinau	4,464	0.3	343
Monaco, Monaco	32	0.74	1,076
Mongolia, Ulan Bator	2,497	1.8	4
Morocco, Rabat	29,779	2.1	173

Country and Capital	1996 Population (1,000)	Annual Growth (%)	People per Square Mile
Mozambique, Maputo	17,878	3.4	59
Myanmar, Yangon	45,976	1.8	181
Namibia, Windhoek	1,677	2.9	5
Nauru, Yaren	10	1.3	1,267
Nepal, Kathmandu	22,094	2.4	418
Netherlands, Amsterdam	15,568	0.6	1,188
New Zealand, Wellington	3,548	1.1	34
Nicaragua, Managua	4,272	2.8	92
Niger, Niamey	9,113	2.9	19
Nigeria, Lagos and Abuja	103,912	3.0	296
Norway, Oslo	4,384	0.5	37
Oman, Muscat	2,187	3.6	27
Pakistan, Islamabad	129,276	2.1	430
Palau, Koror and Melekeok	17	1.7	96
Panama, Panama City	2,655	1.7	90
Papua New Guinea, Port Moresby	4,395	2.3	25
Paraguay, Asuncion	5,504	2.7	36
Peru, Lima	24,523	1.8	50
Philippines, Manila	74,481	2.2	647
Poland, Warsaw	38,643	0.2	329
Portugal, Lisbon	9,865	0	279
Qatar, Doha	548	2.6	129
Romania, Bucharest	21,657	–0.8	244
Russia, Moscow	148,178	0	22
Rwanda, Kigaki	6,853	2.2	711
St. Kitts and Nevis, Basseterre	41	0.8	298
St. Lucia, Castries	158	1.2	669
St. Vincent and Grenadines, Kingstown	121	0.7	904
San Marino, San Marino	25	0.9	1,058
Sao Tome and Principe, São Tomé	144	2.6	389
Saudi Arabia, Riyadh	19,409	3.4	23
Senegal, Dakar	9,093	3.4	123
Seychelles, Victoria	78	0.8	441
Sierra Leone, Freetown	4,793	2.6	173

continues

continued

Country and Capital	1996 Population (1,000)	Annual Growth (%)	People per Square Mile
Singapore, Singapore	3,397	1.8	14,099
Slovakia, Bratislava	5,374	0.4	111
Slovenia, Ljubljana	1,951	−0.2	250
Solomon Islands, Honiara	413	3.4	39
Somalia, Mogadishu	9,639	2.7	40
South Africa, Pretoria	41,743	1.8	89
Spain, Madrid	39,181	0.2	203
Sri Lanka, Colombo	18,553	1.2	742
Sudan, Khartoum	31,548	2.9	34
Suriname, Paramaribo	436	1.6	7
Swaziland, Mbabane	999	2.9	150
Sweden, Stockholm	8,901	0.6	56
Switzerland, Bern	7,207	0.8	469
Syria, Damascus	15,609	3.4	220
Taiwan, Taipei	21,466	0.9	1,723
Tajikistan, Dushanbe	5,916	1.8	107
Tanzania, Dar es Salaam	29,058	2.2	85
Thailand, Bangkok	58,851	1.1	298
Togo, Lomé	4,571	3.6	218
Tonga, Nukualofa	106	0.8	385
Trinidad and Tobago, Port–of–Spain	1,272	0.1	642
Tunisia, Tunis	9,020	1.8	150
Turkey, Ankara	62,484	1.7	210
Turkmenistan, Ashkhabad	4,149	2.0	22
Tuvalu, Funafuti	10	1.6	1,011
Uganda, Kampala	20,158	2.5	261
Ukraine, Kyiv	50,864	−0.2	218
United Arab Emirates, Abu Dhabi	3,057	4.6	95
United Kingdom, London	58,490	0.3	627
United States, Washington, D.C.	265,563	1.0	75
Uruguay, Montevideo	3,239	0.7	48
Uzbekistan, Tashkent	23,418	2.0	136
Vanuatu, Port Vila	178	2.2	31
Venezuela, Caracas	21,983	2.0	65

Country and Capital	1996 Population (1,000)	Annual Growth (%)	People per Square Mile
Vietnam, Hanoi	73,977	1.7	589
Western Samoa, Apia	214	2.3	195
Yemen, San'a	13,483	3.9	66
Yugoslavia, Belgrade	10,612	1.0	269
Zambia, Lusaka	9,159	2.1	32
Zimbabwe, Harare	11,271	1.5	75

Source: U.S. Bureau of the Census, No. 1325
World Population Profile: 1996 (WP/96)

Select Bibliography

American Annual 1996. Encyclopedia of the Events of 1995, Yearbook of the Encyclopedia Americana. Canada: Grolier, 1996.

American Annual 1997. Encyclopedia of the Events of 1996, Yearbook of the Encyclopedia Americana. Canada: Grolier, 1997.

Asia, Australia, New Zealand, Oceania. Land and Peoples: Vol. 2. Danbury, CT: Grolier, 1993.

Australia and New Zealand. Land and Peoples: Vol. 2. Danbury, CT: Grolier, 1993.

Australia. Time–Life Books, eds. London: Time-Life Books, 1985.

Bander, Bruce, Mary Ann Hartwell, and Hector Holthouse. *Australia.* Washington, DC: National Geographic Society, 1968.

Benchley, Peter. "French Polynesia." *National Geographic* 191, no. 6 (June 1997).

Benchley, Peter. "New Zealand's Bountiful South Island." *National Geographic* 141, no. 6 (Jan. 1972).

Botkin, Daniel, and Edward Keller. *Environmental Science: Earth As a Living Planet.* New York: John Wiley & Sons, Inc., 1995.

Busch, Harold, and H. Breidenstein., eds. *Germany: Countryside, Cities, Villages and People.* Frankfurt am/Main: Bronners Druckerei, 1956.

Carpenter, Allan, and Carl Provorse. *The World Almanac of the U.S.A.* Mahwah, NJ: World Almanac Books, 1996.

Compton's Interactive Encyclopedia, *Your Family's Encyclopedia,* 1996 Edition. CD-ROM for Windows 95.

Davies, Norman. *Europe: A History.* Oxford and New York: Oxford Press, 1996.

De Blij, Harm J., and Peter O. Muller. *Geography: Realms, Regions and Concepts.* Seventh Edition. New York: John Wiley & Sons, 1994.

De Blij, Harm J., and Peter O. Muller. *Human Geography: Culture, Society, and Space.* New York: John Wiley & Sons, 1986.

DeMallie, Raymond J., and Douglas. R. Parks. *Sioux Indian Religion*. Norman: University of Oklahoma Press, 1987.

Edwards, Mike. "A Broken Empire: Russia, Kazakhstan, Ukraine." *National Geographic* 183, no. 3 (Mar. 1993).

Edwards, Mike. "Chernobyl." *National Geographic* 186, no. 2 (Aug. 1994).

Edwards, Mike. "Genghis Khan." *National Geographic* 190, no. 6 (Dec. 1996).

Edwards, Mike. "The Great Khans." *National Geographic* 191, no. 2 (Feb. 1997).

Encyclopedia Americana, The International Reference Work. New York: Americana Corporation, 1956.

Esplanshade, Edward B. *Rand McNally Goode's World Atlas*. 19th Edition. Rand McNally & Co., 1995.

Europe. Land and Peoples: Vol. 3. Danbury, CT: Grolier, 1993.

Europe. Land and Peoples: Vol. 4. Danbury, CT: Grolier, 1993.

Fellmann, Jerome, Arthur Getis, and Judith Getis. *Human Geography: Landscapes of Human Activities*. Fifth Edition. Madison: Brown & Benchmark, 1996.

Getis, Arthur, Judith Getis, and Jerome Fellmann. *Introduction to Geography*. Fourth Edition. Dubuque, Iowa: Wm. C. Brown Publishers, 1994.

Greece. Time–Life Books, eds. London: Time–Life Books, 1986.

Hammond Comparative World Atlas. Revised Edition. Maplewood, NJ: Hammond, 1993.

Hammond Standard World Atlas. Deluxe Edition. Maplewood, NJ: Hammond, 1974.

Hepner, George F., and Jesse O. McKee. *World Regional Geography: A Global Approach*. St. Paul: West Publishing Co., 1992.

Indians of South America. Washington, DC: National Geographic Society, 1982.

Italy. Time–Life Books, eds. London: Time–Life Books, 1986.

Jackson, Richard H., and Lloyd E. Hudman. *Cultural Geography: People, Places, and Environment*. St. Paul: West Publishing Co., 1990.

Johnson, Otto. *1997 Information Please Almanac: The International Authority*. Boston, MA: Houghton Mifflin Co., 1996.

Jordan, Robert Paul. "New Zealand: The Last Utopia?" *National Geographic* 171, no. 5 (May 1987).

Jordan, Terry G., and Lester Rowntree. *The Human Mosaic: A Thematic Introduction to Cultural Geography*. Fourth Edition. New York: Harper & Row: 1986.

Kovacs, Kit. "Bearded Seals." *National Geographic* 191, no. 3 (Mar. 1997).

Mairson, Alan. "The Three Faces of Jerusalem." *National Geographic* 189, no. 4 (Apr. 1996).

Map Art. Geopolitical Series. Lambertville, NJ: Cartesia Software, 1994.

McKnight, Tom L. *Essentials of Physical Geography.* Englewood Cliffs, NJ: Prentice Hall, 1992.

McKnight, Tom L. *Physical Geography: A Landscape Appreciation.* Englewood Cliffs, NJ: Prentice Hall, 1987.

Microsoft Encarta '95, Complete Interactive Multimedia Encyclopedia. 1995 Edition. Microsoft Corporation. CD-ROM for Windows 95.

Moffett, Mark W. "Tree Giants of North America." *National Geographic* 191, no. 1 (Jan. 1997).

Momatiuk, Yva, and John Eastcott. "Maoris: At Home in Two Worlds." *National Geographic* 166, no. 4 (Oct. 1984).

Momatiuk, Yva, and John Eastcott. "New Zealand's High Country." *National Geographic* 154, no. 2 (Aug. 1978).

National Geographic Atlas of the World. Revised Sixth Edition. Washington, DC: National Geographic Society, 1996.

Newman, Cathy. "The Light at the End of the Chunnel." *National Geographic* 185, no. 5 (May 1994).

Newton, Douglas. "Maoris: Treasures of the Tradition." *National Geographic* 166 no. 4 (Oct. 1984).

Norris, Robert E. *World Regional Geography.* St. Paul: West Publishing Company, 1990.

Oeland, Glenn. "Emperors of the Ice." *National Geographic* 189, no. 3 (Mar. 1996).

Palmer, R.R. *A History of the Modern World.* New York: Alfred A. Knopf, 1952.

Parfit, Michael. "Mexico City: Pushing the Limits." *National Geographic* 190, no. 2 (Aug. 1996).

Patterson, Carolyn Bennett. "A Walk on the Wild Side." *National Geographic* 163, no. 5 (May 1983).

Random House College Dictionary, Revised Edition. New York: Random House, 1980.

Renwick, William H., and James M. Rubenstein. *An Introduction to Geography: People, Places, and Environment.* Englewood Cliffs, NJ: Prentice Hall, 1995.

Roberts, David. "Egypt's Old Kingdom." *National Geographic* 187, no. 1 (Jan. 1995).

Scandinavia. Time–Life Books, eds. London: Time–Life Books, 1985.

375

Shadbolt, Maurice, and Olaf Ruhen. *Isles of the South Pacific*. Washington, DC: National Geographic Society, 1968.

Spain. Time–Life Books, eds. London: Time–Life Books, 1986.

Stansfield, Charles A., Jr., and Chester E. Zimolzak. *Global Perspectives: A World Regional Geography*. Second Edition. Columbus: Merrill Publishing Co., 1990.

Steger, Will. "Arctic Ocean Traverse." *National Geographic* 189, no. 1 (Jan. 1996).

Summerhay, Soames. "A Marine Park is Born." *National Geographic* 159, no. 5 (May 1981).

Taylor, Ron, and Valerie Taylor. "Paradise Beneath the Sea." *National Geographic* 159, no. 5 (May 1981).

Tourtello, Jonathan B., Mary B. Dickenson, eds. *Discovering Britain and Ireland*. Washington, DC: National Geographic Society, 1985.

United Kingdom. Land and Peoples: Vol. 3. Danbury, CT: Grolier, 1993.

Ward, Andrew. "Scotland: Plaid to the Bone." *National Geographic* 190 , no. 3 (Sept. 1996).

Ward, Geoffrey C. "India." *National Geographic* 191, no. 5 (May 1997).

Webster's Concise World Atlas and Almanac. New York: Chatham River Press, 1991.

Webster's Ninth New Collegiate Dictionary. Springfield, MA: Merriam–Webster, 1988.

Wilson, Barbara Ker. *Australia: Wonderland Down Under*. New York: Dodd, Mead & Co., 1969.

World Almanac and Book of Facts 1997. Mahwah, NJ: World Almanac Books, 1996.

World Population Profile: 1996. Washington, DC: United States Bureau of the Census, 1996.

Index

C

D

383

T

U